Complexity Theory and Law

This collection of essays explores the different ways the insights from complexity theory can be applied to law. Complexity theory – a form of systems theory – views law as an emergent, complex, self-organising system composed of an interactive network of actors and systems that operate with no overall guiding hand, giving rise to complex, collective behaviour in law communications and actions. Addressing such issues as the unpredictability of legal systems, the ability of legal systems to adapt to changes in society, the importance of context, and the nature of law, the essays look to the implications of a complexity theory analysis for the study of public policy and administrative law, international law and human rights, regulatory practices in business and finance, and the practice of law and legal ethics. These are areas where law, which craves certainty, encounters unending, irresolvable complexity. This collection shows the many ways complexity theory thinking can reshape and clarify our understanding of the various problems relating to the theory and practice of law.

Jamie Murray is Senior Lecturer in Law at Liverpool Hope University.

Thomas E. Webb is Lecturer in Law at the University of Lancaster.

Steven Wheatley is Professor of International Law at the University of Lancaster.

Part of The Law, Science And Society series
Series Editors
John B Paterson
University of Aberdeen, United Kingdom
Julian Webb
University of Melbourne, Australia

A GlassHouse Book

for information about the series and details of previous and forthcoming titles,
see www.routledge.com/Law-Science-and-Society/book-series/CAV16

Complexity Theory and Law

Mapping an Emergent Jurisprudence

**Edited by Jamie Murray,
Thomas E. Webb and Steven Wheatley**

Routledge
Taylor & Francis Group

LONDON AND NEW YORK

First published 2019 by Routledge

2 Park Square, Milton Park, Abingdon, Oxfordshire OX14 4RN
52 Vanderbilt Avenue, New York, NY 10017

Routledge is an imprint of the Taylor & Francis Group, an informa business

First issued in paperback 2019

British Library Cataloguing-in-Publication Data
A catalogue record for this book is available from the British Library

Library of Congress Cataloging-in-Publication Data
A catalog record for this book has been requested

ISBN: 978-0-415-78609-6 (hbk)
ISBN: 978-0-367-89525-9 (pbk)

Typeset in Galliard
by Apex CoVantage, LLC

Contents

Contributors

Anna Marie Brennan is a lecturer at the University of Liverpool. She undertook her PhD at University College Cork on 'The Accountability of Transnational Armed Groups under International Law'.

Mark Chinen is a professor of law at the Seattle University School of Law and a fellow of the Fred T. Korematsu Center for Law and Equality. He is the author of several papers on complexity theory and international law.

Lucy Finchett-Maddock is a senior lecturer in law and art at the University of Sussex. Her work focuses on the interaction of property within law and resistance interrogating the spatio-temporality and aesthetics of formal and informal laws, property, commons and protest. She is author of *Protest, Property and the Commons: Performances of Law and Resistance* (Routledge, 2016).

Neville Harris is a professor of law at the University of Manchester. He is the author of *Law in a Complex State* (Hart, 2013).

Daniel M. Katz is associate professor of law at the Chicago-Kent College of Law in the Illinois Institute of Technology. He is the author of numerous publications integrating law, complexity theory, science and technology, including 'Measuring and Modeling the US Regulatory Ecosystem' (2017) 168(5) *Journal of Statistical Physics* 1125 (with Michael J Bommarito II).

Michael Leach is a Ph.D. candidate in Law at Tilburg University researching the promotion of the 'Rule of Law' in international development. Prior to this he undertook an MSt. at the University of Oxford on the evolutionary dynamics of post-crisis banking and financial regulatory change in Indonesia.

Jamie Murray is a senior lecturer at Liverpool Hope University. His articles include 'Complexity Theory & Socio-Legal Studies' (2008) in the *Liverpool Law Review*. He is the author of *Deleuze & Guattari: Emergent Law* (Routledge, 2013).

J.B. Ruhl is David Daniels Allen Distinguished Chair at Vanderbilt University, USA. He is the author of several leading articles on complexity theory and law, including 'Complexity Theory as a Paradigm for the Dynamical Law-and-Society System' (1995–6) in the *Duke Law Journal* and 'Law's Complexity: A Primer' (2007–2008) in the *Georgia State University Law Review*.

Dimitrios Tsarapatsanis is a lecturer at the University of Sheffield. He is the author of a number of articles on legal theory, bioethics and human rights and of *Les fondements éthiques des discours juridiques sur le statut de la vie humaine anténatale* (University of Nanterre Press, 2010).

Julian Webb is a professor of law at the University of Melbourne. His work includes two early attempts to apply complexity thinking, first, to theories of the legal profession in 'Turf Wars and Market Control: Competition and Complexity in the Market for Legal Services' (2004) and then to legal ethics in 'Law, Ethics, and Complexity' (2005).

Thomas E. Webb is a lecturer in law at the University of Lancaster. He undertook his doctorate on aspects of complexity theory and constitutional law writings and is the author of several papers on law and systems theory thinking, including 'Tracing an Outline of Legal Complexity' (2014) in *Ratio Juris*.

Steven Wheatley is a professor of international law at the University of Lancaster. He is the author of *The Idea of International Human Rights Law* (Oxford University Press, 2018), which applies the insights from complexity theory to the study human rights.

Minka Woermann is a senior lecturer in the philosophy department at Stellenbosch University, South Africa. She is the author of *On the (Im)Possibility of Business Ethics: Critical Complexity, Deconstruction, and Implications for Understanding the Ethics of Business* (Springer, 2013) and *Bridging Complexity and Post-Structuralism: Insights and Implications* (Springer, 2016).

Section I
Law's complexity

1 Encountering law's complexity

Jamie Murray, Thomas E. Webb and Steven Wheatley

This collection introduces the reader to the ways that scholars are using complexity theory to make sense of law. Complexity presents a more productive language for legal theory and a revolutionary way of addressing the problems of descriptive, normative and critical jurisprudence as well as understanding the interconnected operations of law and other social activities. Complexity theory developed in the natural sciences as a way of explaining the ways in which order could arise without the need for a guiding hand or central controller. In a complex system, the structure emerges spontaneously as the result of the interactions of the component elements in the system as they encounter new information. Complexity theory has been used, *inter alia*, to explain the workings of insect colonies and the relationship between the mind and the brain (Waldrop, 1994, p. 145). It has also been relied on by certain social scientists (for example, Geyer and Rihani, 2010; Sawyer, 2005; Urry, 2003; Walby, 2007), and there is now a significant, albeit disparate, body of scholarly writing that seeks to apply the insights from complexity theory to law (for example, Hathaway, 2001; Murray, 2006, 2008; Ruhl, 1996a, 1996b, 1997, 2008; Vermeule, 2012; Webb, 2013, 2014, 2015; Webb, 2005; Wheatley, 2016).

Complexity theory understands law as an emergent, self-organising system in which an interactive network of many parts – actors, institutions and 'systems' – operate with no overall guiding hand, giving rise to complex collective behaviours that can be observed in patterns of law communications. The contributions in this volume explore the different ways in which the insights from complexity can be applied to law – addressing such questions as how we understand the idea of law, the role of law as a regulatory tool and the advantages of an approach to legal questions from complexity, including the academic function of critique. The collection focuses on public policy and administrative law, international law and human rights, business and finance and the practice of law and legal ethics because these are the areas in which law has, thus far, been seen most clearly to encounter complexity. The objectives of this chapter are to introduce the reader to the science of complexity, to explain the basic idea of complexity theory, to outline the ways complexity has been used in the academic discipline of law, and to provide an outline of the chapters in the collection.

Why complexity?

There are no shortages of possible approaches to legal theory. *Lloyd's Introduction to Jurisprudence* includes, for example, natural law, positivism, sociological jurisprudence, realism, critical legal studies, feminist jurisprudence, postmodernist jurisprudence and critical race theory (Freeman, 2014). The basic claims of this collection are, first, that complexity theory offers something qualitatively different to these now-traditional approaches and so should be added to the list and, second, that legal complexity is a fact of the world and the tools we currently possess are, on their own, inadequate to the task of making sense of law's complexity, or at least insufficient to understanding the limits of our knowledge about law. The argument for complexity is that law systems are complex systems, and to make better sense of the law we must look to the insights from complexity to develop models that explain what law is and how we should think about the very nature and purpose of law. Simply put, if a research question involves interconnectedness, systemic properties, unpredictability, porous boundaries, some element of bottom-up organisation and rapid innovations in law and regulation, we are concerned with legal complexity, and to make sense of law's complexity, we must engage with complexity theory. This is the case whether we are examining the legal complexity of the governance of global financial markets (Sornette, 2017), the regulation of on-street sex work (Carline and Murray, 2018), or anything in between.

The first challenge lawyers face with the application of complexity theory to law is the apparent diversity of understandings available. Indeed, the anthropologist and computational social scientist John Murphy introduces the subject as follows:

> Complexity theory is a collection of theories and approaches that began to grow to prominence in the 1990s, that attempt to address the behavior of systems not readily understood using traditional approaches . . . Complexity theory addresses highly nonlinear systems and systems that exhibit emergent, self-organized, and adaptive behavior. Domains include virtually every field of study, from economics, to cosmology, to genetic evolution, to cognition and artificial intelligence. Its appeal is that it proposes that common principles guide the dynamics and evolution of systems across all of these domains, and that these principles reflect a deeper order that profoundly structures the physical and social world in which we live.
>
> (Murphy, 2017)

There is then no general science or philosophy of complexity and no agreed-on final definition of the concept. This has not prevented scholars in both the natural and social sciences from looking to the language of complexity to explain the world we inhabit. Physics and chemistry point to the existence of complex systems like the Great Red Spot vortex of Jupiter (Kaufman, 1996, p. 20) and Benard cells, Belasov-Zhabokinski reactions and chemical clocks (Kaufman,

1996, p. 53); biology to the existence of complex adaptive systems like ant colonies and immune systems (Waldrop, 1994, p. 145). The social world is seen to be composed of complex adaptive systems such as those of language and 'symbolic interactions' (Sawyer, 2005, pp. 4–5, 24–25), as well as political systems (Vermeule, 2012, p. 50).

Wherever they look, complexity theorists see complex systems, presenting an exciting picture of ceaseless creativity, transformation, order out of chaos, strange attractors, far-from-equilibrium processes, spontaneous self-organisation, nonlinearity, emergence, adaptation and evolution.

Perhaps because of the diversity of contexts to which complexity theory has been applied, many complexity theorists often refer to the approach as a set of tools (Byrne, 1998, p. 34; Geyer and Rihani, 2010, chapter 3; Webb, 2005, p. 232), but it is perhaps more accurately described as 'a conceptual framework, a way of thinking, and a way of seeing the world' (Mitleton-Kelly, 2003, p. 26). That way of seeing the world is predicated, especially in the social sciences, on a concern that the attempt to emulate the natural scientific method of analytical reductionism closes off an expanse of social experience and interaction – the source of complexity – and presents social existence as a quantifiable, essentially knowable, phenomenon. A view of the world as complex regards our models and descriptions as incomplete by virtue of the tension between our own localness and the scale of that which we seek to explain (Cilliers, 1998, p. 95; Webb, 2005, p. 235). It is opposed to final destinations and only provides descriptions and analysis suitable for the moment (Cilliers, 1998, p. 4, 2001, p. 141; Richardson et al., 2001; see also Webb, 2005, p. 237 and n.43 p. 237). In exposing the deeply interconnected, perpetually interacting, reiterative nature of social existence, complexity theory requires observers to be more precise in their definition of the scope of their investigation and the contingency arising from their spatio-temporal context. Consequently, whilst there is some variation in language, and while some approaches place greater weight on a particular concept or device of complexity theory than others, the essence of complexity is to be found in the modesty it engenders in the observer of society.

Complexity theory has revolutionised many areas of the natural sciences, and its core insights have been adopted by social sciences to provide a better way of thinking about human social existence, emphasising the importance of connectivity and dynamic network organisation, unpredictability, systemic instabilities and rapid change. Complexity theory came into existence following the recognition that the Newtonian model of a clockwork universe that could be taken apart and subjected to reductionist analysis was unable to explain the workings of certain (complex) systems (Capra and Mattei, 2015). Byrne and Callaghan explain the point this way: 'the implications of [complexity] is not that the law focused on Newtonian science is wrong but rather that it is *limited in its rightness*' (Byrne and Callaghan, 2013, p. 19, *emphasis* added). Once we recognise that much of the physical and social world is made up of complex systems, we must accept that these can only be studied through a new complexity paradigm focused on notions of interconnectedness, relationality, nonlinearity, self-organisation, dissipative structures, emergence, systemic openness, adaptation, evolution and transformation. Simply put, complexity and complexity thinking involve a significant modification in how we see and understand our world.

Yet, despite all this, complexity theory remains largely absent from legal thinking, with systems thinking in legal scholarship dominated by autopoietic systems theory. Given the influence of complexity theory in the natural and social sciences it is strange that law as a discipline has remained largely indifferent to it. It is even more strange because law systems exhibit all of the features of complex systems, emerging from the actions and interactions of law actors in a networked relationship, but with different characteristics from those law actors. Complexity gives us, then, the possibility of a way of thinking about law and a language to describe law systems as never before.

Our position is that complexity presents a view of law and society which is qualitatively different from that of autopoietic theory and, we argue, significantly enhances the value of systems theory thinking in law. This is the case for four reasons.

First, complexity theory is better science. In the natural sciences, little reliance is placed on autopoiesis beyond the narrow discipline of cell biology; by way of contrast, there are numerous references to complexity theory across all scientific disciplines. Moreover, the literature on autopoiesis draws narrowly on the work of (Maturana and Varela, 1987), notwithstanding that Niklas Luhmann and those after him have developed a highly sophisticated, internally coherent theory of autopoiesis (Luhmann, 2004; Teubner, 1993, 2012). Furthermore, while autopoiesis is presented as an approach founded in the sciences, it does not acknowledge the narrowness of its foundations, nor does it represent the conclusions it draws about the proper order of a functionally differentiated society as being founded in that science. Complexity, on the other hand, is well established in physics, chemistry and biology, and the literature has drawn on a wide range of sources in these disciplines to produce *inter alia* socially influential metaphors, such as 'butterfly effects' and 'tipping points' (see Lewin, 1992, p. 11; Lorenz, 1993).

Second, autopoiesis asks us to think in terms of communication systems we cannot see, touch, or hear; we must accept, as a matter of faith, their existence. Indeed, the notion of autopoietic, functionally differentiated subsystems is an artefact of autopoietic thinking, not of social observation (Webb, 2013, pp. 135–139). There is no particular reason why in autopoietic thought certain definitions and boundaries of system communication, such as law, politics, health and education, are to be preferred, other than that they simply *are* preferred. For autopoiesis, this means that, although individual events can have multiple meanings across different systems (see King, 1993, pp. 223–226; also Luhmann, 1992a, p. 1432), there is no opportunity for those meanings to directly engage one another to create new logics for the system (they may only structurally couple, King, 1993 p. 225) – functional differentiation perpetuates the status quo and increases the risk of entropy. Complexity, on the other hand, though it conceptualises law as an emergent property of the communication acts of law actors like parliaments and courts that we can easily perceive, does not require that communications be framed according to a predetermined list of social functions. And, more important, it anticipates that the confluence of communications amongst different actors, institutions and systems – the interface of their respective descriptions strategies, their boundaries of understanding – is the most important aspect of social behaviour to observe for law.

Third, the dehumanised nature of autopoiesis is highly problematic (Wheatley, 2018). There is already an established critique and counter-critique to the removal of the person from autopoiesis (Bankowski, 1996; Paterson, 1996) that addresses this question on autopoiesis' own terms. Similarly, autopoiesis has also been challenged on the exclusion of the physical, corporeal existence of humanity from autopoiesis both from the compassionate perspective of the concept of vulnerability and in terms of the implications for the longer-term stability of autopoietic social systems (Phillipopoulos-Mihalopoulos and Webb, 2015). For complexity theory, the setting aside of the importance of human agency, and the ascription of volition and the construction of meaning principally to social not – as autopoiesis would say – psychic systems, closes off great swathes of activity which are neither anticipated nor understood at a systemic level. The operational context of any source of meaning, be that an individual, an institution, or a 'whole' system, has an impact on the subject matter with which that source of meaning is concerned. Complex systems operate within and across many different scales, producing models of understanding according both to their operational context and to the scale to which that context is addressed.

The final problem relates to the assumptions which autopoiesis makes about regulation and the reasons for regulatory failure (see Luhmann, 1992b, p. 397). First, autopoietic identity (ego) is tied up with self-reference, thus reference to the other (alter) ruptures that relationship. The interdependence of system identity and the processes of self-reference that sustain it means that autopoiesis finds it difficult to countenance using 'law as a means of direct intervention in social systems . . . as a means for purposeful intervention in adaptive, open systems' (Teubner, 1988, p. 219), because this would entail external reference. The second reason is that the autopoietic identity relies upon a binary code. The ability of the system to distinguish itself from its environment, and thus to recognise communications as being part of the system, as having a meaning which it can understand, is wholly reliant on the perpetuation of this functionally derived code. The difficulty with both these explanations for why regulatory failure occurs is that they rest on the assumption that, were perfect communication somehow possible, regulatory failure would not occur. Yet, complexity theory shows us there is no perfect form of regulation available. This demonstrates the qualified nature of autopoietic accounts of regulatory failure. They are useful in that they demonstrate the challenges of communicating, but they do not grasp the fact it is not that regulation seeks to remedy a known problem with a quantifiable solution but, instead, that regulatory space is forever being destabilised by events, new actors and new interactions (see further Geyer and Rihani, 2010; Ruhl and Salzman, 2002, 2003). A complex version of law will be flexible and adaptable, but it will not provide a *solution* to regulation, only the possibility of failing less frequently and reacting more adeptly.

What is complexity?

The origins of complexity theory can be traced back to early work on cybernetics and information theory (Waldrop, 1994; Woermann, 2016), but the notion of a distinctive theory of complexity is normally credited to the Belgian physical chemist Ilya Prigogine, who won the Nobel Prize for Chemistry in 1977 for his work on far from equilibrium systems, a type of complex system. Prigogine introduced the notion of 'order out of chaos' (Prigogine and Stengers, 1984), which can be taken as the first aphorism of a theory of complex systems. The story then shifts to the activities of the US Santa Fe Institute (established 1986), which played host to some of the leading thinkers on complexity theory, including Kauffman and Holland, who each worked on biological complexity (Holland, 1995, 1998; Kauffman, 1993, 1996), and Arthur, who worked on economic complexity (1994, 2014). Complexity theory was, at the time, mostly based in the academic disciples of physics, mathematics and computer sciences, but there was also an interest in the subject in the continental philosophy of Deleuze and Guattari (1987) and Morin (2007).

Whilst there is no agreed-on definition of a complex system, there is some consensus in the literature on the characteristics of complex systems. First, complex systems are self-organising. There is no controlling power or central control in a complex system, which is the result of the actions and interactions of micro-level component elements. Second, complex systems have a meso-level of creative organisation in which the interconnections and interrelations of micro-molecular elements result in another level of complexity. Third, the actions and interactions of component elements at the micro- and meso-levels result in the emergence of macro-system-level characteristics with different properties or capacities from the lower-level elements. This is normally explained in terms that, in a complex system, 'the whole is greater than the sum of its parts'. Fourth, complex systems change over time with the flow of new matter, energy or information into the system, generating novel emergent properties. Fifth, complex systems not only interact with agents and elements in the external environment but also with other complex systems, leading to the possibility of even higher-level emergent properties. Furthermore, complex systems interconnecting and interacting with other complex systems will become nested, with increased complexity. Finally, whilst complex system may remain stable for long periods, their nonlinearity means that radical change can happen quickly and unexpectedly, with complex systems existing somewhere between entropy (where the system decays over time) and chaos (where too much activity makes stable structures impossible to maintain) (Capra, 2016; Coveney and Highfield, 1996; Heylighen et al., 2007; Lewin, 1992; Richardson and Cilliers, 2001; Waldrop, 1994, *passim*).

On this view, complexity theory might be a postmodern theory – in the most extreme sense – because of the central importance it accords to contingency and emergence, and thus the empirical unpredictability of social life, that is, the possibility that 'anything goes' (Cilliers, 1995, 1998, *viii*), that there are no real structures or boundaries to speak of. However, from our perspective this misunderstands the purpose of these insights in the context of complexity theory. Complexity is not a postmodern theory because it is not concerned with doing away with, or otherwise transcending boundaries. It is concerned with the means by which those boundaries – of actors' understanding, of institutions, of systems, of concepts – are constructed, with their justifications and with their responses to stimulation by other boundaries. Without being so hubristic as to believe that we can fully comprehend the nature and implications of the notion of emergence for law and society more generally, a complexity view of the law should nonetheless be committed to the aspiration of attempting to understand the idea of emergence and to grasp what it means for law in both specific and general contexts.

Thinking about complexity

The philosopher and sociologist Edgar Morin introduced a well-known distinction in the discussions on complexity between (what he calls) restricted (or modern) writings that look to discover mathematically formulated laws of complexity, and general (or postmodern) scholarship which regards all attempts to produce laws of complexity as a negation of the insight that some systems cannot be modelled perfectly because they are complex systems (Morin, 2007, p. 10). The difference can be seen, for example, in the divergent methodologies of the Santa Fe Institute, which aim to formalise the laws of general complexity, and Morin's own project, which stayed faithful to the more open and philosophical concerns of general systems theory. Morin argues that the search for the laws of complexity is influenced by the paradigm of classical science, of the need for simplification, but that when the principle of reduction is applied to complex systems some important elements will inevitably and necessarily be missed, meaning that predictions of the future shape and form of the system become impossible to make with any certainty. By way of contrast, general complexity concludes that it is not possible to uncover general laws of complexity and tries instead to make sense of the relationships between the whole and the parts by focusing on notions of order and disorder (Morin, 2007, p. 10).

For Morin and others, the alternative to restrictive complexity and attempts to develop laws of complexity, and indeed the correct way of engaging with complexity, is to develop a philosophy of complexity that is epistemologically modest and ethically embedded (Cilliers, 1998; Morin, 2007; Byrne and Callaghan, 2013; and Woermann, 2016). The general (postmodern) accounts focus on the philosophical insights that result from the realisation that complex systems cannot be described or explained because they are the result of the interactions between the system and its component elements, the interactions between the component elements, and their interactions with elements outside of the system, with the result that all descriptions and predictions about the workings of complex systems inevitably involve the exercise of subjective judgment. Woermann and Cilliers explain the point this way: 'As soon as we engage with complexity, we have to make certain modelling choices when describing phenomena . . . [in other words] our modelling choices are based on subjective judgements about what matters' (Woermann and Cilliers, 2012, p. 448; also de Villiers and Cilliers, 2010). The aim of writings within the general (postmodern) complexity school is not to work out the laws of complexity but to emphasise what we do not know, indeed cannot know, and thereafter to focus on our own ethical responsibilities as thinkers about complexity theory in law (Cilliers, 2004). To the restricted (or modern) complexity theorists, the writings of the general (postmodern) complexity scholars can be seen 'as pure chattering, pure philosophy' (Morin, 2007, p. 27).

Heylighen, Cilliers and Gershenson argue that, given their scientific backgrounds, most complexity researchers 'still implicitly cling to the Newtonian paradigm, hoping to discover mathematically formulated "laws of complexity" that would restore some form of absolute order or determinism to the very uncertain world they are trying to understand' (Heylighen et al., 2007, p. 124). The emergence of computational complexity, associated with the work of the Santa Fe Institute, is the clearest evidence of the tendency to try to capture physical and social complexity with rule-based models, and we see efforts to develop laws of complexity in Holland's analysis of emergence (1998), Mitchell's general analysis of complexity theory (2009), and Miller and Page's work on social complexity (2007). For the postmodern complexity theorist, these efforts are both futile and a rejection of the very notion of a complex system, which is defined by its incompressibility and unpredictability (Richardson and Cilliers, 2001, pp. 8–9), with the consequence that any description of a complex system will fail to capture its full complexity and adaptability, meaning that predictions of the future shape and form of the system become impossible to make with any certainty.

The difficulty for those who argue for a postmodern ethic of complexity (or a postmodern reading of complexity theory) is that one of the central lessons of complexity is that, whilst the functioning of complex systems cannot be predicted with absolute certainty, neither are complex systems completely unpredictable. Complexity theory involves a rejection of both the ambition of modernity to understand and explain everything and the claim of postmodernity (or at least its characterisation) that everything is contingent and nothing can be explained (which appears paradoxically to be another grand narrative). Marais Kobus expresses the point this way: complexity theory refuses to follow either the claims of modernism to explain everything or the argument from postmodernism that everything is contingent and context-dependent and instead regards 'the universal and the contingent, consistency and change as constituent factors of reality [and] through this stance, it hopes to do justice to the wholeness and interrelatedness of reality' (Kobus, 2014, p. 17).

Much of the mainstream literature in complexity attempts to navigate this tension between restrictive and general complexity, rejecting the reductionist paradigm of classical modern science, that is, of the need to explain everything but without feeling the need to refrain from telling us something about complex systems. Most writings on complexity theory take the view that there is nothing inherently problematic in trying to *better understand* complex systems, but they also recognise it would be a mistake to think we could ever fully understand complex systems because of the limits of our knowledge of their workings and the unpredictable consequences of seemingly small events on the system – from the removal of a keystone species, like the sea star that keeps populations of mussels and barnacles in check; to the attempted coup against Mikhail Gorbachev and collapse of the Soviet Empire; to the selling of sub-prime mortgages in Florida and the 2008 financial crisis. The central lesson from restricted and general complexity theory is that whilst we can know some things, we can never know everything, and we should not delegate ethical and political decision-making to computation models of complex systems that must be, by definition, and in ways of which we are unaware of, wrong, limited and thus imperfect. If there is one lesson from complexity theory, it is the need for epistemic humility: the certain knowledge that we can never be certain when dealing with complex systems, including the complex systems of law.

Complexity theory and law

We are now in a position to reflect on some of the ways that complexity theory may help us to answer some of the questions facing the academic discipline of law and the arena of legal practice. One initial difficulty, as this collection of essays makes clear, is that there is no standard approach, no jurisprudence of complexity that runs throughout the various contributions, or indeed the wider literature. While there is no paradigmatic 'jurisprudence of complexity', there are a number of insights from complexity that can be applied to law and which might influence the way in which legal theory addresses the central questions of jurisprudence. These relate to the unpredictability of legal systems; the idea of the law system as emergent, the result of the interactions between law actors; the ability of law to adapt to changes in its external environment and the functioning of other law systems; the importance of context to understanding the law; the unclear, contested and open nature of law system boundaries and the way they interface with society; and the fact that practitioners and scholars cannot avoid ethical responsibility in their work.

To understand the possibilities of complexity theory, consider the following. A legal researcher is tasked with writing a report on the law on the protection of wild animals. She draws up 3 lists of the legislative provisions, court decisions, and governmental regulations. But, she knows she must also take into account judicial cases on the interpretation of legislation, the way government regulations have implemented acts of parliament, and any reliance by the courts on executive orders. Now she can see a networked relationship between the rules adopted by the legislature, courts and executive, evidencing a more complicated picture than suggested by her lists, and, being a talented computer scientist, she develops a programme to model those relationships. But, still, she sees only part of the picture, as there will be relevant legislative provisions and judicial decisions in property law and human rights law, etc. When writing her report and reflecting on the legal rules and networks of relationships, she begins to see patterns in the rules, a body of "Wildlife Law", albeit its content is not always clear, and she must make choices when filling in the gaps in this Wildlife Law, and in the exercise of that discretion, we would expect (and hope) she would use her professional and ethical judgement.

The scenario highlights the different ways complexity theory can be used in legal research: in observing the complicated nature of law communications; in suggesting the possible use of new technologies; in emphasizing the organizing properties of emergent phenomena; and in pointing out the ethical responsibilities of the legal researcher. The first approach equates complexity with *complicatedness* – the notion that the law system is simply too complicated, or complex, for any mortal lawyer to understand. The literature here highlights the difficulties of capturing every combination and permutation of legal rules and practices. Peter Schuck, for example, argues that a legal system is complex 'to the extent that its rules, processes, institutions, and supporting culture possess four features: density, technicality, differentiation, and indeterminacy or uncertainty' (Schuck, 1992, p. 3). Second, and related to complicatedness, is the idea of *computational complexity*, which draws on the mathematical theory of complexity outlined by computer scientists to develop computational algorithms to model law systems (Kades, 1996–1997, p. 403). Third, there is the approach that sees emergence – the idea of 'the whole being more than the sum of the parts' – as the distinguishing feature of complexity. *Emergence* describes phenomena that arise from and depend on the interaction between underlying phenomena that are at the same time autonomous from those phenomena: something novel emerges from 'below'. Understood as a philosophical method, emergence complexity is concerned with understanding and explaining the ways that novel properties emerge from the actions and interactions of the component parts (Humphreys, 2016). Finally, there is the *general (or postmodern)* approach to complexity first identified by the philosopher and sociologist Edgar Morin, which regards all attempts to produce laws of complexity as a negation of the central insight that some systems cannot be modelled perfectly because they are complex systems.

To make sense of the literature in the emerging literature on complexity theory in law, we need to ask two questions. First, does the argument in the work under consideration depend fundamentally on the presence of emergent phenomena? The literature that equates complexity with complicatedness and the related computational models of complexity are not concerned with emergence because they are not looking to explain the novel phenomena that emerge from the actions and interactions of component agents but to make better sense of the networks of connections between agents and communications in the law system. Second, is the scholar trying to better explain the subject of the research, or is their central insight the unknowability of certain phenomena? Any approach that looks to make sense of the workings of the law system, or to propose reform of the system, looks to restricted (or modern) writings on complexity, whereas the literature that points to the limits of our knowledge, and the ethical responsibilities of lawyers and law academics, looks to the general (or postmodern) scholarship.

Depending on the answers we get to these two questions, we can position scholarly materials on complexity theory and law relative to one another on a plot (Figure 1.1). This approach allows us to account for the tendency of individual contributions to contain characteristics of multiple typologies and avoids the reductionist, closed nature of a grid.

Publications in the first (Modern/ Emergence) quadrant understand the law as an abstract (but real) entity that results from the interactions of law agents and other actors and looks to explain the workings of the system and how it can be improved, with a particular focus on law reform in relation to the regulation of other complex systems like the environment. Work in the second (Modern/Non-Emergence) quadrant is interested in explaining the networks of relationships between law actors or law communications, such as court judgments and legislative acts, but is not concerned with emergent properties (such as the abstract notion of the 'the law' system) that develop through the interactions of lawmaking actors. The third (Post-Modern/Non-Emergence) part of the literature focuses on the lack of certainty in our knowledge of complicated systems of networked relationships, and what they tell us about epistemic humility. It is essentially an argument for accepting the limits of our knowledge about the workings of law. Finally, there is the (Post-Modern/Emergence) work, which is interested in our ethical responsibilities as practitioners, regulators and scholars when studying the law system. Whilst all writings (and all four approaches) use the term *complexity*, it is clear they are using it in different ways, for different purposes, and those engaging with the literature on complexity should not make the mistake of trying to read the corpus of material on complexity theory and law (including that reflected in this collection) as a unified body of work of scholarship that shares

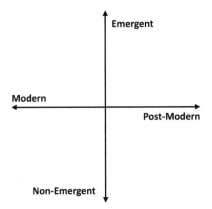

Figure 1.1 A plot-based typology of complexity

the same methodology. Instead, we recommend the reader utilise the typology of complexity theory thinking implied by the plot-based typology of complexity to anchor their engagement with the collection of approaches contained within complexity theory thinking.

The structure of the book

Following this initial chapter is an essay by J. B. Ruhl and Dan Katz on 'Mapping Law's Complexity with "Legal Maps"'. Ruhl and Katz argue that whilst scholars have begun to use complexity science to examine descriptive and normative questions about the law system, there is a need to move to a more empirical phase to influence the practice of law. They argue that law scholars can draw on the wealth of legal materials, such as court judgments, regulations and statutes, to analyse the effects of introducing and removing materials, allowing policymakers to speculate about the implications of proposed initiatives. In effect, Ruhl and Katz propose that lawyers create a mechanism for monitoring legal complexity through the device of 'Legal Maps' to deal with the ever-increasing complexity in the law.

The remaining essays in the collection are organised along traditional subject lines within the discipline of law. The reason for this is that most law scholars work within disciplinary subjects and *inter alia* identify themselves primarily by discipline – as constitutional and administrative lawyers, international lawyers, business lawyers and legal philosophers operating from a particular theoretical perspective.

Section II looks at Complexity and the State: Public Law and Policy. The chapters here examine the implications from complexity theory for our understanding of the formulation and implementation of public policy, and the utility of complexity theory as a tool to understand discussions concerning constitutions.

Neville Harris's chapter, 'Complexity: Knowing It, Measuring It, Assessing It', examines the efforts to simplify the taxation and social security systems in the United Kingdom. The perception of complexity has prompted efforts at simplification that aim to counter complexity either by managing it better, for example, through information technology systems, or by reducing complexity, by streamlining the legal and administrative frameworks. Harris argues that complexity is something that can be tested and that the taxation and social security systems are empirically complex, but he is sceptical about the degree to which complexity can be mapped, arguing that there are limits to how much we can know about law's complexity. His chapter concludes by shifting the focus to the legitimacy deficit that flows from the fact of complicated regulatory frameworks that are difficult for non-experts to understand and navigate.

Thomas Webb's chapter, 'Asylum and Complexity: The Vulnerable Identity of Law as a Complex System', examines the how complexity asks us to make sense of the administrative justice processes around refugee applications through the notion of systemic vulnerability and identity. Vulnerability is not only a feature of human existence but is also the existence of social systems. Just as crossing the critical juncture of vulnerability, between the ever-present possibility of vulnerability and the realisation of that possibility, will be severely detrimental to a human, so, too, will it harm systems. The harm to systems lies in the risk to their identity. Social assemblages (DeLanda, 2006) – a term which encompasses both humans and larger collections of individuals and processes – only exist because they can define themselves relative to the environment through a process of relational differentiation. Anything which undermines this ability to define, to establish an identity, produces exclusion, and, thus, dedifferentiation, which is tantamount to system death. By exploring how complexity theory reacts to the concepts of identity and vulnerability, the chapter asks the reader to consider the implications for individuals and larger social assemblages of procedures and policies that exclude individuals from social processes, such as asylum application procedures, and thus prevent them from establishing an identity. It is suggested that, without remedial measures, the risk to law is a loss of legitimacy and thus identity.

Section III, Complexity Beyond the State: Human Rights and International Law, examines the ways in which complexity theory can inform our understanding of the nature and function of the international law system, with a focus on the way in which human rights function in the complex environment of world politics. The three essays here all look, in different ways, to emergence to enhance our understanding of the doctrine and practice of international law, including human rights.

The opening essay by Steven Wheatley, 'Explaining Change in the United Nations System: The Curious Status of Security Council Resolution 80 (1950)', looks to complexity theory to explain change in the international law system. Taking as its case study the alteration in the voting procedures in the Security Council that occurred without formal amendment of the UN Charter, the chapter relies on the central insights from complexity – of emergence and evolution, path dependency and change and the power of events – to explain the change in UN law. The UN system evolved as member states responded to the 'empty chair' policy of the Soviet Union. Whilst a change in the plain meaning of Charter provisions can be explained by the role of subsequent practice in the interpretation of treaties, there remains the problem of the status of the first resolution adopted under the 'new' procedure, here Resolution 80 (1950), which was not adopted in accordance with the old (literal) rule, requiring the positive support of the 'P5', but nor could there be a new pattern of practice, meaning that only the absence of a veto was needed. Wheatley argues that to make sense of innovations in regulatory practices of complex systems, like the United Nations, we have to foreground the factor of time. By looking to explain change within a timeframe, we can explain how an innovation practice like Security Council resolution 80 (1950) can result in a change of understanding in a regulatory system, as part of a new pattern of practice.

Dimitrios Tsarapatsanis's 'The "Consensus Approach" of the European Court of Human Rights as a Rational Response to Complexity' relies on complexity theory to provide a defence for the consensus approach of the European Court of Human Rights, whereby the court looks to the practice of states to explain the meaning of the convention rights. Tsarapatsanis defends his practice-based account by relying on an argument for practical reasoning in the real-world conditions faced by judges on the Court of Human Rights and the limited amount of time a judge can give to a single case. Rather than substitute her own subjective position on human rights, Tsarapatsanis argues that the judge should look to the actual human rights practice of states as a defensible reasoning strategy in non-ideal conditions, concluding that the emergent position of states parties, reflected in the consensus of states parties, gives a non-ideal interpretation of the human rights treaty. He concludes that we should see the 'consensus approach' as a collective intelligence device that exploits the presence of patterns of emergent solutions.

Anna Marie Brennan looks to complexity theory to make the case for holding non-state actors responsible for crimes committed by terrorist groups under international criminal law, given the legal paradigm of individual responsibility. In her chapter, 'Prospects for Prosecuting Non-State Armed Groups under International Criminal Law: Perspectives from Complexity Theory', she argues that if the whole group commits an international crime, the focus of responsibility should be the group and not the individual members. The work challenges the 'command and control' paradigm of the laws of war and examines the ways we can think of the actions of non-state actors as an emergent property of the activities of the group. Given that the commission of international crimes by non-state actors is often the result of the policies and practices of the group, it makes no sense, she argues, to focus on the individual and proposes that we align the moral responsibility of the group with the practice of international criminal law.

The fourth section on Complexity and Business and Finance Regulation examines the way in which complexity theory can inform our understandings of the task of regulating dynamic and complex business and financial activities at national, regional and global scales. The focus here is often on establishing better models of regulation following the financial crisis of 2008, with policymakers looking to complexity to make sense of the requirements for effective regulation of the highly interdependent and interconnected architecture of the global financial system.

Mark Chinen's chapter, 'Governing Complexity', turns to complexity theory to understand financial regulation in the aftermath of the 2008 financial crisis. Chinen sees international finance and the global economy as complex systems, with networks of heterogeneous agents acting and interacting within a regulatory space. The adaptive nature of these systems makes legal regulation difficult, given the limitations in forecasting the future behaviours of financial systems, or the implications of regulatory interventions, but, he contends, these challenges should not prevent policymakers from engaging in financial governance planning, forecasting or attempting to control systems of international finance and the global economy. Chinen argues for the development of new strategies in which legal regulation co-adapts and co-evolves with the complex systems of international finance and the global economy to promote systemic stability and resilience in international financial systems. But it remains for policymakers to establish the rules for international finance and the global economy. Complexity theory, for Chenin, does not dictate any set of procedural and substantive values for the governance of complex adaptive systems but focuses on participation, accountability, effectiveness, responsiveness and fairness.

The chapter by Michael Leach, 'Complex Regulatory Space and Banking', examines the implications for law and regulation of understanding banking as a complex system. For Leach, banking is quintessentially a complex system, even in its most basic form of deposit collecting and lending, and his primary concern is the ways that law can regulate complex systems like banking and finance. This chapter outlines a sketch of the regulatory space of a simplified, but still complex, banking system to highlight how we might understand the role of complexity in legal regulation. Leach concludes with a preliminary evaluation of the utility of blending complexity theory with regulatory theory to explore the complex space of banking regulation.

Jamie Murray's chapter, 'Regulating for Ecological Resilience: A New Agenda for Financial Regulation', outlines the fundamental shifts that have taken place in the financial system in the years since the global financial crisis, with a transformed understanding of risk and the development of a new 'macroprudential' approach to regulating systemic risk in complex financial systems. The work argues against a financial regulation centred on systemic risk and consequent regulation for engineering resilience, and in favour of a complexity theory understanding of ecological resilience should now become central to regulating complex financial systems. The chapter explores the concept of ecological resilience and how financial regulation could seek to regulate complex financial systems. In doing so, it draws on established complexity jurisprudence that has developed in relation to the problematics of governing for ecological resilience in adaptive management, assisted self-organisation and reflexive regulation self-management. Taking both an understanding of ecological resilience to complex financial systems and an understanding of the complexity jurisprudence for regulating complex systems for ecological resilience, the chapter sets out a new agenda for financial regulation.

The final section on Complexity and the Ethics of Law and Legal Practice discusses the challenges facing legal practitioners, academics and students in view of their complex operative context.

Lucy Finchett-Maddock's chapter, 'Nonlinearity, Autonomy and Resistant Law', focuses on social centres – an emergent corner of social organisation that exists outside of formal legal structures that are reliant on an absolute understanding of time and space as essentially linear. Finchett-Maddock takes the example of these centres as a case study to examine the implications of nonlinearity for how lawyers, and society in general, think about and respond to law and legal complexity. The chapter is especially concerned with the tension between the reality of social existence as essentially complex, nonlinear and emergent, as compared with the constructed reality of law as largely determinate and thus predictable. She argues that it has not always been the case that legal and social realities were so at odds with one another. Instead, Finchett-Maddock contends that, whereas law originally emerged as a product of communal interaction, the increasing influence of private property and the protection of rights in it over time, forced law – and by implication, our understanding of law – into a more linear, spatially limited form. The consequences of the dominance of an absolute conception of linear time have been to deny the existence of alternative forms of social organisation, for example decentralised, leaderless, emergent networks. By demonstrating the viability of social centres as a form of organisation not defined by this linearity, but nonetheless quite capable of existing coherently, Finchett-Maddock seeks to return legitimacy to these alternative, autonomously emerging forms of social organisation.

Minka Woermann's chapter, 'Complexity and the Normativity of Law', commences with an argument for why ethics is critical to any serious engagement with complexity – how models are indispensable for rendering complex systems meaningful. However, these models are necessarily limited, exclusionary and the outcome of normative evaluations. Ethical considerations are, therefore, ever present. Acknowledging the ethics of complexity requires constant and critical engagement with the status and implications of our models. The implications of these observations are explored in the second part of the chapter in terms of the constant reinterpretation, establishment, implementation, policing and transformation of law. For Woermann, law engages in this reinterpretation as an organisationally open, yet operationally closed system – though this is not a simplistic dichotomy. The law maintains its own processes, while also actively engaging in the life of the social environment. In this way, the complex legal system witnesses the entry of the environment into the system as an integral part of its sustained existence, rather than the mere re-entry of the system's own internal construction of the environment as seen in autopoietic constructions. This process, Woermann reasons, means that we can never achieve a perfect model of law once we accept that the world is inherently, irreducibly and uncontrollably complex, and that law's efforts to understand it are co-produced by that environment. Instead, the codification of law is needed to produce legally useful understandings in the moment, accepting that it will not, indeed cannot, account for all the complexities of the social interactions which law regulates and can aspire only to producing 'just' outcomes

In the final chapter 'Regulating the Practise of Practice: On Agency and Entropy in Legal Ethics', Julian Webb, argues that the regulation of legal ethics operates through a relatively stable and clearly bounded system of rules and principles. It thus provides an interesting and potentially useful model for considering the nature of 'ruleness' and the ways in which a system is at risk of regulative 'entropy', that is, the decay of rule-described behaviour and hence of the predictive value of the rules ascribed. He argues that legal formalist and legal positivist accounts remain insufficiently sensitive to the significant and complex operation of agency in multi-agent systems. Webb offers an alternative representation in his chapter arguing that we should see 'ethical' practice as a process of agentic (i.e. self-organising and self-regulating) 'playing with the rules', which serves to normalise minimal consistency and perhaps even inconsistency, between informal (cultural) and formal norms of practice. The chapter concludes by considering ways in which insights from complex systems theory might help us design systems of regulation that are negatively entropic and better able to impede ethical fading and creative compliance.

Bibliography

Arthur, W. Brian (1994), *Increasing Returns and Path Dependence in the Economy* (Ann Arbor: University of Michigan Press).

Arthur, W. Brian (2014), *Complexity & the Economy* (Oxford: Oxford University Press).

Bankowski, Zenon (1996), 'How Does It Feel to Be on Your Own? The Person in the Sight of Autopoiesis', in David Nelken (ed.) *Law as Communication* (Aldershot: Dartmouth Publishing).

Byrne, David (1998), *Complexity Theory and the Social Sciences: An Introduction* (London: Routledge).

Byrne, David and Gillian Callaghan (2013), *Complexity Theory & the Social Sciences* (Abingdon: Routledge).

Capra, Fritjof (2016), *The Systems View of Life: A Unifying Vision* (Cambridge: Cambridge University Press).

Capri, Fritjof and Ugo Mattei (2015), *An Ecology of Law: Towards a Legal System in Tune with Nature and Community* (Oakland: Berratt Koehler).

Carline, Anna and Jamie Murray (2018), 'Reconceptualising On-Street Sex Work as a Complex Affective Social Assemblage', in Sharron A. FitzGerald and Kathryn McGarry (eds.) *Realising Justice for Sex Workers: An Agenda for Change* (London: Rowman & Littlefield).

Cilliers, Paul (1995), 'Postmodern Knowledge and Complexity (or Why Anything Does Not Go)', 14(3) *South African Journal of Philosophy* 124.

Cilliers, Paul (1998), *Complexity and Postmodernism: Understanding Complex Systems* (London: Routledge).

Cilliers, Paul (2001), 'Boundaries, Hierarchies and Networks in Complex Systems', 5(2) *International Journal of Innovation Management* 135.

Cilliers, Paul (2004), 'Complexity, Ethics and Justice', 5 *Tijdschrift voor Humanistiek* 19.

Coveney, Peter and Roger Highfield (1996), *Frontiers of Complexity: The Search for Order in a Chaotic World* (London: Faber & Faber).

de Villiers, Tanya and Paul Cilliers (2010), 'The Complex "I": The Formation of Identity in Complex Systems', in Paul Cilliers and Rika Preiser (eds.) *Complexity, Difference and Identity (Issues in Business Ethics, Volume 26)* (London: Springer).

DeLanda, Manuel (2006), *A New Philosophy of Society: Assemblage Theory and Social Complexity* (London: Continuum).

Deleuze, Gilles and Felix Guattari (1987), *A Thousand Plateaus* (Minneapolis, MI: University of Minnesota Press).

Freeman, Michael (2014), *Lloyd's Introduction to Jurisprudence* (London: Sweet & Maxwell, 9th ed.).

Geyer, Robert and Samir Rihani (2010), *Complexity and Public Policy: A New Approach to Twenty-First Century Policy and Society* (London: Routledge).

Hathaway, Oona A. (2001), 'Path Dependence in the Law: The Course and Pattern of Legal Change in a Common Law System', 86(January) *Iowa Law Review* 601.

Heylighen, Francis, Paul Cilliers and Carlos Gershenson (2007), 'Philosophy and Complexity', in Jan Bogg and Robert Geyer (eds.) *Complexity, Science and Society* (Oxford: Radcliffe Pub.), 117.

Holland, John H. (1995), *Hidden Order: How Adaptation Builds Complexity* (Reading, MA: Helix Books).

Holland, John H. (1998), *Emergence: From Chaos to Order* (Cambridge, MA: Perseus Books).

Humphreys, Paul (2016), *Emergence: A Philosophical Account* (Oxford: Oxford University Press).

Kades, Eric (1996–1997), 'The Laws of Complexity and the Complexity of Laws: The Implications of Computational Complexity Theory for the Law', 49 *Rutgers Law Review* 403.

Kauffman, Stuart (1993), *The Origins of Order: Self-Organization and Selection in Evolution* (New York: Oxford University Press).

Kauffman, Stuart (1996), *At Home in the Universe* (New York: Oxford University Press).

King, Michael (1993), 'The "Truth" About Autopoiesis', 20(2) *Journal of Law and Society* 218.

Kobus, Maraism (2014), *Translation Theory and Development Studies: A Complexity Approach* (New York: Routledge).

Lewin, Roger (1992), *Complexity: Life at the Edge of Chaos* (Oxford: Macmillan Maxwell International).

Lorenz, Edward N. (1993), *The Essence of Chaos* (Seattle: Washington University Press).

Luhmann, Niklas (1992a), 'Operational Closure and Structural Coupling: The Differentiation of the Legal System', 13 *Cardozo Law Review* 1419.

Luhmann, Niklas (1992b) 'Some Problems With <<Reflexive Law>>', in Gunther Teubner and Alberto Febbrajo (eds.) *State, Law, and Economy as Autopoietic Systems: Regulation and Autonomy in a New Perspective (European Yearbook in the Sociology of Law, Double Issue)* (Milan: Giuffrè).

Luhmann, Niklas (2004), *Law as a Social System*, trans. Klaus A. Ziegert (Oxford: Oxford University Press).

Maturana, Humberto R. and Francisco J. Varela (1987), *The Tree of Knowledge: The Biological Roots of Human Understanding* (Boston: Shambhala Publications).

Miller, John H. and Scott E. Page (2007), *Complex Adaptive Systems: An Introduction to Computational Models of Social Life* (Princeton, NJ: Princeton University Press).

Mitchell, Melanie (2009), *Complexity: A Guided Tour* (Oxford: Oxford University Press).

Mitleton-Kelly, Eve (2003), 'Ten Principles of Complexity and Enabling Infrastructures', in Eve Mitleton-Kelly (ed.) *Complex Systems and Evolutionary Perspectives on Organisations: The Application of Complexity Theory to Organisations* (London: Pergamon Press), 23.

Morin, Edgar (2007), 'Restricted Complexity, General Complexity', in Carlos Gershenson, Diederik Aerts and Bruce Edmonds (eds.) *Worldviews, Science and Us: Philosophy and Complexity* (Singapore: World Scientific), Chapter 5.

Murphy, John T. (2017), 'Complexity Theory', *Oxford Bibliographies*. <www.oxfordbibliographies.com>, accessed 23 Novermber 2017.

Murray, Jamie (2006), 'Nome Law: Deleuze and Guattari on the Emergence of Law', 19 *International Journal for the Semiotics of Law* 127.

Murray, Jamie (2008), 'Complexity Theory and Socio-Legal Studies, Coda: Liverpool Law', 29 *Liverpool Law Review* 227.

Paterson, John (1996), 'Who Is Zenon Bankowski Talking to? The Person in Sight of Autopoiesis', in David Nelken (ed.) *Law as Communication* (Aldershot: Dartmouth Publishing).

Philippopoulos-Mihalopoulos, Andreas and Thomas E. Webb (2015), 'Vulnerable Bodies, Vulnerable Systems', 11(4) *International Journal of Law in Context* 444.

Prigogine, Ilya and Isabelle Stengers (1984), *Order out of Chaos: Man's New Dialogue with Nature* (London: Flamingo).

Richardson, Kurt and Paul Cilliers (2001), 'What Is Complexity Science? A View from Different Directions', 3(1) *Emergence* 5.

Richardson, Kurt, Paul Cilliers, and Michael Lissack (2001), 'Complexity Science: A "Gray" Science for the "Stuff in Between"', 3(2) *Emergence* 6.

Ruhl, J. B. (1996a), 'The Fitness of Law: Using Complexity Theory to Describe the Evolution of Law and Society and Its Practical Meaning For Democracy', 49 *Vanderbilt Law Review* 1407.

Ruhl, J. B. (1996b), 'Complexity Theory as a Paradigm for the Dynamical Law-and-Society System: A Wake-Up Call for Legal Reductionism and the Modern Administrative State', 45(March) *Duke Law Journal* 849.

Ruhl, J. B. (1997), 'Thinking of Environmental Law as a Complex Adaptive System: How to Clean Up the Environment by Making a Mess of Environmental Law', 34(Winter) *Houston Law Review* 933.

Ruhl, J. B. (2008), 'Law's Complexity: A Primer', 24 *Georgia State University Law Review* 885.

Ruhl, J. B. and James Salzman (2002), 'Regulatory Traffic Jams', 2(2) *Wyoming Law Review* 253.

Ruhl, J. B. and James Salzman (2003), 'Mozart and the Red Queen: The Problem of Regulatory Accretion in the Administrative State', 91 *Georgetown Law Journal* 757.

Sawyer, Robert K. (2005), *Social Emergence: Societies as Complex Systems* (Cambridge: Cambridge University Press).

Schuck, Peter H. (1992), 'Legal Complexity: Some Causes, Consequences, and Cures', 42 *Duke Law Journal* 1.

Sornette, Didier (2017), *Why Stock Markets Crash: Critical Events in Complex Financial Systems* (Oxford: Princeton University Press).

Teubner, Gunther (1988), 'Evolution of Autopoietic Law', in Gunther Teubner (ed.) *Autopoietic Law: A New Approach to Law and Society* (New York and Berlin: Walter de Gruyter Publishing), 217.

Teubner, Gunther (1993), *Law as an Autopoietic System* (Oxford: Wiley-Blackwell).

Teubner, Gunther (2012), *Constitutional Fragments: Societal Constitutionalism and Globalization*, trans. Gareth Norbury (Oxford: Oxford University Press).

Urry, John (2003), *Global Complexity* (Oxford: Polity Press).

Vermeule, Adrian (2012), *The System of the Constitution* (New York: Oxford University Press).

Walby, Sylvia (2007), 'Complexity Theory, Systems Theory, and Multiple Intersecting Social Inequalities', 37 *Philosophy of the Social Sciences* 449.

Waldrop, Mitchell M. (1994), *Complexity: The Emerging Science at the Edge of Order and Chaos* (London: Penguin Books, 2nd ed.).

Webb, Julian (2005), 'Law, Ethics, and Complexity: Complexity Theory and the Normative Reconstruction of Law', 52 *Cleveland State Law Review* 227.

Webb, Thomas E. (2013), 'Exploring System Boundaries', 24 *Law and Critique* 131.

Webb, Thomas E. (2014), 'Tracing an Outline of Legal Complexity', 27(4) *Ratio Juris* 477.

Webb, Thomas E. (2015), 'Critical Legal Studies and a Complexity Approach: Some Initial Observations for Law and Policy', in Robert Geyer and Paul Cairney (eds.) *Handbook on Complexity and Public Policy* (Cheltenham: Edward Elgard).

Wheatley, Steven M. (2016), 'The Emergence of New States in International Law: The Insights from Complexity Theory', 15(3) *Chinese Journal of International Law* 579.

Wheatley, Steven (2018), *The Idea of International Human Rights Law* (Oxford: Oxford University Press).

Woermann, Minka (2016), *Bridging Complexity & Post-Structuralism: Insights & implications* (Berne: Springer).

Woermann, Minka and Paul Cilliers (2012), 'The Ethics of Complexity and the Complexity of Ethics', 31 *South African Journal of Philosophy* 447.

2 Mapping law's complexity with "Legal Maps"

J. B. Ruhl and Daniel M. Katz*

Introduction

As intuitive as it is to any lawyer that the law and the legal systems administering it are complex, getting a handle on exactly what that means and what to do about it is no simple matter. First, one needs a theoretical foundation to describe complexity in terms relevant to law and legal systems, however we define them. What is *legal* complexity, and what attributes go into making legal systems complex? Then one must develop metrics and methods to measure and monitor those attributes in the legal system, to determine how complex the legal system *is*. Armed with such data and findings, legal theorists, politicians, and citizens can begin an evidence-based debate regarding how complex the law *ought* to be. And, if it were determined that the law is too complex or not complex enough, it would be useful to have the means to adjust and manage the law's complexity. Of course, none of these undertakings is a small task.

Legal scholars have begun to employ complexity science (also known as complexity theory) as one lens through which to probe these descriptive and normative questions about law's complexity (Ruhl et al., 2017; Ruhl, 2008). The focus of complexity science is complex adaptive systems, systems "in which large networks of components with no central control and simple rules of operation give rise to complex collective behaviour, sophisticated information processing, and adaptation via learning or evolution" (Mitchell, 2009, p. 13). Legal scholars using this discipline to study law's complexity have thus far focused primarily on describing legal systems as complex adaptive systems to understand the origins of legal complexity and explore its theoretical and normative implications. But the theory of legal complexity will remain stuck in theory until it moves to the empirical phase of study. In short, we cannot put the theory of legal complexity to work without robust empirical tools.

To put this problem in practical terms, consider the US Tax Code, which is widely considered to be notoriously complex. But exactly how complex *is* the Tax Code, and how complex *ought* it be compared to, say, securities laws or environmental protection laws? One reason it is difficult to approach these questions is that the metrics often used for claiming the Tax Code is complex turn the problem on its head. The Tax Code is not complex because of its costs of compliance, difficult readability, number of rates and special provisions; the complexity of tax compliance software; or so on. Rather, the Tax Code imposes costly compliance burdens, is difficult to read, has lots of rates and special provisions, and poses a challenge to software developers *because it is complex.* These attributes are *consequences* of Tax Code complexity, not its *causes.*

The Tax Code in this respect is a microcosm of legal complexity in general and an example of how little we understand its causes, consequences, and cures. The same questions could be asked of environmental law, securities law, health law, and dozens of other legal fields; answers would be wanting in those fields, as well (McGarity, 2013). In short, there is very little empirically robust understanding of the causes of legal complexity, which reduces the normative debate over legal complexity and how to "simplify" law largely to scholarly theory and political rhetoric (Katz and Bommarito, 2014).

Our claim is that complexity science, with its origins in physics and ecology, provides a useful framework for studying *legal* complexity. Most lawyers are likely unfamiliar with complexity science, yet complexity science has had tremendous influence in other social science disciplines, such as economics (Beinhocker, 2006) and political science (Page, 2006), and has been applied in the study of a wide variety of policy challenges including terrorist networks (Bousquet, 2012) and health care (Bar-Yam et al., 2012). This does not mean complexity science will necessarily have the same utility when applied to legal systems, but if one believes legal complexity is a concern, it is probably worth exploring whether anything can be gained from applying a scientific discipline singularly devoted to the study of complexity in social and physical systems.

At the outset, we appreciate that it is impossible to open the door to the question of legal system complexity without confronting the age-old puzzle – what is the legal system? We do not have any intention, however, of going down that jurisprudential rabbit hole. Indeed, as we explore in the following, bringing complexity science to law makes more apparent than ever that there is no definitive answer to that question. Legal systems are designed at their core to regulate and interact with other social systems, such as the financial system and healthcare system. Those systems, in turn, interact with each other and with the legal system – they all co-evolve as a "system of systems" (Ruhl et al., 2017). To facilitate political and social discourse, it makes sense for societies to classify social systems and assign different actors to particular systems, such as banks to the financial system, hospitals to the health care system, and courts to the legal system, based on what might be called "centres of gravity." But it is not as if banks have nothing to do with the legal system, or that courts have nothing to do with the financial system – tentacles of influence reach across the artificially constructed boundaries. And some actors have no obvious centre of gravity – is a bail bond company part of the legal system or the financial system? But the point of bringing complexity science to law is not to obfuscate conceptions of the legal system or other social systems but, rather, to illuminate their respective centres of gravity and pathways of cross-influence.

Building on that theme, this chapter explores the theoretical and empirical dimensions of legal complexity in terms we hope are accessible and of practical value to lawyers and legal scholars not already familiar with complexity science. We begin by reviewing the core concepts of complexity science and legal scholars' application of these theories to the law. There have been three major themes in this body of scholarship. First, a descriptive body of work has focused on mapping complexity science concepts onto legal systems to enable explanation of legal systems as complex adaptive systems. Second, a prescriptive thrust has moved from mapping concepts towards developing principles for structural design and normatively acceptable operation of legal systems given their complex adaptive system properties. Finally, an ethical focus in the literature explores what it means to be an actor in a complex legal system. The chapter then shifts to the empirical front, identifying potentially useful metrics and methods for studying legal complexity.

The chapter closes with a proposal for monitoring legal complexity over time by conceptualising what we call Legal Maps – a multilayered, Google Maps–style active representation of the legal system network at work. Legal Maps would link together layers of legal domains horizontally and vertically, displaying cross-references within and between different layers. For example, all cross-references between a statute's provisions would be linked, then all references between that statute's provisions and provisions of other statutes would be linked, and then all references to those provisions made in agency regulations and court decisions would be linked, and so on to the edges of the defined "legal system." Once constructed, new cross-references and new provisions (as well as repeals and revisions) could be integrated into the network in real time, thus allowing observation of the network as it evolves. Hypothetical changes to the system, such as a proposed repeal of a major law, then could be tested ex ante to gain a fuller understanding of the impact to the system as a whole. Legal Maps could also be linked to other social system models, such as of the financial system, to explore how the systems co-evolve. To be sure, no such representation of the legal system exists today, but by all means the data and computational techniques needed to build it do exist and are used extensively in similar applications (Ruhl et al., 2017). The chapter thus establishes an agenda for identifying the empirical questions and methodological approaches ripe for studying complexity in legal systems through Legal Maps.

The complexity science theory of legal complexity

The key premise in applying complexity science to legal systems is that there is a difference between complexity in the sense of "complicatedness" and complexity in the sense of system structure and behaviour. That distinction, which goes to the essence of complexity science theory, is aptly described in a leading text by John Miller and Scott Page (2007, p. 9):

> In a complicated world, the various elements that make up the system maintain a degree of independence from one another. Thus, removing one such element (which reduces the level of complication) does not fundamentally

alter the system's behavior apart from that which directly resulted from the piece that was removed. Complexity arises when the dependencies among the elements become important. In such a system, removing one such element destroys system behavior to an extent that goes well beyond what is embodied by the particular element that is removed.

Few dispute that law is complicated; whether it is complex in the systems context is another matter. To be sure, the complicatedness of law should not be discounted. Law can be vast, dense, vague, and intricate, making compliance a daunting undertaking. Complexity as used for our purposes, however, is getting at something different. Complexity science emphasises the systems effects, studying inter-agent connections and the system-wide effects they produce. In the context of social systems, complexity science offers a different approach from that taken in small-number agent models (such as in bilateral game theory) and large-number agent models (such as the rational actor in law and economics). The problem with these inter-agent modelling approaches is that most economic, political, and social interactions involve moderate numbers of people. Again, Miller and Page offer a concise take on the problem (2007, p. 33):

Most social science models require either very few (typically two) or very many (often an infinity) agents to be tractable. When an agent interacts with only a few other agents, we can usually trace all the potential actions and reactions. When an agent faces an infinity of other agents, we can average out . . . the behaviour of the masses and again find ourselves back in a world that can be easily traced. It is in between these two extremes – when an agent interacts with a moderate number of others – that our traditional analytic tools break down.

In other words, traditional models of inter-agent behaviour do not work well when there are too many interacting agents to fit neatly into bilateral models but not enough agents to ignore idiosyncratic behaviour by averaging out to an infinite-numbers "rational actor" model. Throughout the legal system, agents in legal institutions and instruments interact in ways suggesting that the differences between agents matter. Thus, mean-field approximations do not always capture useful or relevant dynamics. The number of judges, lawyers, agencies, laws, or regulations is neither small nor infinite, and we can find no legal scholarship claiming that the differences between, say, judges or regulations, do not matter.

Complexity science thus is about building models for contexts in which agent heterogeneity and interrelatedness can and usually do influence outcomes. Legal scholars have developed descriptive, prescriptive, and ethical models of what this approach means for law.

Descriptive theories

Legal scholars applying complexity science to legal systems have thus far focused primarily on mapping key concepts of complexity science onto legal systems (Ruhl, 2014). Consider the general definition of a complex adaptive system mentioned earlier: a large network of components, with no central control and simple rules of operation, giving rise to complex collective behaviour, sophisticated information processing, and adaptation via learning or evolution. Anyone with training in law can easily map this framework onto the legal system. The legal system's components comprise a broad diversity of institutions (the organisations of people who make, interpret, and enforce laws) and of instruments (the laws, regulations, cases, and related legal content the institutions produce). These components are interconnected and interactive. Institutions are interconnected through structures and rules such as hierarchies of courts and legislative creation and oversight of agencies; institutions interact in forums such as judicial trials, legislative hearings and debates, and agency rule-makings. The instruments also are interconnected through mechanisms such as code structures, which, in turn, interact through cross-references and other devices.

The highly interconnected architecture of such a system drives the way it behaves over time. An agency adopts a rule, which prompts another agency to enforce a different rule, which leads to litigation before a judge, who issues an opinion overruled by a higher court, which prompts a legislature to enact a new statute, which requires another agency to adopt a rule, and so on. The institutional agents follow procedural rules (e.g., opportunity for public comment), and even the instrumental agents have rules for rules (e.g., canons of statutory construction), but there is no central controller pulling all the strings. There are hierarchies for various institutions (e.g., courts) and instruments (e.g., statues can pre-empt common law). Yet there is no master agent controlling *the system.*

The descriptive branch of legal complexity theory has focused on this kind of mapping exercise to demonstrate the legal system's complexity by examining how each attribute of complex adaptive systems described in complexity science research finds close parallels in legal system structure and behaviour (Ruhl, 2008, pp. 898–901; Cilliers, 1998, pp. 119–23; Webb, 2014). Indeed, Stuart Kauffman, one of the leading thinkers in complexity science since its early development in the 1990s, used the common law as an example of complex adaptive system behaviour (Kauffman, 1995, p. 169). The judiciary's hierarchical structure and practice of stare decisis link courts with courts and opinions with opinions in ways that produce complicated *and* complex (as complexity science defines it) feedback connections (Mitchell, 2011). The "substantive jurisprudence" emerges from this system through a process of gradual development and evolution of doctrine based on bedrock principles, some of which were set down centuries ago. Although one must read the cases to know the common law of, say, property, the common law of property is something more than just the sum of the cases. In the United States, for example, the *Restatement of Property* is more than a case reporter – it is the product of tremendous effort by property law experts working over many decades to synthesise and compress case law into emergent, macro-scale doctrinal themes and structures, as well specific micro-scale rules and principles.

There have been numerous such accounts of complex adaptive system attributes in a broad range of legal systems including administrative law, mediation and alternative dispute resolution, bankruptcy law, environmental law, business law, international law, land-use regulation law, intellectual property law, international development law, regulation of the internet, the law of war, health law, and telecommunications regulation, as well as in more general accounts of legal systems (summarised in Ruhl and Katz, 2015). It is beyond this chapter's scope to articulate all such examples – the point is that these scholarly contributions have established a robust descriptive model of legal systems as complex adaptive systems.

It is appropriate to pause here and ask the critic's question: So what? Accepting for now that the attributes of complex adaptive systems map well onto legal systems, what is the value of having a robust descriptive model of legal systems as complex adaptive systems? The value of such a model is that it changes perspective and leads to new questions. To borrow from how Brian Arthur (2015, p. 2), a leading thinker in applying complexity science to economics, described the impact of complexity science in economics (we replace economics with law):

> [T]his new approach is not just an extension of standard [legal theory], nor does it consist of adding agent-based behavior to standard models. It is a different way of seeing the [legal system]. It gives a different view, one where actions and strategies constantly evolve, where time becomes important, where structures constantly form and re-form, where phenomena appear that are not visible to standard equilibrium analysis, and where a meso-layer between the micro and the macro becomes important.

In other words, the descriptive model of legal systems as complex adaptive systems provides a different perspective on legal systems. Admittedly, thus far the model has been constructed based on intuition, analogy, and example, but that by no means makes it unusual in the world of legal theory. Either you are persuaded on that basis or not, but we will proceed for now on the assumption that there is theoretical coherence to the model. The obvious next question is what to do with it.

Prescriptive theories

If the legal system is a complex adaptive system, how should legal agents and society at large act in such a system? An important point – one that cannot be overemphasised – is that describing the legal system as a complex adaptive system assumes no normative position about complex adaptive systems or legal systems. Instead, describing the legal system as a complex adaptive system is merely an observation about the way the legal system is constructed and behaves. Assuming that as a given, however, the nature of the legal system as a social system means that, unlike complex physical and biological systems, humans have a say in how it is designed and operated. Hence, as legal theorists constructed the descriptive model of legal complexity, they also turned to normative questions about the model's implications for legal system structure and performance.

This inquiry is distinct from the separate but related question of how to design legal systems given that their target regulatory subject is often a complex adaptive system. Therefore, it makes sense to think that the design of legal regimes intended to manage human interaction with ecosystems should consider that property (Cherry, 2007, p. 371). But both sides of the equation must be taken into account. Law itself is a complex adaptive system, and it necessarily influences and is influenced by the systems it is intended to regulate or manage. Hence, a principal concern of legal theorists interested in legal complexity has been to develop some sense of how best to respond to the legal system's complexity, considering that the legal system is just one member of a "system of systems" (Ruhl et al., 2017).

Legal complexity theory has worked on designing legal institutions and instruments that seem to fit well with complex adaptive system attributes. The theoretical premise is not that complexity is necessarily normatively good and should be promoted, but that some structural designs are less likely to disrupt the complexity dynamics of the system and are more likely to work well within the system as a whole and, perhaps as important, to facilitate the legal system's interaction with other complex social systems. The operative principle is that the legal system should be designed with its complexity in mind.

The main thrust of this prescriptive branch of legal complexity theory is a deep scepticism that top-down, centralised regulation can avoid unintended consequences or keep up with the co-evolving systems and that more flexible, decentralised forms of governance fit better with the legal complexity model. For example, administrative law expert Donald Hornstein argues that understanding regulatory law as a complex adaptive system counsels in favour of relying more on the distributed power of states for policy formulation and for making federal administrative agency governance more experimental, adaptive, and collaborative Hornstein, 2005). Similarly, telecommunications law expert Barbara Cherry argues that rapid technological, social, and economic change – systems co-evolving with law – demand a more adaptive governance structure (Cherry, 2008). Cherry also argues that wholesale deregulation as a means of "simplifying" legal regimes can lead to disastrous results due to complex system cascade effects. Instead, building regulatory resilience – the capacity to withstand shocks from technology and other systems – should be the priority.

The thrust of this prescriptive branch of legal complexity theory is to build adaptability and resilience into legal systems to keep pace with co-evolving social, technological, physical, and biological systems. The predominant view among legal complexity theorists is that law cannot deregulate its way there, nor can it command and control its way there (Adler, 2012). There are no easy answers – how to put law's complexity to work will be quite the challenge for legal design, particularly if there are no reliable metrics for assessing how the legal system performs as a complex adaptive system.

Ethical theories

Some legal complexity theorists have gone beyond descriptive and prescriptive accounts of legal system design to examine the ethical implications of viewing law as a complex adaptive system (Cilliers, 1998). As Julian Webb (2005) suggests, one might conclude from the descriptive and prescriptive theories that "we have little choice but to accept that the system will organise and adapt itself in the manner most likely to ensure its survival," and thus, "resistance to law is likely to achieve little or no immediate gain." But Webb offers an alternative to this pessimistic view, arguing that "[c]omplexity . . . emphasizes the distributed nature of power; the inability of any person (or institution) to claim that it exerts control over society." The upshot of this is that "we have to take responsibility for the effects of all our decisions."

Exercising that responsibility, argues Webb, implicates three overarching principles. First, an appreciation of legal complexity confirms not only why "the law delivers justice as much by accident as by design" but also "encourage[s] emancipatory movements to embrace the uncertainty this provides." Second, complexity science reveals the interconnectedness of seemingly self-referentially closed social systems, meaning that "a failure to achieve normative consistency between systems will generate system-conflicts." Last, Webb argues that activating certain ethical values consistent with complex adaptive system behaviour, such as altruism, pluralism, and interdependence, will support the maintenance and development of the legal system. In short, Webb's take on the ethical implications of legal complexity calls for polity-wide responsibility and participation in the legal system and a deep re-examination of fundamental ethical notions of power, rights, and rules. But the question remains: How, exactly, should such ethical principles be operationalised in concert with legal complexity if there is no reliable way of measuring legal complexity?

Measuring legal complexity

The descriptive, prescriptive, and ethical theories of legal complexity rely largely on intuition, analogy, and example for their persuasion. This approach has taken the legal complexity project far, but the path has come to an end. What else is there to say about legal complexity that derives from intuition, analogy, and example? Not much. Now that it is developed, the core theory of legal complexity can be used as a lens to examine different fields of law or legal problems, but this leads to little theoretical advancement. Rather, this technique maps the theory onto author-selected contexts and elaborates on why legal complexity is a useful model for understanding how the discrete legal context is operating. To be sure, it is essential when working out a theory to compare hypotheses to the real world by intuition, analogy, and example. If a theory does not cohere at that level, it is probably not worth pursuing. But we believe it is fair to conclude that the theory of legal complexity has been sufficiently tested at this level to confirm it is worth pursuing further. So, what is the next step in that cause?

As with any posited theory, the next step for legal-complexity theory is to respond to the critic's demand for empirical proof (e.g., prove the Tax Code is too complex). Asking that a theory withstand empirical testing is not an obstructionist demand. Particularly when normative claims are based on a theory, those making the claims should be expected to offer support beyond the mere elegance or intuitive appeal of the theory. If one believes legal complexity imposes constraints on the legal system or, conversely, that it opens up tremendous opportunities, one should want to know when, where, and by how much the complexity activates those conditions. And if one believes legal complexity justifies using adaptive approaches to respond to those constraints, one must have answers for when, where, and through what means should the law be adaptive. If the quality and quantity of legal complexity matters for either of those questions, how are the quality and quantity of legal complexity measured and described? These are questions one should naturally ask of legal-complexity theorists making normative claims about what the theory means for legal-system design and behaviour. If the theory is to produce answers to such questions, legal-complexity theorists must initiate an empirical phase of study.

The first step in such an undertaking is to design and field test a set of relevant system metrics and methods to measure legal complexity. Unfortunately, complexity science has arrived at no standard toolbox of metrics or methods but a synthesis of various accounts by complexity scientists, and by the few legal scholars that have explored legal complexity empirics, suggests several dominant themes we believe will be most useful for studying legal complexity. We divide these into a system-structure set and a system-behaviour set.

Complexity and system structure

Agents and agent sets: composition, classification, and diversity

The "ecosystem" archetype provides a useful descriptor of the rich and complex dynamics underlying law's evolution. However, in order to advance such statements beyond mere metaphor, it is necessary to retrofit and apply rigorous tools from appropriate intellectual domains such as systems ecology, physics, biology, and complex systems. One threshold step in the process of characterising the broader landscape is to identify all potential agents whose individual behaviour might impact the collective behaviour of the broader system.

The law, like other complex adaptive systems, exhibits a diversity of agents and agent sets. The set of all potential agents is vast and includes institutions (i.e., courts, legislatures, administrative agencies, corporations, public interest organisations, etc.), individual actors (i.e., judges, legislators, lobbyist, bureaucrats, etc.), and the law itself (i.e., rules, adjudications, decisions, etc.). Individual agents often belong to agent sets and those agent sets can themselves be nested within broader agent sets. The nested nature of agent sets is an important complication that must be confronted in the process of deconstructing and measuring legal complexity. At the same time, such theoretic representations of the respective agent sets can be a useful manner through which to begin exploring the operation and dynamics of the respective complex adaptive system.

With the basic identification step in place, an agent-centric metric thus would classify the respective legal agent sets by segmenting them and placing them in a broader taxonomy of agents and agent sets. This classification step is itself complex because it requires the development of categories whose boundaries are typically difficult to cleanly segment. The set of agents may be (and often is) quite diverse. A variety of measures can assess the diversity of a particular set of agents and agent sets. Both an absolute and comparative question, an agent-centric diversity measure could illuminate a variety of interesting research questions.

Formal architecture: trees and other formal hierarchies

The sheer number of agents offers just a partial characterisation of the overall complexity of a given complex adaptive system. Agents are connected in a variety of ways, including by formal architectures that serve specific purposes and functions. Formal architecture is an important default proposition for any complex system, helping set some contours of its performance and offering a partial description of its behaviour and topology.

Formal hierarchical architecture is typically represented in a structure known as a "tree." Trees are a well-studied mathematical structure composed of nodes connected by branches. Conceptualised as a graph, a tree is a connected, undirected graph with no simple circuits. Direction only flows one way; each node and branch is associated with a level, with levels starting at the root node and terminating at the leaf nodes. These are important features that distinguish a tree from other graphs (such as those typically studied in network science).

Typically the byproduct of system designers or instantiated by formal rules, tree-based architecture is designed to serve important functions. Those functions might be institutional, or they might serve as a means to help make sense of a given system's complexity. For an institutional example, consider the American federal judiciary. The federal judiciary features a formal hierarchy of judges and judicial staff whose collective behaviour help shape "the path of the law" (Holmes, 1897). The basic formal hierarchy is memorialised in the formal multi-tier structure that begins with federal magistrate judges and terminates with the U.S. Supreme Court. Hierarchies of this sort permeate legal systems.

Network architecture: emergent hierarchies

Complex system architectures take on a variety of forms and complex adaptive systems exhibit multi-scalar hierarchies, organisations, and other structural forms within which the agents are distributed (Boccaletti et al., 2006). As a matter of system evolution, there are two forces typically in constant operation – forces building up hierarchies and forces operating to tear those very hierarchies down. At any given moment, these countervailing dynamics operate to yield different kinds of observed structures.

The tree conception has some important limitations, but many limitations are overcome by considering the interconnection networks that exist between respective objects. As recently noted, "hierarchies emerge and occur widely in self-organising and evolutionary systems, such as food webs (ecological), neural networks (biological), open-source software (technological), and industrial production networks (economic)" (Luo and Magee, 2011, p. 51). This is equally applicable to describing legal systems. Hierarchy is a fundamental feature of legal systems, but the nature of that hierarchy is likely to vary across particular agents and agent sets.

Hierarchies are typically not the byproduct of a random process. Quite the opposite, their forms are the consequence of specific underlying generating dynamics. While hierarchies can be the byproduct of choices by system designers, they more commonly emerge as a result of actions undertaken by agents. Thus, observed system architecture is usually not the function of top-down choices made by a system designer but, rather, is the aggregate byproduct of bottom-up decisions offered by various agents and agent sets. Thus, in addition to the formal legal hierarchies discussed earlier, there are emergent legal hierarchies that develop through a series of micro-choices made by the respective actors. Such emergent hierarchies can operate alongside any nominal hierarchy that might also exist, thereby confounding one's ability to understand the dynamics and predict the behaviour of a given system.

One way to formalise this emergent architecture is through the tools of network science (also known as applied graph theory; Barabasi, 2003; Lazer, 2009; Watts, 2003). Networks consist of nodes that can, for example, in the simple case represent actors, institutions, and documents. The connections between these nodes are represented by edges (bidirectional) or arcs (unidirectional). Such connections can memorialise simple binary {0, 1} connections or can be weighted to represent far more sophisticated types of relationships. Network science is among the fastest-growing fields in all of science and includes scholarship in wide-ranging disciplines including, more recently, law (Fower, 2007; Smith, 2007; Strandburg et al., 2009).

Information storage and computation

Complex adaptive systems store and process information (Haken, 2010). The agents and architecture described earlier play an important role in characterising the operation and flow of information undertaken therein. If trees and networks represent the architecture and agents are the nodes, then information would be the "electrical current" that flows across the respective institutional circuitry. The act of processing the information is computation (broadly construed). This is true whether the complex system is one's cognitive architecture or the operation of a biological or physical system. In this application, legal systems can be conceptualised as computational complex systems – systems that store and process information. As it concerns this storage and processing task, not all computational complex systems are equally complex. Even among otherwise complex systems, there is a spectrum.

Law's complexity is a long-standing social and political issue, and various technologies help lawyers and laypersons confront the sheer volume of information and overall attendant complexity of legal systems. As the saying goes, necessity is the mother of all invention, and the complexity of law has necessitated the development of legal information technology as a rational and necessary response to law's complexity. Even in a pre-computing era, the tradition of compilation and synthesis of legal doctrine in legal treatises can be thought of as an early form of legal information technology allowing various end users to better understand the law in a given area. In addition, various indexing systems and other legal taxonomies – such as the West Key Number System discussed previously – also represent early forms of legal technology. Again, their use allowed an end user to more quickly assemble the relevant information content contained therein.

Complexity and system behaviour

Interest in legal complexity is in part motivated by interest in the behaviour of the legal system and its *predictability*. Some basic level of predictability is an obvious and straightforward normative goal for any legal system. The difficulty arises in instances where predictability conflicts with other normative goals, such as fairness and various efforts to ensure that the law evolves to take account of changes in broader society. In the aggregate, various efforts to particularise the law to better distinguish various classes of conduct are one important source of legal complexity. In this context, complexity arguably serves a positive normative purpose. However, each increase in complexity can have unintended consequences, including making the overall legal system less transparent and less understandable to laypersons.

As highlighted herein and across the literature, legal systems are complex adaptive systems. Our desire to predict system-level behaviour must be tempered by the realities that are attendant in working with complex adaptive systems. There are real limits in our ability to make forecasts. In the general case discussing the relationship between system complexity and prediction, scholars highlight the distinction between two famous complex systems – tides and weather. Both feature fairly complex dynamics, but from a prediction standpoint, tides are easy and weather is hard (in some cases, perhaps, impossible).

Taken as a whole, and in many specific instances, legal systems exhibit properties that make them behave more like weather and less like tides, which run on well-known schedules. However, this is merely conjecture (albeit, perhaps, well-founded and intuitive conjecture). To evaluate that proposition more robustly requires greater scientific exploration, characterisation, and measurement of legal systems and their complexity using appropriate tools.

Despite these real limitations along a variety of dimensions, it is possible to make forecasts about the future behaviour of complex systems. Indeed, a core portion of lawyers' professional judgment includes forecasting uncertain legal environments. In certain instances, complexity makes this task more challenging. The tools used by complex-systems scholars such as networks, trees, and computation and terms such as *emergence*, *path-dependence*, *feedback*, and *diffusion* can help those embedded in an environment better understand (and hopefully predict) relevant behaviour.

Networks, trees, diffusion, and system behaviour

Complex adaptive systems exhibit information processing, feedback, and feedforward mechanisms producing structural interconnectedness and interdependence between agents throughout the system itself. As discussed earlier, a structural metric would construct a model of the legal system's networked agents and structure, showing all interconnections and interdependencies, and measure the strengths and directions of information feedback and feedforward channels.

In the context of law, we are interested in the social spread of ideas and paradigms. The development of the common law, for example, is a distributed process. No individual jurist, academic, or lawyer is able to unilaterally impose his or her specific vision of what the law is or what the law ought to be. It is a process of prestige and persuasion – where prestige is a function of one's structural position within a network and persuasion is about one's ability to use legal argumentation to convince his or her colleagues of the merit of their argument (Baum, 2006). We are interested in the origin, persistence, and ultimate success of various legal ideas, doctrines, and paradigms. In law, as in many other pursuits, there exists a marketplace for ideas where most ideas do not persist. However, some do. An important question is, Why do some persist and others fade?

With a reasonable understanding of the current and future structure in place, it is possible to study the flow, spread, and success or failure of legal ideas and paradigms using the tools of social epidemiology. Among other things, social epidemiology and social physics is the study of how various social structures impact the spread or persistence of various ideas (Pentland, 2014). Like very contagious pathogens, transformative ideas tend to win out while poor ideas rarely catch fire (Blackmore, 2000; Cotter, 2005). However, there is a large intermediate class of ideas whose fate can be said to be contingent. If those without social authority do not embrace the idea, it will not persist (even if it is superior to its alternatives). For those classes of ideas, structure matters.

Emergence, feedback, and system prediction

With all we currently know and all we might know about the operation of any given system, it is all too tempting to overstate our ability to forecast its behaviour. In his book *A Philosophical Essay on Probabilities*, the renowned French mathematician Pierre-Simon Laplace fell victim to the trap of determinism, or what has been called by many "Laplace's Demon" (Shermer, 1995). Loosely speaking, Laplace argued that, if someone knew the precise location and momentum of every atom in the universe, their past and future values for any given time could be precisely calculated using the laws of classical mechanics. Of course, this specific line of thinking has been thoroughly discredited.

In a sense, Laplace was offering a strong case of modernist thinking. By contrast, complexity science is a discipline anchored to postmodern thinking. However, unlike much of the work done under the umbrella of postmodernism, it is actually rigorous. The discipline also has the benefit of actually building positive knowledge (as opposed to merely demonstrating what we do not know). Among other things, complex-systems scholars have identified two major dynamics that frustrate our ability to predict system behaviour: (1) feedback and (2) emergence.

FEEDBACK

There are two basic forms of feedback every system generates. *Negative feedback* systems tend toward stability over time as a change in the variable being considered brings about some sort of contrary response that moves that variable in the opposing direction. For example, heat applied to a cup of coffee is not stable because it will slowly cool through a process of negative feedback until it reaches equilibrium at room temperature. Standard models of social, economic, and political sciences tend to emphasise the equilibrium properties of a given phenomenon. As a first-order description of the relevant dynamics, such characterisations tend to perform fairly well. However, they are missing an important source of system behaviour – positive feedback.

In systems that display *positive feedback*, small changes get amplified because they run in the direction that the systems are already moving (or they are able to permanently push the system in that direction). Positive feedback systems are sensitive to initial conditions where small changes get amplified. Herds, bubbles, avalanches, cascades, and network effects are empirically observable phenomena whose theoretical origins are linked to various forms of positive feedback. Understanding this dynamic informs future predictions of system behaviour. On average, across all its respective agents, the law is a system rapidly moving in the direction of social authority inequality. Without a significant change in the underlying dynamics, law is a complex system that currently features (and will continue to feature) positive feedback and large amounts of social authority inequality among cases, judges, law reviews, law schools, and other related social institutions (Katz and Stafford, 2010).

EMERGENCE

Another important source of frustration is the tendency of complex systems to display emergent behaviour. Complex adaptive systems produce emergent-scale behaviour – behaviour sometimes incapable of being understood except through system-wide study (Johnson, 2002). There is not complete agreement about the conditions giving rise to emergent phenomena. In general, however, systems display emergence when the micro-study of individual actors in a given system yields incomplete information about the entirety of the organisation (De Wolf and Holvoet, 2005). Instead, interactions between the components help structure the outputs of the given system. These themes are well articulated in classic treatments such as *Micromotives and Macrobehavior* by Thomas Schelling (1978) and *Emergence: From Chaos to Order* by John Holland (1998). As Peter Corning describes (2002, p. 21), "[a]mong other things, complexity theory gave mathematical legitimacy to the idea that processes involving the interactions among many parts may be at once deterministic yet for various reasons unpredictable."

There is a variety of examples of emergent behaviour in social and physical systems, including ecosystems, where order emerges from the interspecies interactions, the phase transition of various chemicals, and the rise of fads and other cultural cascades. Some such systems seem mundane, such as a traffic jam, but are nonetheless emergent (Mendes et al., 2012). So, too, it is also the case for legal systems. The study of emergence in legal systems would help us better quantify the magnitude of the legal system's irreducibility and incompressibility. This would, in part, provide a representation of how much we are *unable* to know and predict about the system through the construction of theoretical and empirical models that might include various structural and performance metrics.

Mapping legal complexity

Applying the measurement metrics outlined earlier to a legal system or subsystem would provide a snapshot of the system's complexity. But important questions would remain: Compared to what? How much is too much? In which direction is the system moving? One way of enriching knowledge in this regard would be by repeating the measurements over time and over many subsystems to gain a deeper understanding of comparative complexity (e.g., tax versus environmental law) and complexity trends. But still, such exercises would provide a sense only of how the different metrics behave over time, not of how the system as a whole behaves over time.

Monitoring legal system complexity thus should operate at two levels. On the surface, comparative and trend analyses like those just described, including of user features such as compliance burdens, provide real-time assessments of how complex a legal system is and whether relative complexity is increasing or decreasing (Kim, 2013). Extreme shifts in these metrics could raise red flags as to system performance. At a deeper level, however, monitoring changes in network interconnection and synchronisation would allow more direct evaluation of system-wide behaviour and a platform for testing system performance. This part outlines a platform and the methods for doing so.

Designing "Legal Maps" for network behaviour monitoring

Measuring system content, structure, information, and computation is necessary to construct a network model of the system, but once that model is constructed, another set of metrics is necessary to assess what is happening inside the system. For example, a metric of network growth would measure how the network expands or contracts, and a metric of system intensity could measure the rate of information flow and orientation along different feedback and feedforward channels. To make such evaluations requires a platform representing the networked system in real time – what we call Legal Maps.

Legal Maps is the legal system equivalent of more familiar applications for geographic navigation, such as Google Maps and similar map applications. The building block of such map tools is Geographic Information Systems (GIS) technology, which is a computer system for capturing, storing, checking, and displaying data related to positions on Earth's surface. By integrating many "layers" of data, such as rainfall, vegetation, roads, and so on, GIS technology can show many different kinds of data on one map, thus enabling people to more easily see, analyse, and understand patterns and relationships. Assembling GIS maps requires data capture, conversion, and digitisation of data from many sources into compatible formats, metrics, and scales; integration of the multi-sourced data into one projection; and manipulation of the data structures to allow mapping, modelling, and other methods to extract information about patterns and relationships, such as the effect of rainfall levels on vegetation near roadways.

Mapping tools combine a highly layered GIS map of geographic and other details with sophisticated algorithms, allowing the user to search the map for directions, distances, points of interest, and so on. In addition, these applications can feed data from smartphones and other sources into the map on a continuous basis to provide a dynamic, real-time user interface to communicate useful information, such as traffic density. For example, if a highway is closed due to an accident, drivers can use the map tool to identify where traffic is at its worst and alternate routes carrying less traffic.

Legal Maps would be built on the same kind of platform as GIS mapping applications, starting with layers of data relevant to the legal system network. For example, the hierarchy network of statutory codes could be represented as a discrete layer, as would the hierarchies of agency rules, court systems, and so on. Then the network's architecture would be represented. Cross-references within each layer, such as between sections of the statute code (nodes), would be represented as connections (edges) representing directionality and strength (e.g., one provision references another provision three times). Citation network visualisations like this already exist for judicial opinions in search engines, such as Ravel Law. Then references between layers, such as a regulation referencing a statutory provision and a court referencing the regulation, would also be mapped. Additional layers relevant to the system behaviour could be added – such as provisions in the Constitution, citations in attorney briefs, administrative rulings, and so on – and the interconnections within and between each layer could be mapped. Search algorithms can then be devised to identify patterns such as clusters of tightly connected statutory and regulatory provisions, particular courts' and agencies' decisions, and so on (Garbarino, 2010).

Legal Maps, like Google Maps, would also operate as a real-time (or nearly real-time) representation of the legal system's dynamics. Events such as promulgation or repeal of a regulation or a new judicial opinion can be streamed into the map system with appropriate representations of cross-references and citations, and the system's information flow paths and rates could be observed (e.g., are certain regulations strong gatekeeper nodes between the statutory provisions they reference and judicial opinions referencing the regulation?). Streams from news and social media could also be fed into Legal Maps to observe how the legal system responds to rising social interest in a policy topic (e.g., how long before courts mention the trend and new regulations are promulgated around the trend?). Like

Google Maps, layers could be selected or excluded to allow analysis of paired layers, and over time, a user's search history could be tracked to provide tailored maps such that a practitioner of tax or environmental law could work within the sector and layers of the system most relevant to his or her practice. The end result would be as close a representation of a map of the legal system network as one could attain using current technology.

Indeed, what we describe can be achieved today. All the data described are already available in digital form. Capturing them and converting them into compatible digital representations would be no more complicated than what Google Maps accomplishes for geographic data. Indeed, Koniaris et al. (2015) recently constructed a partial representation of such a model for European Union (EU) legislation, plotting more than 250,000 legal documents (nodes) spanning 60 years of EU legislation. Their model linked three layers – treaties, statutes, and judicial opinions – yielding almost 1 million connections (edges) within the network. Using this network representation, they performed a temporal analysis of the evolution of the legislation network, as well as a robust resilience test to assess its vulnerability under specific cases that may lead to possible breakdowns. Similarly, the search algorithms we anticipate Legal Maps using are no more sophisticated than those used in Google Maps. The only constraints to further development of such models for the legal system are time and money. But, even assuming the time and money were available, why build Legal Maps? What would we do with it?

Using "Legal Maps"

Smartphone mapping applications have obvious valuable uses, not the least of which is providing directions between two points. Of even greater value when out on the road are the traffic density and trip rerouting functions. And Google now provides an application program interface – a set of routines, protocols, and tools for building software applications – allowing other application builders to integrate Google Maps into their user interfaces. Legal Maps could provide all these functions as well, several of which would greatly enhance the capacity for monitoring legal system complexity and behaviour. We provide a few examples in the following.

Synchronisation monitoring

The feedback mechanisms characteristic of a complex adaptive system are the source of both system resilience and systemic risk. The term *systemic risk* has become closely associated with the financial system collapse of 2008 (Anabtawi and Schwarcz, 2011), but the concept of systemic risk is not limited to financial systems – it applies to all complex systems. Dirk Helbing (2013) of the Swiss Federal Institute of Technology defines systemic risk as

> the risk of having not just statistically independent failures, but interdependent, so-called "cascading" failures in a network of N interconnected system

components. . . . Even higher risks are implied by networks of networks, that is, by the coupling of different kinds of systems.

Helbing argues that this global environment is a "hyper-connected" world exposed to massive systemic risks driven by systemic instability (such as tipping points, positive feedback, and complexity). The upshot is that catastrophic damage scenarios are increasingly realistic, yet our political and economic systems simply are not wired with the incentives needed to imagine and guard against these outlier events.

Quite simply, we need to build systemic risk into our scenarios of the future, including for the legal system. The legal system must (1) not only anticipate systemic failures in the systems it is designed to regulate but also (2) anticipate systemic risk in the legal system as well. Legal Maps could provide a platform for the second type of monitoring. Complexity scientists have identified a strong marker for systemic risk in the form of highly synchronised positive-feedback systems that give rise to networked risks. When all feedback in the system has harmonised in the same self-reinforcing direction, a small, seemingly noncausal disruption to the system can lead to massive failure. As econo-physicist Didier Sornette (2009) puts it, "[t]he collapse is fundamentally due to the unstable position; the instantaneous cause of the collapse is secondary." His assessment of the financial crash, for example, is that, like other financial bubbles, over time "the expectation of future earnings rather than present economic reality . . . motivate[d] the average investor." What pops the bubble might seem like an inconsequential event in isolation, but it is enough to set the collapse in motion once the system is ripe.

By tracking information flow and structure in the legal system over time, including the conduits across which it moves and their direction, strength, and timing, Legal Maps could help monitor for the build-up of highly synchronised information pathways that could open the door to cascade failures. For example, if financial, environmental, and other regulators receive information along a tightly synchronised set of pathways and then move in the same direction based on information input (e.g., increase monitoring if information indicates a certain trend), interruption in the information flow or a surge in unreliable information can set the legal system up for a cascade of failures. Analyses of both the financial collapse of 2008 and the BP Deep Horizon oil spill suggest such forces were at play in the relevant regulatory systems and contributed to the cascade of failures within and outside the legal system (Financial Crisis Inquiry Commission, 2011; National Commission on the BP Deepwater Horizon Oil Spill & Offshore Drilling, 2011).

Stress tests

Financial system models allow an introduction of perturbations to assess what happens when the system is put under stress (Weber, 2014). Similar stress testing could be applied in legal system models. For example, the rate of information flow (e.g., rate of variation in financial instruments or number of pollution violations) could be manipulated in the network Legal Maps model to see how the legal system handles high-flow rates and where flow jams occur under different stress conditions. Or, as Koniaris et al. (2015) performed on their European Union legislation network model, pieces of network structure could be deleted (as in a proposed major deregulatory event) or added (as in a proposed enactment of a major new regulatory regime) to test how network structure and behaviour would respond in terms of reconfigured synchronisation patterns and information flow jams. While there are always unforeseen circumstances (i.e., unknown unknowns) which any legal system must confront on a constantly evolving landscape, it is still possible to stress test a legal system model against a range of known or proposed scenarios in the effort to determine its robustness.

Interdependent systems analysis

Transportation disaster planning and assessment are turning to interdependent systems analysis (ISA) to move beyond single-disaster assessment (which usually focuses on identifying human error) to understand why disasters happen in general (Burton and Egan, 2011). ISA uses network analytics and stress testing to link the system under study to its co-evolving systems over relevant time scales. ISA improves the ability of planners to identify the endogenous and exogenous conditions leading to systemic risk and, ultimately, to failure cascades. The legal system, in coevolution with the social systems it is intended to regulate and protect, is a perfect medium for ISA. From the sectors law regulates and protects, Legal Maps could be built out to include data feeds such as financial, manufacturing, environmental, and demographic data. When one system experiences failure, such as a financial crash, retrospective analysis of how the legal system responded prior to, during, and after the event can help identify where stress and failure were rising or falling in the legal system, such as by excessive synchronisation or information flow jams.

Comparative design studies

In a world where all the preceding could be accomplished, it would then be useful to conduct cross-system comparisons, such as between common law and civil code systems, or across different bodies of law (e.g., tax versus environmental) and national law systems (e.g., the United Kingdom compared to China). Particularly as economic and social phenomena occur increasingly at global scales, nations' legal systems are increasingly interdependent, thus supporting the case for building out and linking Legal Maps for all nations.

Conclusion

Much of the foregoing would have seemed like sheer fantasy as recently as just a few years ago. But that was before massive advances in data storage and computational capacity and their use to promote robust complexity and network sciences.

These advances challenge traditionalists' claims that the legal system is so exceptional or impenetrable that it cannot, in some substantial degree, be measured and modelled through computational methods applied in other disciplines (Ruhl et al., 2017). The legal system, a phrase used ubiquitously in legal scholarship, is just that – a system. As such, its description and assessment are open to the empirical approaches we have suggested. This goes well beyond mere extrapolation from the empirical techniques already in play, such as citation databases and judicial voting studies. This is about building computational models of the legal system, its complexity, and its systemic risks.

Early attempts to develop legal complexity metrics, build out a system model through Legal Maps, conduct stress tests, locate systemic risk, seed the system with machine learning sensors, and propose new legal designs will be rudimentary, coarse, and often wrong and will be criticised for that. But succumbing to such critiques would have kept economists using the abacus and ecologists counting tree rings. It is time for lawyers to move beyond case studies, rhetoric, and conventional statistical methods – it is time to study the deep structure of legal complexity through the empirical and technological methods of complexity science.

Note

* This chapter is, with permission, an extensively condensed and edited version of an article the authors published in Volume 101 of the *Iowa Law Journal* (Ruhl and Katz, 2015).

Bibliography

Adler, J. H. (2012), 'Rights, Markets, and Changing Ecological Conditions', 42 *Environmental Law* 93.

Anabtawi, Iman and Steven L. Schwarcz (2011), 'Regulating Systemic Risk: Towards an Analytical Framework', 86 *Notre Dame Law Review* 1349.

Arthur, W. Brian (2015), 'Complexity Economics: A Different Framework for Economic Thought', in *Complexity and the Economy 1* (Oxford: Oxford University Press).

Barabási, Albert-László (2003), *Linked: How Everything Is Connected to Everything Else and What it Means For Business, Science, and Everyday Life* (New York: Basic Books).

Bar-Yam, Yaneer et al. (2012), *A Complex Systems Science Approach to Healthcare Costs and Quality* (Cambridge: New England Complex Systems Institute).

Baum, Lawrence (2006), *Judges and Their Audiences: A Perspective on Judicial Behavior* (Princeton, NJ: Princeton University Press).

Beinhocker, Eric D. (2006), *The Origin of Wealth: Evolution, Complexity, and the Radical Remaking of Economics* (Boston: Harvard Business Review Press).

Bettencourt, Luis M. A. et al. (2006), 'The Power of a Good Idea: Quantitative Modeling of the Spread of Ideas from Epidemiological Models', 364 *Physica A* 513.

Blackmore, Susan (2000), 'The Power of Memes', 283 *Scientific American* 64.

Boccaletti, Stefano et al. (2006), 'Complex Networks: Structure and Dynamics', 424 *Physics Reports* 175.

Bousquet, Antoine (2012), 'Complexity Theory and the War on Terror: Understanding the Self-Organising Dynamics of Leaderless Jihad', 15 *Journal of International Relations & Development* 345.

Burton, Lloyd and M. Jude Egan (2011), 'Courting Disaster: Systemic Failures and Reactive Responses in Railway Safety Regulation', 20 *Cornell Journal of Law & Public Policy* 533.

Cherry, Barbara A. (2007), 'The Telecommunications Economy and Regulation as Coevolving Complex Adaptive Systems: Implications for Federalism', 59 *Federal Communications Law Journal* 369.

Cherry, Barbara A. (2008), "Institutional Governance for Essential Industries Under Complexity: Providing Resilience Within the Rule of Law', 17 *CommLaw Conspectus Journal of Communications Law & Policy* 1.

Cilliers, Paul (1998), *Complexity and Postmodernism: Understanding Complex Systems* (New York: Routledge).

Corning, Peter A. (2002), 'The Re-Emergence of "Emergence": A Venerable Concept in Search of a Theory', 7 *Complexity* 18.

Cotter, Thomas F. (2005), 'Memes and Copyright', 80 *Tulane Law Review* 331.

De Wolf, Tom and Tom Holvoet (2005), 'Emergence Versus Self-Organisation: Different Concepts But Promising When Combined', in Sven A. Brueckner et al. (eds.) *Engineering Self-Organizing Systems: Methodologies and Applications* (Berlin and Heidelberg: Springer-Verlag), 3.

Epstein, Richard A. (1995), *Simple Rules for a Complex World* (Boston: Harvard University Press).

Financial Crisis Inquiry Commission (2011), *The Financial Crisis Inquiry Report* (Washington, DC: U.S. Government Printing Office).

Fowler, James et al. (2007), 'Network Analysis and the Law: Measuring the Legal Importance of Precedents at the U.S. Supreme Court', 15 *Political Analysis* 324.

Garbarino, Carlo (2010), 'A Model of Legal Systems as Evolutionary Networks: Normative Complexity and Self-Organization of Clusters of Rules' (Boston Univ., Legal Studies Research Paper No. 1601338) <http://papers.ssrn.com/sol3/papers.cfm?abstract_id=1601338>.

Haken, Hermann (2010), *Information and Self-Organization: A Macroscopic Approach to Complex Systems* (Berlin and Heidelberg: Springer-Verlag).

Helbing, Dirk (2013), 'Globally Networked Risks and How to Respond', 497 *Nature* 51.

Holland, John H. (1998), *Emergence: From Chaos to Order* (New York: Basic Books).

Holmes, Oliver Wendell, Jr. (1897), 'The Path of the Law', 10 *Harvard Law Review* 457.

Hornstein, Donald T. (2005), 'Complexity Theory, Adaptation, and Administrative Law', 54 *Duke Law Journal* 913.

Johnson, Steven (2002), *Emergence: The Connected Lives of Ants, Brains, Cities, and Software* (New York: Scribner).

Katz, Daniel M. and Derek K. Stafford (2010), 'Hustle and Flow: A Social Network Analysis of the American Federal Judiciary', 71 *Ohio State Law Journal* 457.

Kauffman, Stuart (1995), *At Home in the Universe: The Search for the Laws of Self-Organization and Complexity* (Oxford: Oxford University Press).

Kim, Rakhyun E. (2013), 'The Emergent Network Structure of the Multilateral Environmental Agreement System', 23 *Global Environmental Change* 980.

Koniaris, M. et al. (2015), 'Network Analysis in the Legal Domain: A Complex Model for European Union Legal Sources', *arXiv* (21 January 2015) <http://arxiv.org/pdf/1501.05237.pdf>.

Lazer, David et al. (2009), 'Life Computational Social Science', 323 *Science* 721.

Luo, Jianxi and Christopher L. Magee (2011), 'Detecting Evolving Patterns of Self-Organizing Networks by Flow Hierarchy Measurement', 16 *Complexity* 53.

Martin Katz, Daniel M. and Michael J. Bommarito II (2014), 'Measuring the Complexity of the Law: The United States Code', 22 *Artificial Intelligence & Law* 337.

McGarity, Thomas O. (2013), 'When Strong Enforcement Works Better Than Weak Regulation: The EPA/DOJ New Source Review Enforcement Initiative', 72 *Maryland Law Review* 1204.

Mendes, G. A., L. R. da Silva, and H. J. Herrmann (2012), 'Traffic Gridlock on Complex Networks', 391 *Physica A* 362.

Miller, John H. and Scott E. Page (2007), *Complex Adaptive Systems: An Introduction to Computational Models of Social Life* (Princeton, NJ: Princeton University Press).

Mitchell, Jonathan F. (2011), 'Stare Decisis and Constitutional Text', 110 *Michigan Law Review* 1.

Mitchell, Melanie (2009), *Complexity: A Guided Tour* (Oxford: Oxford University Press).

National Commission on the BP Deepwater Horizon Oil Spill & Offshore Drilling. (2011), *Deep Water: The Gulf Oil Disaster and the Future of Offshore Drilling* (Washington, DC: U.S. Government Printing Office).

Page, Scott E. (2006), 'Path Dependence', 1 *Quarterly Journal of Political Science* 87.

Pentland, Alex (2014), *Social Physics: How Good Ideas Spread – The Lessons from a New Science* (London: Penguin Books).

Ruhl, J. B. (2008), 'Law's Complexity: A Primer', 24 *Georgia State Law Review* 885.

Ruhl, J. B. (2014), 'Managing Systemic Risk in Legal Systems', 89 *Indiana Law Journal* 559.

Ruhl, J. B. and Daniel M. Katz (2015), 'Measuring, Monitoring, and Managing Legal Complexity', 101 *Iowa Law Review* 191.

Ruhl, J. B., Daniel M. Katz and Michael J. Bommarito II (2017), 'Harnessing Legal Complexity', 355 *Science* 1377.

Schelling, Thomas C. (1978), *Micromotives and Macrobehavior* (New York: W. W. Norton & Company).

Shermer, Michael (1995), 'Exorcising Laplace's Demon: Chaos and Antichaos, History and Metahistory', 34 *History & Theory* 59.

Smith, Thomas A. (2007), 'The Web of the Law', 44 *San Diego Law Review* 309.

Sornette, Didier (2009), 'Dragon-Kings, Black Swans and the Prediction of Crises', 2 *International Journal of Terraspace Science & Engineering* 1.

Strandburg, Katherine J. et al. (2009), 'Patent Citation Networks Revisited: Signs of a Twenty-First Century Change?' 87 *North Carolina Law Review* 1657.

Watts, Duncan J. (2003), *Six Degrees: The Science of a Connected Age* (New York: W. W. Norton & Company).

Webb, Julian (2005), Law, Ethics, and Complexity: Complexity Theory and the Normative reconstruction of Law', 52 *Cleveland State Law Review* 227.

Webb, Thomas E. (2014), 'Tracing an Outline of Legal Complexity', 27 *Ratio Juris* 477.

Weber, Robert F. (2014), 'A Theory for Deliberation-Oriented Stress Testing Regulation', 98 *Minnesota Law Review* 2236.

Section II

Complexity and the state: Public law and policy

3 Complexity
Knowing it, measuring it, assessing it

Neville Harris

Introduction

References to complexity in public discourse often present it as a pervasive and negative characteristic of regulatory and governance systems. The assumed normative position is that, where laws, structures and processes are complex, administrative and regulatory authorities may struggle to meet a legitimating goal of applying rules accurately (Mashaw, 1983) and will incur greater administrative costs: so, '[c]omplex rules hurt the government' (Kerwin, 2003, p. 97). Furthermore, compliance, knowledge and enforcement of rights can be undermined. More generally, complexity is seen as having implications for the status of law on the basis that if a rule's complexity leads to its misapplication or to sub-optimal compliance, this can weaken the law's legitimacy, while by causing legal uncertainty it could even be prejudicial to the Rule of Law (Craig, 1997, p. 469). National and international initiatives have been focused not only on reducing complexity but also managing it better. Some of them, such as the European Union's (EU's) Better Regulation initiative (see European Commission, 2016), favour legislative changes through codification and the removal of duplication, inconsistencies and obsolete provisions. Others look to structural features of law-based systems, as in the UK's recent welfare reforms aiming to make systems easier and less costly to operate.

Manifestations of complexity in law and governance are the concern of this chapter. The discussion rests on a dual conception of complexity in relation to laws, processes and systems: first, denoting those that are highly elaborate and complicated with many interconnected and interdependent elements or features and, second, in the way associated with complexity theory, as autonomous and intricate systems of law and governance which are complex in the sense that the interactions within them do not always operate in a linear cause-and-effect way but can also be marked by unpredictability. The chapter views complexity as a largely empirical phenomenon and considers its features and how far measurement is possible. It also assesses the role of simplification as a policy response. In looking at the functioning of law-based systems, the chapter focuses, in particular, on the key public law areas of social security and taxation, which have an impact on a majority of citizens, are acknowledged to be particularly complex areas of governance and are ones in respect of which simplification remains a policy goal. Another complex area that is highlighted is 'Brexit' – the process of UK withdrawal from the EU – and its surrounding legal context. The chapter also considers the relationship between complexity and rights.

Knowing it

Complexity as a driver for reform

Complexity has become a driver for public policy reform across a number of fields.[1] For example, the 2010–15 Coalition government's case for welfare reform in the UK involving the introduction of two major new benefits, Universal Credit[2] and Personal Independence Payment,[3] included that 'the system is too complex: for claimants . . . and to administer' (Department for Work and Pensions (DWP), 2010a, p. 7). Complexity had contributed to the 'rising costs of state support – including waste through unproductive administration, error and fraud' (ibid). Universal Credit is a new generic benefit replacing six separate means-tested benefits (see DWP, 2010b). Benefit is now administered by one government agency, the Department for Word and Pensions (DWP), rather than three separate agencies (the DWP, Her Majesty's Revenue and Customs and local authorities), saving in operational costs an estimated £0.5 billion and £0.9 billion per annum.[4] The new single-tier pension under the Pension Act 2014 also represents a simplification reform (DWP, 2013, paras 32–34).

The field of welfare benefits is one where the more problematic consequences of complexity are not merely administrative burdens and costs but also errors in day-to-day decision-making, difficulties in managing scheme transitions, underutilisation (take up) of benefit, failures of conditionality compliance among recipients and claimant difficulties in planning for life events such as retirement or take up of employment (National Audit Office [NAO], 2005). Greater simplicity is also part of the case for schemes of universal basic income, which have been under active consideration both in the UK (Roberts, 2017) and elsewhere, such as in Finland (Kalliomaa-Puha et al., 2016), France (Martin, 2017), the Netherlands (Boffey, 2015) and Ontario, Canada (Kassam, 2017). Universal basic income would be paid to all citizens, regardless of their employment status. It can be expensive (increased taxation of the better off may be needed to help to pay for it: van Parijs and Vanderborght, 2017), but, assuming payment is always at a standard rate (although age or disability variables may be allowed), it is simpler and more easily administered than means-tested benefit (Spicker, 2011, pp. 121–124; Martinelli, 2017).[5]

In the UK tax system, complexity 'has increased to the point where the burden for taxpayers in complying with the law has become a real issue' (Bowler, 2009, para.2.1). Nonetheless, an assumption that legal simplification alone can remedy complexity has been challenged (James and Edwards, 2008). In particular, both the underlying tax policy and the tax system as a whole are considered contributory factors to complexity (Arnold, 2009, p. 1; Lee, 2003, p. 41). Thus, although a Tax Law Rewrite Project was in operation from the mid-1990s to 2010, it apparently achieved only limited success in reducing complexity (Salter, 2010, 2011). The establishment of the Office of Tax Simplification (OTS) in 2010 has, however, offered some momentum towards simplification. More importantly, perhaps, the OTS has made strident efforts to define and measure complexity, as discussed later. Tax simplification was also a theme in the 2016 budget, in response to concerns about the impact of complexity on 'micro-entrepreneurs' (HM Treasury, 2016, para.1.170). A whole category of National Insurance Contributions (Class 2, applicable to the self-employed), 'an outdated and complex feature', was proposed for abolition (Ibid para 1.166). Meanwhile, the plan to place the OTS on a statutory footing (Ibid, para 2.213) has materialised, and the Finance Act 2016, s.186(1) now requires the OTS, on the Chancellor's request, to 'conduct a review of any aspect of the tax system for the purpose of identifying whether, and if so how, that aspect of the tax system could be simplified'.

Features of complexity

Much of policymakers' attention towards complexity is directed at the law. Governance systems are founded on law and have decision-making functions which involve the application of legal rules. While the design of rules will be a key factor in complexity – with length, denseness, technicality, precision/openness and interdependency considered relevant characteristics (Schuck, 1992)[6] – the institutional context in which the rules operate is also important. Within many governance systems, the interactions between different parts and agencies generate what has been coined 'extrinsic complexity' (Spicker, 2011) but is actually an internal characteristic. For example, different parts of the social security system are closely interlinked. Many benefits are 'passported': an entitlement to one opens a gateway to another, but this system is 'complex to understand, establish entitlement, and administer' (DWP, 2012b, ch.2 para.7). System interconnections can be particularly problematic for claimants who need to report a change in personal circumstances since an inability to understand them may lead to a failure to disclose relevant changes to the correct agency. Consequently, benefit may be underpaid or overpaid (and be recoverable, causing hardship; Harris, 2013, pp. 61–66).

Complexity also arises from adaptation or adjustment of systems in line with changing policy goals and technical initiatives or in response to judicial rulings.[7] Such changes typify the small piecemeal amendments frequently made to an already-complex legal framework. Dynamic environments of this kind are strongly associated with complexity. Part of the significance of the Universal Credit reform, mentioned earlier, is in representing an attempt to redesign a critical part of the welfare state to ensure a simpler framework than the very complex one that evolved over many years.

The complex design of many systems of governance is associated with their multiple roles and diverse policy goals within politically and financially sensitive fields. In relation to social security, for example, Ghai (2002, p. 4), identifies three common objectives: 'reducing destitution; providing for social contingencies; and promoting greater income and consumption equality'. Each of the distinct models of welfare state (see Esping-Andersen, 1990; Arts and Gelissen, 2002) through which there would be an attempt to realise these goals tends to involve highly complex schemes. British social security law also reflects 'the underlying complexity of the policies it expresses' (Social Security Advisory Committee, 2007, para.1.19). They include discouraging dependency and encouraging the take-up of employment, as reflected in various welfare-to-work schemes (e.g. the Work Programme and work-for-your-benefit provisions),[8] rules requiring lone parents to be available for employment and actively seek it (see Haux, 2011), the work capability assessment for incapacity benefits[9] and the benefit cap[10] restricting overall household entitlement.[11] Another policy, of more recent origin, has been localisation. It has affected the Social Fund, which was a national scheme but now each area makes its own provision (see Grover, 2012), an arrangement which 'potentially add[s] further complexity' (House of Commons Work and Pensions Committee, 2012, para.209). Similarly, for council tax relief, a national scheme has been replaced by a multiplicity of local schemes (Harris, 2013, pp. 65–66) – a policy which, the Institute for Fiscal Studies says, 'undermines the drive towards simplicity' (Hood and Phillips, 2015, p. 2; see also Finch et al., 2014). Other policy pushes involving measures increasing complexity include those of reducing the supposed magnetic pull of welfare benefits to EU migrants and attempting to curb levels of health care 'tourism' (Harris, 2016).

The tax system offers similar examples of specific policy aims within discrete areas, such as the specific tax reliefs aimed at environmental protection and job creation. In the field of taxation 'a complex policy invariably requires complex legislation, which may need to be interpreted into a complex administrative process' (OTS, 2015, p. 6). The implication for attempts to simplify the system is therefore that 'without policy changes the benefits from rewriting legislation are limited' (Tax Law Review Committee, 1996, para.6.12), although there is still a case for simpler rules, as discussed later.

Various aspects of the management of systems are also associated with complexity (Spicker, 2011, p. 141). A system's design reflects not only its intended role (see the following discussion) but also the forms of administrative control over decision-making that government deems necessary or desirable. Copious but tightly drawn 'bright-line' rules tend to be used to delineate and define, and in 'rule-based governance regimes' such as in welfare, we see rules of some complexity developed as a mechanism to set 'goals and expectations' for the managing agencies (Scott, 2017, p. 269). Changes to systems also have to be managed, and in an era of austerity, government departments are 'trying to tackle complex reforms with fewer staff and smaller budgets' (NAO, 2017, p. 7, para.1).

Two specific areas of management identified as net contributors to complexity are the contracting-out or commissioning of functions and the use of information technology (IT). Contracting-out has been used extensively by the welfare system in areas concerned with medical assessment for work incapacity and disability and in activation strategies (notably the Work Programme; NAO, 2016; House of Commons Work and Pensions Committee, 2011a). There were 5.5 million contracted-out health and disability assessments in the five years to 2015; a further 7 million were anticipated up to 2018, at a cost of £1.6 billion (NAO, 2016). There has been 'a transition from a highly centralised bureaucracy providing standard services to a more complex public-private delivery network' (Finn, 2011, p. 132). Both the nature of the contracting regime itself and the need for state management of the arrangements, such as monitoring contractor performance, enforcing contractual terms and disciplining for underperformance, contribute to complexity (HCCPA, 2013). A further element is the impact of contracting out on the state–citizen relationship: as a third party, the contractor is accountable only to the state (contractually) and not the claimant. Such arrangements 'partially decouple the citizen from the state' (Carmel and Papadopoulos, 2009, p. 102). A claimant could, for example, be reported by the contractor for failing to attend a medical examination but sanctioned for it by the DWP, which establishes the reason for the failure.[12]

The contribution of IT systems to complexity relates in part to the scale and intricacy of their functions. Aside from their capacity to store and process huge amounts of data, they use algorithms in applying rules and criteria to individual circumstances. Kades (1996, p. 409) highlights the argument that 'tax law seems precisely like the kind of formal, mechanical set of rules for which computers are ideally suited, to manage more and more complexity with fewer and fewer (human) headaches'. One identifiable benefit derived from applying IT programmes in the determination of individual claims or assessments is that officials may require less detailed knowledge of the actual rules (NAO, 2005). IT systems also facilitate the administration of individual cases because officials can easily track them and share relevant information within or outside the agency. Nevertheless, the complexity of the schemes themselves necessitates careful IT design and technical maintenance. Moreover, policy formulation does not always take sufficient account of how IT will deliver it. A further issue is that changes to IT systems that are needed when policy reforms are introduced may contribute to delays in implementing the new policy (Hudson, 2009, p. 298). IT systems, in any event, become outdated and upgrading can have significant cost implications. The robustness of interfaces between systems, such as those between local authority and central systems, can also be uncertain (House of Commons Work and Pensions Committee, 2012, para.232).

It is also important to consider unpredictability. It is central to complexity theory, as noted earlier, but also has an empirical basis. Theory posits the idea that what characterises a system as complex is the propensity for unpredictable outcomes to arise from the operation of its internal dynamics, as distinct from the predictable consequences of interactions within linear processes (Geyer and Rihani, 2010). The theory is applied to various social systems, including the welfare state (Geyer, 2003; Rhodes et al., 2011). Unpredictable nonlinear interactions may hinder the realisation of underlying objectives (Geyer and Rihani, 2010, p. 4). Geyer has noted how both the operation and the resourcing of the welfare state are predicated on linear assumptions, but it is 'an evolving complex adaptive system' with 'complex dynamics' in forging solutions to policy questions, including resource allocations, and is beset with unpredictability in a somewhat uncertain social, political and economic environment (Geyer, 2003, p. 40).

There may, however, be scope for learning lessons from a system's operation with a view to managing the effects of unpredictability. For example, the NAO (2015, para.2.1) advises the Department for Work and Pensions to develop 'a strategic approach to managing uncertainty' arising from welfare reforms, which 'have uncertain impacts both in isolation and in how they affect people cumulatively'. The department should 'set realistic expectations for how processes will work, understand its own capacity to manage reforms, and anticipate how to re-shape programmes to achieve its core objectives when problems arise' (Ibid, para.2.2). Modelling the impact of assumptions to counter the risks of 'optimism bias' (that is, where proponents are unrealistically optimistic about the likelihood of a reform's success) is recommended by the National Audit Office (NAO) (Ibid, para.2.10). Failure to predict various consequences can undermine policy initiatives. For example, the intended simplification of the state retirement pension[13] left it 'still riddled with complexities' and, contrary to public expectation, resulted in a majority of pensioners receiving much less than the standard rate (Baroness Altman, quoted in Mikhailova, 2017).

Legal change and unpredictability are also central to the perceived complexity of 'Brexit'. As an organisation the EU is complex but so is its legal, constitutional and economic relationship with the UK and each of the other 27 Member States (House of Commons Exiting the European Union Committee, 2016, para.76). From a UK perspective a normative goal of Brexit negotiation and planning is certainty, not least in order to ease the transition for trade, commerce, governance and regulation as withdrawal from the EU occurs (HM Government, 2017). Yet the 'complexity of ensuring legal certainty in the UK on the day after Brexit must not be underestimated' (House of Commons Exiting the European Union Committee, 2017, para.17). Through what was originally termed the 'Great Repeal Bill' and has subsequently been published as the European Union (Withdrawal) Bill, the Government aims to offer certainty through the conversion of the body of EU law – the Acquis Communautaire – into UK domestic law, allowing 'businesses to continue trading in the knowledge that the rules will not change significantly overnight' and providing 'fairness to individuals whose rights and obligations will not be subject to sudden change' (HM Government, 2017, para.1.2). But the Acquis is not all in one place and includes some jurisprudence – itself a complicating factor because of uncertainty surrounding the weight to be attached to rulings by the Court of Justice of the EU post-Brexit, which has prompted the president of the UK Supreme Court to call for government clarification (Wright, 2017). So, this will be 'a complex task for both the Government and Parliament' (House of Commons Exiting the European Union Committee, 2017, para.17).

The legal changes needed in consequence of Brexit represent 'potentially one of the largest legislative projects ever undertaken in the UK' (Simpson Caird, 2017, p. 4). An estimated 15 new statutes will be needed in addition to many regulations (White and Rutter, 2017, p. 10). Under the Bill as presented, the use of 'Henry VIII clauses', giving power to amend primary legislation via statutory instrument, would be utilised 'to correct the statute book, where necessary, to rectify problems occurring as a consequence of leaving the EU' (Department for Exiting the European Union, 2017, para.1.15). The EU (Withdrawal) Bill would create a ministerial power to, 'by regulations make such provision as the Minister considers appropriate to prevent, remedy or mitigate' any failure of EU law as retained in domestic law to 'operate effectively' or any other deficiency in the retained EU law.[14] There is considerable democratic concern about this far-reaching power (House of Lords Select Committee on the Constitution, 2017, paras 42 and 44). The white paper on the bill concludes that 'it is not possible to predict at this stage how every law is to be corrected, as in some areas of policy the solution may depend on the outcome of negotiations' (Department for Exiting the European Union, 2017, para.3.12). Therefore, the question of predictability arises again. The risks and uncertainties as the UK disentangles itself from the EU are contributing to the complexity inherent in the process.

Measuring it

Why?

There are many reasons why being able to measure degrees of complexity may be important. First, it can enable policymakers to identify a case for simplification – particularly by comparing complexity levels across a system. It has been argued that without a suitable measurement method, how would legislators know which provisions to 'tweak to weed out the complexity' (Ruhl and Katz, 2015, p. 196)? Second, enabling the complexity impact of specific policy changes to be measured can 'give policy makers the chance to track changes in relative complexity . . . over time' (OTS, 2015, p. 3). Third, there may be benefits in being able to compare complexity levels across different systems. Such comparison could facilitate assessment of the potential for international harmonisation of laws. Harmonisation is said to be hindered by disparities in complexity levels (Crettez et al., 2009). Comparison of complexity levels across different types of system may also inform system design. For example, the NAO (2011, p. 9) has compared complexity levels across different methods of means-testing.

A fourth reason for measurement, identified by Kades (1996), could be to identify areas of conflict where the level of complexity makes disputes intractable, enabling expensive litigation to be avoided. Other reasons for measuring complexity concern the issue of optimal legal complexity, centred on 'what set of legal rules will bring out the best in human beings, by some measure of social welfare' (Epstein, 2004, p. 2) and on the question, '[H]ow complex should the legal system be to get its job done without undue risk of systemic failure?' (Ruhl and Katz, 2015, p. 240). Ruhl and Katz (Ibid) argue that 'improved empirical understanding of legal complexity can inform its management', on the assumption that since legal complexity may not be intrinsically bad, the goal may not necessarily be to remove or reduce it but rather to maximise the 'good' within a complex system 'while minimizing the "bad" '.

So the case for finding a suitable method, or suitable methods, for measuring complexity is clearly very strong. The next question is therefore how to measure it effectively.

How?

The question of how to measure complexity is not really separable from the question of why. One would expect a measurement method to reflect its underlying purpose. Thus, to find ways of making a complex system easier to administer, or more accessible, the methodology may focus on the impact that particular features of it have on these areas. This contrasts with a more theoretical orientation towards metrically based methodologies (e.g. Ruhl and Katz, 2015), albeit that theory application may be directed ultimately at facilitating the establishment of empirical proof. Kades (1996) refers to computational complexity theory and a mathematical theoretical model with algorithms to determine complexity levels in disputes. My focus, however, is on practical attempts to develop a complexity index, which have occurred within two of the more complex areas of governance affecting individual citizens: social security and taxation.

Social security

In the social security field, more than 70 years ago Beveridge (1942, para.397) recognised that a reformed system would be 'still be a machine with many parts and complications to deal with all the complexities of need and variety of persons'. However, the quest for a workable method of measuring social security complexity did not materialise until the 2000s. The House of Commons Committee of Public Accounts (2006, para.6) found the system to be 'getting more complex, but there is no way of measuring the degree of complexity' and it called for an 'easy to understand' basis of measurement, linked to factors such as customer satisfaction levels, error and take-up rates and the length of, and linkages within legislation. The DWP's Benefit Simplification Unit, established in 2005, had a stated priority in 2007 to complete its analysis of the scope for devising a 'complexity index' (DWP, 2007). Despite having few staff, the unit published until 2009–10 a series of annual 'simplification plans' and an inter-departmental *Simplification Guide to Best Practice*. The guide aimed to ensure that policies sought to avoid 'unnecessary complexity' and that simplification was pursued (see Harris, 2013, pp. 18–19). It called on policy designers to explain why any changes increasing complexity were unavoidable. However, the index did not materialise, as it was concluded that 'no single metric that could act as a suitable measure of complexity' (House of Commons Work and Pensions Committee, 2007b, paras 11–12).

There was also no systematic plan of simplification (House of Commons Work and Pensions Committee, 2007a, para.79), but the DWP decided to focus on the impact of complexity on the 'customer journey' – the claimant's interactions with the benefit system. Different journeys of diverse levels of complexity were selected as models in commissioned research (Royston, 2007). Other approaches referred to the 'burdens' placed on claimants, or 'compliance costs' – financial (such as telephone or postal charges and travel costs), time-related (completing forms or attending appointments) and psychological (factors such as stress, fear and depression; Bennett et al., 2009). Also considered relevant to how complexity was experienced were variable characteristics of claimants such as their level of intelligence and knowledge/education (see NAO, 2005). But there was no parallel attempt to base measurement of complexity on the burdens on the state, despite this being quite extensively analysed elsewhere (see, in particular, NAO, 2005 and House of Commons Work and Pensions Committee, 2007a). Nevertheless, the Labour government began to consider structural reform – in particular, the rationalisation of working age benefits – to ensure not only greater clarity and certainty for claimants but also improved administrative efficiency (see Freud, 2007 and Sainsbury and Stanley, 2007). However, this idea stalled in this administration's final years (Timmins, 2016, p. 21).

The dual focus on improved efficiency and greater certainty, combined with an emphasis on ensuring work incentives and easing transitions to employment, nevertheless continued under the Coalition government post-2010.[15] Coalition reforms included the introduction of Universal Credit and Personal Independence Payment, outlined earlier. While the creation of a simpler system, both for claimants and administrators, would be a measure of these reforms' success, no scientific method of calculating baseline and post-reform complexity was applied, making it difficult to evaluate the government's claim to be 'sweeping away the complexities of the current benefit system' (DWP, 2012a). Although the Institute for Fiscal Studies judges there to have been a 'significant simplification of the benefits system as a whole' (Browne and Roantree, 2013, p. 5), this conclusion seems based mostly on the structural changes rather than evidence pertaining to the operation of the system. Moreover, some structural elements, such as the localisation of council tax support, have undermined the simplification goal, as noted previously. Evidence suggests that the complexity of the welfare reform process itself was not fully worked out by the DWP – part of a more general lack of strategic insight regarding change (NAO, 2015). One might expect the DWP to heed the message to learn from the experience and remove 'unnecessary complexity' on its discovery (see Finch et al., 2014, p. 40), but its current policy objectives make no explicit reference to complexity reduction. It is at most implicit in its general objective of efficient delivery (DWP, 2017).

Taxation

Measurement of complexity has received the most attention in the area of taxation. The OTS's Complexity Index is 'designed to indicate the relative complexity of different parts of tax legislation' (OTS, 2013, para.1.2.1). The OTS believes that the index could also be utilised in other governance fields. The index aims to facilitate prioritisation and targeting of simplification efforts and 'give policy makers a tool to track the relative complexity of their policy changes' (OTS, 2015, p. 3). The OTS considers that once areas of high complexity are identified the underlying reasons for their complexity can be determined and thus so can a basis for simplification. Tracking changes in complexity over time enables the success of any simplification reforms to be gauged. In the accountancy age, this also becomes an auditing concern since estimates of cost savings from reforms must be published as part of policy rationale.

The OTS Complexity Index bases measurement on a total of ten indicators across two complexity heads (see Figure 3.1; OTS, 2015, pp. 4–9), with a scoring system (of up to 1 per indicator and an aggregate of 0–10) based on a standardisation formula. The first head is **underlying complexity**: the structural complexity of a taxation measure as reflected in its policy, legislation and administration. The key measurement criteria for *policy complexity* are the number of exemptions and reliefs within the particular area and the number of Finance Acts since 2010, on the basis that 'change is a significant contributor to complexity' (OTS, 2015, p. 7). In the interests of relative simplicity, the OTS is, however, ignoring the magnitude of change. On *legislative complexity* the criteria relate to the legislation's readability (based on the 'Gunning-fog readability index')[16] and its length, whilst acknowledging that short legislation can also be complex and that sometimes legislation is lengthy in order to avoid complexity – such as where it defines many technical terms or limits cross-referencing by repeating provisions. On *operational complexity*, relevant criteria are the complexity of official guidance and the information required from a taxpayer.

The second head is the **impact of complexity,** as judged by the associated costs for the taxpayer and the tax authority (Her Majesty's Revenue and Customs). Account is taken of the *number of taxpayers* (the greater the number, the higher the complexity rating), the *compliance burden* for taxpayers and the tax authority, the average *ability of taxpayers* (the impact of complexity will be greatest among the least able) and the measurable *impact of error, failure to take reasonable care or avoidance* (since complexity may be conducive to the occurrence

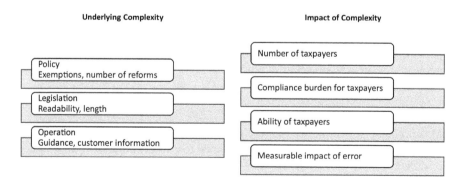

Figure 3.1 The OTS Complexity Index (outline)

of each of these phenomena). Error, lack of care and avoidance resulted in £10.2 billion of uncollected tax in 2012–13 (cited in OTS, 2015, p. 9).

The OTS provides worked examples covering air passenger duty, inheritance tax and landfill tax; see Table 3.1. Inheritance tax scores as the most complex of the three, its high score reflecting, in relation to **underlying complexity**, the number of exemptions or reliefs and the length of the relevant legislation, and on the **impact of complexity**, the ability of taxpayers to understand and deal with the tax and the degree of risk of avoidance action. Air passenger duty is the least complex.

In seeking to measure underlying complexity with reference to the burden on individual taxpayers lacking relevant expertise or experience, the OTS's approach is similar to that applied to the social security 'customer journey' mentioned earlier. As we have seen, the regulatory burdens or 'compliance costs' on individuals will to a degree reflect the underlying complexity of the system and its rules. As Epstein (1995, p. 29) says, legal complexity is as much a measure of 'how deeply' rules 'cut into the fabric of ordinary life' as of their formal properties. Underlying complexity generates an impact – as Katz and Bommarito (2014, p. 346, original emphasis) say in relation to the US Code, its '*structure, language* and *interdependence* . . . collectively impact the complexity of the law as experienced by an end user seeking to determine whether certain conduct is covered by a particular legal rule'. The OTS demonstrates that the impact will also be affected by a range of separately measurable factors that are not *directly* related to the inherent qualities of the laws, policies or processes at issue.

By reflecting the notion of relative complexity, the Complexity Index serves its underlying objective of identifying the most complex areas to target for reform. Moreover, the conclusion that viewing complexity other than in relative terms may not be possible is reinforced; the OTS (2015, p. 2) had 'difficulties in establishing an objective definition of complexity'. Similarly, Katz and Bommarito (2014, p. 370) acknowledge that their framework for empirical measurement of complexity, applied to the US Code, is for scoring the *relative* complexity of the Code's titles. The OTS's Complexity Index appears to have utility as a policy/governance tool, although one yet to be applied beyond the area of taxation. Where it fails, however, is in ignoring unpredictability in the collective operation of rules and processes, save that over time the scale of unpredictable occurrences may be reflected in the complexity impact score. This may in part be a consequence of the Index's development as 'a diagnostic tool rather than as [part of] a rigorous academic analysis of complexity' (OTS, 2015, p. 2).

Table 3.1 OTS Complexity Index: Three Tax Areas[17]

	Air Passenger Duty	Inheritance Tax	Landfill Tax
Total Underlying Complexity	2.0	6.5	4.4
Total Impact of Complexity	0.6	5.6	1.3

Assessing it

Understanding of complexity, and insights into methods of measuring it, may guide plans for its reduction. However, complexity has a purpose or value in itself since it is often a manifestation of arrangements designed to achieve various social benefits. Moreover, complexity may be necessary for the regulation of a complex field of activity. The OTS (2015, p. 10) refers to 'necessary complexity' arising in taxation areas concerned with complex transactions or financial arrangements – 'real-world commercial complexity'. Where complexity is unavoidable, the Complexity Index may have less utility, unless one can measure the complexity of the tax's policy objectives, which the OTS considers would require 'an entirely different index' (Ibid). We can also see areas of policy, such as the social security policies of discouraging dependency and incentivising entry to work, which in order to be implemented to maximum effect require elaborate rules – the use of discretion in the welfare benefits field having been largely abandoned (a process of juridification: Zacker, 1987, p. 404) in the interests of ensuring tighter control over decision-making (see Harris, 2000). For example, housing benefit rules aim to target rental support on those in the greatest financial need but to link entitlement to objectively determined accommodation requirements (as reflected in the much-litigated 'bedroom tax' restrictions).[18] They also aim to avoid exploitation by landlords (inflated rents) and to prevent fraud. So the necessarily complex rules[19] span, for example, means-testing, calculating eligible rent and defining 'contrived tenancies'. Complexity therefore follows the policy objectives.

Even if there *is* scope for reducing scheme or system complexity, there are issues to balance. Epstein refers to the 'justice' realisable via complex rules. If the rules can

> identify enough factors, can indicate ways in which they can be taken into account, can specify the appropriate burden of proof, and can provide for the exhaustive collection of evidence, then maybe . . . the legal system will reach the heady level of perfection to which it aspires.
>
> (Epstein, 1995, p. 38)

There may be a trade-off between fairness and simplicity. For example, the Personal Independence Payment's 'daily living' component has two rates whereas there were three under the equivalent element of the Disability Living Allowance for the over-16s that this benefit replaced. The new benefit's simpler framework is intended to be easier to understand and administer (DWP, 2011, p. 18), but it means there is a higher entry threshold of disability. Similarly, when the distinction between householder and non-householder rates of income support was replaced by one based on age alone, with the divide at age 25, the targeting of extra support on householders, a status requiring a complex assessment, was replaced by a much simpler but cruder and less accurate test of need based on the criterion of age (see Harris, 1989, pp. 74–76).

Complexity also has an ambiguous relationship with rights. A more personalised entitlement is likely to require a more complex set of rules to target resources on individuals' circumstances and needs (Kaplow, 1995, p. 161; see also Jewell, 2007, p. 381). Thus, for example, in the design of means-tests, the level of complexity increases with the range of factors requiring consideration (NAO, 2011, p. 9). In this way, there is less risk that an individual will be denied an appropriate level of provision. On the other hand, if complexity hinders the capacity for accurate decision-making or adversely impacts on the 'customer journey', it can undermine the realisation of individual rights. Simplification can be a double-edged sword, improving access to the system and aiding administration (although simple rules entailing the exercise of discretion and judgment can be difficult to administer) but resulting in starker, less sophisticated distinctions, with sharper cut-offs. Simplification tends mostly to serve administrative goals and expenditure savings. Improved efficiency may ultimately, through its economic benefits, yield wider public benefit but work against social justice.

Conclusion

Complexity in the regulatory functioning of the state may be a mark of societal advancement in reflecting a closer responsiveness to increasingly diverse needs and interactions within multilayered and interdependent social and economic spheres. Yet it presents various challenges for government and citizen. The quest for better understanding of the causes, features and ways of measuring complexity is partly a matter of academic endeavour, but it is also driven at governmental level by a desire for simplification to reduce cost and counter complexity's other negative effects on governance, such as hindering accurate and consistent rule-based decision-making.

In the context of governance and regulation a better understanding of complexity is, however, also necessary not merely to find ways of reducing it or ameliorating its effects but also to counter its potential impact on the democratic processes which underpin these functions. Overly complex laws which are difficult to administer consistently, efficiently and transparently pose a risk to their democratic legitimacy since respect for them is bound to be undermined by evidence of inconsistency, non-compliance and obscurity. Moreover, as the Brexit referendum debate and process have illustrated, there is a risk that legal realities which are overlaid with complexity are simply ignored or glossed over in public discourse and political processes. There are also risks to the realisation by citizens of their rights, particularly those in the areas of public law where the impact of complexity is most marked. Academic analysis of such rights, the degree of respect paid to them and the extent of their practical enforceability needs to be informed by a greater understanding of complexity so that its actual or potential impact can be better assessed.

There is still some way to go in understanding complexity and especially in measuring it effectively. Furthermore, it is unclear how far complexity indices could be used to reform schemes' or systems' established policy objectives. As we have seen, unless such objectives change significantly it is unlikely that overall system complexity can be addressed fundamentally. Thus, it is necessary to be cautious, even about simplification reforms such as those aimed at addressing complexity in the welfare or tax systems, since they have tended to leave the system's objectives unaltered. Yet if simplification is taken too far it can carry risks to the realisation of inherent programme goals, such as meeting need and targeting policies of support or redistribution.

In whatever way complexity is measured, it remains a relative concept. Simplification therefore involves a shift along a continuum. The benefits of such a shift are easily identified. But while simpler processes will generally benefit all, simpler rules carry risks. They often involve broad distinctions which can operate restrictively against whole categories of people, denying (social) justice to individuals within them whose circumstances may warrant the kind of differential treatment that would be possible under more complex and multi-pointed rules. Not everyone would agree with Epstein's argument that since, in making a single critical distinction, a simple rule can assist in the organisation of 'large areas of social life', we should 'celebrate tests that give us a 95 percent fit to their chosen end, and not bemoan the 5 percent of cases that stand between us and some unattainable perfection' (Epstein, 1995, p. 42).

Notes

1 See for example Bakirtzi (2011), referring at various points to efforts to simplify tax/social insurance contributions administration and interfaces across a range of states.
2 This is intended as the principal means-tested benefit providing assistance towards living costs, including housing costs, for those out of work or in low-paid employment.
3 These changes were made under the Welfare Reform Act 2012. Personal Independence Payment is the main benefit to assist with the costs associated with disability; see Harris (2014).
4 The higher figure was advanced by the Economic Dependency Working Group (2009, p. 301); the lower figure was given in DWP (2010b, ch.7 para 7).
5 In Switzerland, a proposed basic income scheme was rejected by a national referendum in 2016.
6 But see Epstein (1995, p. 28): 'It is . . . a dangerous mistake to think of a rule as simple just because it is short'.
7 See for example the Social Security (Personal Independence Payment) (Amendment) Regulations 2017 (SI 2017/194), seeking to counter the effect of the decisions in *Secretary of State for Work and Pensions v LB (PIP)* [2016] UKUT 0530 (AAC) and *MH v Secretary of State for Work and Pensions (PIP)* [2016] UKUT 0531 (AAC).
8 See *R (Reilly and Wilson) v Secretary of State for Work and Pensions* [2013] EWCA Civ 66, ruling on the Jobseeker's Allowance (Employment, Skills and Enterprise Scheme) Regulations 2011 (SI 2011/917), which were declared *ultra vires* and replaced by the Jobseeker's Allowance (Schemes for Assisting Persons to Obtain Employment) Regulations 2013 (SI 2013/276).
9 Welfare Reform Act 2007 and the Employment and Support Allowance Regulations 2008 (SI 2008/794).
10 See now the Welfare Reform Act 2012 s.96, as amended by the Welfare Reform and Work Act 2016 s.8 (from 7 November 2016). See also the Benefit Cap

(Housing Benefit and Universal Credit) (Amendment) Regulations 2016 (SI 2016/909).

11 Welfare Reform Act 2012, ss.96 (as amended) and 97. Details are added by regulations, such as the Universal Credit Regulations 2013 (SI 2013/376), regs 78–83.

12 As to the potential consequences, see House of Commons Work and Pensions Committee, 2011b, para 65.

13 Pensions Act 2014; State Pension Regulations 2015 (SI 2015/13). See DWP (2013).

14 European Union (Withdrawal) Bill, 2017, as first published, clause 7(1).

15 Influential was Economic Dependency Working Group (2009). This Group was established by the Centre for Social Justice.

16 This index is based on 'a weighted average of the number of words per sentence, and the number of long words per word. An interpretation is that the text can be understood by someone who left full-time education at a later age than the index'; Gunning-fog-index.com (accessed 9 August 2017).

17 Based on OTS (2015), Annex.

18 The bedroom tax rules alone have a specific policy objective of ensuring that the state does not subsidise accommodation the size of which (in terms of the number of bedrooms) exceeds the needs of the claimant; see the Housing Benefit Regulations 2006 (SI 2006/213), reg.B13. On the litigation, see in particular *R (MA & ors) v Secretary of State for Work and Pensions* [2014] EWCA Civ 13; *R (Rutherford & ors.) v Secretary of State for Work and Pensions* [2016] EWCA Civ 29; and *R (Carmichael and Rourke) v Secretary of State for Work and Pensions* [2016] UKSC 58.

19 SI 2006/213 n.18 above.

Bibliography

Arnold, Brian J. (2009), 'Australian Tax Rewrite', 19(1) *Revenue Law* 1.

Arts, W. and J. Gelissen (2002), 'Three Worlds of Welfare Capitalism or More? A State-of-the-Art Report', 12(2) *Journal of European Social Policy* 137.

Bakirtzi, Effrosyni (2011), *Case Study in Merging the Administrations of Social Security Contribution and Taxation* (Washington, DC: IBM Centre for the Business of Government).

Bennett, Fran, Mike Brewer and Jonathan Shaw (2009), *Understanding the Compliance Costs of Benefits and Tax Credits* (London: Institute for Fiscal Studies).

Beveridge, William H. (The Inter-Departmental Committee on Social Insurance and Allied Services) (1942), *Social Insurance and Allied Services* (Cmnd 6404) (London: HMSO).

Boffey, Daniel (2015), 'Dutch City Plans to Pay Citizens a "basic income", and Greens Say It Could Work in the UK', *The Guardian*, 26 December 2015.

Bowler, Tracey (2009), *Countering Tax Avoidance in the UK: Which Way Forward*, TLRC Discussion Paper No.7 (London: Institute for Fiscal Studies).

Browne, James and Barra Roantree (2013), *Universal Credit in Northern Ireland: What Will Its Impact Be, and What Are the Challenges?* (IFS Report R77) (London: Institute for Fiscal Studies).

Carmel, Emma and Theodoros Papadopoulos (2009), 'Governing Social Security: From Protection to Markets', in J. Millar (ed.) *Understanding Social Security: Issues for Policy and Practice* (Bristol: Policy Press, 2nd ed.), 93.

Craig, Paul (1997), 'Formal and Substantive Conceptions of the Rule of Law: An Analytical Framework', *Public Law* 467.

Crettez, Bertrand, Bruno Deffains and Régis Deloche (2009), 'On the Optimal Complexity of Law and Legal Rules Harmonization', 27 *European Journal of Law and Economics* 129.

Department for Exiting the European Union. (2017), *Legislating for the United Kingdom's Withdrawal from the European Union* (Cm 9446) (London: Department for Exiting the European Union).

DWP (Department for Work and Pensions). (2007), 'Further Supplementary Note from DWP', in House of Commons Work and Pensions Committee (ed.) *Benefit Simplification* (HC 463-II) (London: The Stationery Office), Ev 132.

DWP (Department for Work and Pensions). (2010a), *21st Century Welfare* (Cm 7913) (London: The Stationery Office).

DWP (Department for Work and Pensions). (2010b), *Universal Credit: Welfare That Works* (Cm 7957) (London: DWP).

DWP (Department for Work and Pensions). (2011), *Government's Response to the Consultation on Disability Living Allowance Reform* (Cm 8051) (London: DWP).

DWP (Department for Work and Pensions). (2012a) Press Release 'Iain Duncan Smith: Early Roll-Out of Universal Credit to Go Live in Manchester and Cheshire' (24 May 2012) <www.gov.uk/government/news/iain-duncan-smith-early-roll-out-of-universal-credit-to-go-live-in-manchester-and-cheshire>, accessed 22 May 2017.

DWP (Department for Work and Pensions). (2012b), *Universal Credit: The Impact on Passported Benefits: Report by the Social Security Advisory Committee and response by the Secretary of State for Work and Pensions*, Cm 8332 (Norwich: The Stationery Office).

DWP (Department for Work and Pensions). (2013), *The Single-Tier Pension: A simple Foundation for Saving* (Cm 8528) (London: The Stationery Office).

DWP (Department for Work and Pensions). (2017), 'Single Departmental Plan 2015–2020 (Updated)' <www.gov.uk/government/publications/dwp-single-departmental-plan-2015-to-2020/dwp-single-departmental-plan-2015-to-2020>, accessed 22 May 2017.

Economic Dependency Working Group. (2009), *Dynamic Benefits* (London: Centre for Social Justice).

Epstein, Richard A. (1995), *Simple Rules for a Complex World* (Cambridge, MA: Harvard University Press).

Epstein, Richard A. (2004), *The Optimal Complexity of Legal Rules*, John M. Olin Law and Economics Working Paper No.210 (Chicago: The Law School, University of Chicago).

Esping-Andersen, Gøsta (1990), *The Three Worlds of Welfare Capitalism* (Cambridge: Polity Press).

European Commission. (2016), *Better Regulation: Delivering Better Results from a Stronger Union* (COM(2016) 615 final) Brussels 14.9.2016 (Brussels: European Commission).

Finch, David, Adam Corlett and Vidhya Alakeson (2014), *Universal Credit: A Policy Under Review* (London: Resolution Foundation) <www.resolutionfoundation.org/app/uploads/2014/09/Universal-Credit-A-policy-under-review1.pdf>, accessed 22 May 2017.

Finn, Dan (2011), 'Welfare to Work After the Recession: From the New Deals to the Work Programme', in C. Holden, M. Kilkey and G. Ramia (eds.) *Social Policy Review 23: Analysis and Debate in Social Policy, 2011* (Bristol: Policy Press), 127.

Freud, David (2007), *Reducing Dependency Increasing Opportunity: Options for the Future of Welfare to Work* (Leeds: Corporate Document Services).

Geyer, Robert (2003), 'Europeanisation, Complexity and the British Welfare State', Paper presented to the UACES/ESRC Study Group on The Europeanisation of British Politics and Policy-Making, Department of Politics, University of Sheffield, 19 September 2003, at http://aei.pitt.edu/1719/, accessed 21 April 2017.

Geyer, Robert and Samir Rihani (2010), *Complexity and Public Policy: A New Approach to 21st Century Politics, Policy and Society* (London: Routledge).

Ghai, Dharam (2002), *Social Security Priorities and Patterns: A Global Perspective*, DP 141/2002 (Geneva: ILO).

Grover, Chris (2012), 'Abolishing the Discretionary Social Fund: Continuity and Change in Relieving "special expenses" ', 19 *Journal of Social Security Law* 12.

Harris, Neville (1989), *Social Security for Young People* (Aldershot: Avebury).

Harris, Neville (ed.) (2000), *Social Security Law in Context* (Oxford: Oxford University Press).

Harris, Neville (2013), *Law in a Complex State. Complexity in the Law and Structure of Welfare* (Oxford: Hart).

Harris, Neville (2014), 'Welfare Reform and the Shifting Threshold of Support for Disabled People', 77(6) *Modern Law Review* 888.

Harris, Neville (2016), 'Demagnetisation of Social Security and Health Care for Migrants to the UK', 18(2) *European Journal of Social Security* 130.

Haux, Tina (2011), 'Lone Parents and the Conservatives: Anything New?' in C. Holden, M. Killkey and G. Ramia (eds.) *Social Policy Review 23: Analysis and Debate in Social Policy, 2011* (Bristol: Policy Press), 147.

HM Government. (2017), *The United Kingdom's Exit from and New Partnership with the European Union* (Cm 9417) (London: HM Government).

HM Treasury. (2016), *Budget 2016* (HC 901) (London: HM Treasury).

Hood, Andrew and David Phillips (2015), Benefit Spending and Reforms: The Coalition Government's Record (IFS Briefing Note BN160) (London: Institute for Fiscal Studies).

House of Commons Committee of Public Accounts. (2006), *Tackling the Complexity of the Benefits System* (HC 756) (London: The Stationery Office).

House of Commons Committee of Public Accounts. (2013), *Department for Work and Pensions: Contract Management of Medical Services* (HC 744) (London: The Stationery Office).

House of Commons Exiting the European Union Committee. (2016), First Report of Session 2016–17, *The process for exiting the European Union and the Government's Negotiating Objectives* (HC 815) (London: House of Commons).

House of Commons Exiting the European Union Committee. (2017), Third Report of Session 2016–17, *The Government's Negotiating Objectives: The White Paper* (HC 1125) (London: House of Commons).

House of Commons Work and Pensions Committee. (2007a), *Benefits Simplification* (HC 463-I) (London: The Stationery Office).

House of Commons Work and Pensions Committee. (2007b), *Benefit Simplification: Government's Response to the Committee's Seventh Report* (HC 1054) (London: The Stationery Office).

House of Commons Work and Pensions Committee. (2011a), *Work Programme: Providers and Contracting Arrangements* (HC 718) (London: The Stationery Office).

House of Commons Work and Pensions Committee. (2011b), *The Role of Incapacity Benefit Reassessment in Helping Claimants into Employment* (HC 1015-I) (London: The Stationery Office).

House of Commons Work and Pensions Committee. (2012), *Universal Credit Implementation: Meeting the Needs of Vulnerable Claimants* (HC 576-I) (London: House of Commons).

House of Lords Select Committee on the Constitution (2017), 9th Report of Session 2016–17, *The Great Repeal Bill and delegated powers* (HL Paper 123) (London: House of Lords).

Hudson, John (2009), 'Social Security and Information Technology', in J. Miller (ed.) *Understanding Social Security: Issues for policy and practice* (Bristol: Policy Press, 2nd ed.), 295.

James, Simon and Alison Edwards (2008), 'Developing Tax Policy in a Complex and Changing World', 38(1) *Economic Analysis and Policy* 35.

Jewell, Christopher J. (2007), 'Assessing Need in the United States, Germany, and Sweden: The Organization of Welfare Casework and the Potential for Responsiveness in the "Three Worlds"', 29(3) *Law & Policy* 380.

Kades, Eric (1996), 'The Laws of Complexity and the Complexity of Laws: The Implications of Computational Complexity Theory for the Law', 49 *Rutgers Law Review* 403.

Kalliomaa-Puha, Laura, Anna-Kaisa Tuovinen and Olli Kangas (2016), 'The Basic Income Experiment in Finland', 23(2) *Journal Social Security Law* 75.

Kaplow, Louis (1995), 'A Model of the Optimal Complexity of Legal Rules', 11(1) *Journal of Law, Economics and Organization* 150.

Kassam, Ashifa (2017), 'Canadian Cities to Test Universal Basic Income', *The Guardian* 25 April.

Katz, Daniel M. and M. J. Bommarito II (2014), 'Measuring the Complexity of the Law: The United States Code', 22 *Artificial Intelligence and Law* 337.

Kerwin, Cornelius M. (2003), *Rulemaking: How Government Agencies Write Law and Make Policy* (Washington, DC: CQ Press).

Lee, Natalie (2003), 'The New Tax Credits', 10(1) *Journal of Social Security Law* 7.

Martin, Phillipe (2017), 'Ideas, Controversies and Proposals About the Universal Basic Income in France', 24(1) *Journal Social Security Law* 31.

Martinelli, Luke (2017), *The Fiscal and Distributional Implications of Alternative Universal Basic Income Schemes in the UK* (Bath: Institute for Policy Research).

Mashaw, Jerry L. (1983), *Bureaucratic Justice: Managing Social Security Disability Claims* (New Haven and London: Yale University Press).

Mikhailova, Anna (2017), 'Most Denied the New "flat-rate" State Pension', *The Sunday Times (Money)*, 16 April 2017, p. 12.

NAO (National Audit Office). (2005), *Department for Work and Pensions: Dealing with the Complexity of the Benefits System* (HC 592) (London: The Stationery Office).

NAO (National Audit Office). (2011), *Means Testing: Report by the Comptroller and Auditor General* (HC 1464) (London: NAO).

NAO (National Audit Office). (2015), *Department for Work and Pensions: Welfare Reform – Lessons Learned* (London: NAO).

NAO (National Audit Office). (2016), *Department for Work and Pensions: Contracted-Out Health and Disability Assessments* (HC 609) (London: NAO).

NAO (National Audit Office). (2017), *Digital Transformation in Government* (HC 1059) (London: NAO).

OTS (Office of Tax Simplification). (2013), *Definitions in Tax Legislation and Their Contribution to Complexity* (London: OTS).

OTS (Office of Tax Simplification). (2015), *The OTS Complexity Index* (London: OTS). <www.gov.uk/government/uploads/system/uploads/attachment_data/file/438587/OTS_complexity_index_methodology_June_2015.pdf>, accessed 17 March 2017.

Rhodes, Mary Lee, Joanne Murphy, Jenny Muir and John A. Murray (2011), *Public Management and Complexity Theory: Richer Decision-Making in Public Services* (New York: Routledge).

Roberts, Rachel (2017), 'Scotland Set to Pilot Universal Basic Income Scheme in Fife and Glasgow', *The Independent*, 2 January 2017.

Royston, Sue (2007), *Benefits Simplification and the Customer* (London: DWP).

Ruhl, J. B. and Daniel M. Katz (2015), 'Measuring, Monitoring, and Managing Legal Complexity', 101 *Iowa Law Review* 191.

Sainsbury, Roy and Kate Stanley (2007), *One for All: Active Welfare and the Single Working Age Benefit* (London: IPPR).

Salter, David (2010), 'The Tax Law Rewrite in the United Kingdom: Plus ca Change Plus c'est la Meme Chose?' *British Tax Review* 671.

Salter, David (2011), 'The Ipsos Mori Review of Rewritten Income Tax Legislation – Contrasting and Converging Viewpoints', *British Tax Review* 622.

Schuck, Peter H. (1992), 'Legal Complexity: Some Causes, Consequences and Cures', 42(1) *Duke Law Journal* 1.

Scott, Colin (2017), 'The Regulatory State and Beyond', in P. Drahos (ed.) *Regulatory Theory: Foundations and Applications* (Canberra, ANU Press), 265.

Simpson Caird, Jack (2017), House of Commons Library, Briefing Paper No.7793, 23 February 2017, *Legislating for Brexit: the Great Repeal Bill* (London: House of Commons Library).

Social Security Advisory Committee. (2007), *Twentieth Report: 2007* (Leeds: Corporate Document Services).

Spicker, Paul (2011), *How Social Security Works: An Introduction to Benefits in Britain* (Bristol: Policy Press).

Tax Law Review Committee. (1996), *Final Report on Tax Legislation* (London: Tax Law Review Committee).

Timmins, Nicholas (2016), *Universal Credit: From Disaster to Recovery?* (London: Institute for Government).

van Parijs, Phillipe and Yannick Vanderborght (2017), *Basic Income: A Radical Proposal for a Free Economy and Sane Society* (Cambridge, MA: Harvard University Press).

White, Hannah and Jill Rutter (2017), *Legislating Brexit: The Great Repeal Bill and the Wider Legislative Challenge* (London: Institute for Government).

Wright, Oliver (2017), 'Clear Up Repeal Bill or Judges Will Do It for You, May Warned', *The Times*, 9 August 2017, pp. 8–9.

Zacker, Hans (1987), 'Juridification in the Field of Social Law', in G. Teubner (ed.) *Juridification of Social Spheres* (Berlin: Walter de Gruyter), 373.

4 Asylum and complexity

The vulnerable identity of law as a complex system[1]

Thomas E. Webb

Introduction

Fineman argues that the essence of human vulnerability is the possibility of that change will expose vulnerability (Fineman, 2008–09, p. 12). Refugee movement is precipitated by persecution of one kind or another. By definition, those subject to persecution are vulnerable because they are subject to changes in their lives outside of their control. Those forced to flee their homes because of changes either in their immediate locality, or their country, represent the quintessence of vulnerability, a situation where the possibility of irresistible, negative change has become a reality. In the United Kingdom the response to this realisation of vulnerability has been mixed. The humanitarian and administrative justice challenges posed by refugee applications are the subject of intense political debate. In the national media, the credibility of asylum claimants has historically been viewed with some scepticism (Matthews and Brown, 2012, pp. 802–804; Philo, Briant and Donald, 2013, pp. 29–32). The political system, ever sensitive to popular political opinions shaped, in part, at least, by the concerns of the mass media, have echoed these concerns (Khosravinik, 2010, pp. 10–11; Pearce et al., 2009, pp. 152–153). As a consequence, the structures for understanding the concept of asylum, and each instance of asylum claimed, have come under pressure. What, though, are the consequences for the legal processes which have developed alongside these political considerations? Are simple solutions to the eminently complex challenges posed by intersecting vulnerabilities and conflicting national sentiments possible? For example, should the legal system accept political efforts to minimise, exclude or invisibilise the refugee as a solution to this complexity, and the tensions which their presence produces? Or will this reduce the meaning of law in this context, and the legitimacy which attaches to the use of social power through law, into pure politics?

In this chapter I argue that the perception of legitimacy is essential to the legal system's identity, and that identity is vital to the continuation of law's ability to claim to be *the* site for the resolution of legal disputes. To demonstrate this, I consider the extent to which complexity theory requires law to incorporate, rather than exclude, the concept of vulnerability. I argue that if vulnerability is not incorporated this poses risks to law's perceived legitimacy and, thus, law's identity. To show this, I reflect on how the concept of vulnerability is not only relevant to individual humans, or to humanity in general, but – drawing on DeLanda's concept of the social assemblage as a means to permit a more nuanced understanding of the idea of the 'system' (DeLanda, 2006) – social systems and processes too. To better elucidate how complexity theory deals with vulnerability, in places I contrast the complexity approach to that of autopoiesis. However, since my aim is to gain an appreciation of the complexity approach to vulnerability as a vehicle for understanding the importance to questions of legitimacy/identity of recognising vulnerability, these considerations should be treated as incidental (for a more detailed discussion of autopoiesis and vulnerability see Philippopoulos-Mihalopoulos and Webb, 2015). The primary reason for including any discussion of autopoiesis is as a foil to consider the consequences of either including or excluding a concept of human vulnerability from systemic thinking, and the implications of this for the systemic vulnerability of the legal system viewed from a complexity perspective.

This discussion is framed in the context of law's approach to applications for asylum in the United Kingdom, though there are other contexts in which it would also be applicable, most obviously with regards to mental health adjudication, or concerns around human rights more generally. Much of the chapter is dedicated to conceptually unpacking the concepts of identity, difference and vulnerability vis-à-vis complexity theory and legal complexity specifically. The closing part of the chapter brings together this discussion to apply the conclusions drawn to two recent judgments – *R (on the application of Detention Action)* and *R (on the application of Public Law Project)*.[2] I argue that the interpretive action taken by the courts in these cases can be understood through the language of complexity theory and, specifically, how it interprets identity, difference and vulnerability as an attempt to minimise exclusion and maximise integration in order to maintain law's legitimacy and, thus, its identity. This provides one explanation for why the courts responded to the cases in the way that they did. That is, the need to pay attention to the systemic risk posed to law's legitimacy, and as a consequence, law's identity, by invisibilising or otherwise excluding the vulnerable subject from view.

It is evident that by examining the question of vulnerability in administrative justice processes – such as those of asylum – through complexity theory we can encourage the visibilisation of the vulnerable subject in those processes and in the discourses which surround them. This enables two things. First, it causes social systems, and particularly the legal aspects of such systems, to confront the question of their own conceptual and material vulnerability if exclusion – of which vulnerability is a cause and a consequence – is not responded to. Second, and for the individual human experience perhaps the more important point, it demonstrates the reasons why social exclusion is detrimental to both individuals and society (see Neves, 2001; Philippopoulos-Mihalopoulos and Webb, 2015).

Constructing systems and the meaning of difference

With my fellow editors I have given a more detailed view of the mechanics and consequences of complexity theory for law and legal systems elsewhere in this volume. Whereas that earlier discussion was intended as a general account of our broad view of complexity theory and law, in what follows I reflect on how complexity theory views the concept of system and consider how the ideas of emergence and boundaries lead to a productive understanding of difference. It will be shown that the ability to establish difference is essential to developing an identity (difference, albeit on alternative conceptual foundations, is also key to establishing identity in autopoiesis, see Luhmann, 1988, for example p. 16, 1992a, p. 172; Teubner, 1993, p. 9).

The only caveat to add to my observations on complexity theory and, to the extent that I deal with it here, autopoiesis, is that, in accordance with the modesty required of all observations based in complexity theory thinking (Cilliers, 2005, p. 256, 2010, p. 8; Cilliers et al., 2007, p. 130; Preiser and Cilliers, 2010, p. 269), I do not think that a complexity theory approach provides a complete explanation for the behaviour of, *inter alia*, the law. Rather, it provides a shift in analytical perspective which permits access to previously unconsidered reflections on law and legal behaviour. Likewise, although I am sceptical of the analytical utility of autopoiesis when considering what can loosely be called meta-, and perhaps also meso-, level social processes, as a framework for conceptualising individual, observer-defined social processes, autopoiesis has value. With these points in mind, it is now possible to begin a brief exploration of complexity theory thinking.

Systems

There is a risk when discussing complexity theory to tie oneself in knots over definitions of what is meant by 'the system' and thus never get to the substance of applying the theory. Autopoiesis superficially avoids this problem by defining systems according to certain social functions; this approach is called functional differentiation (see Luhmann, 1992b). However, if one steps outside of the reality constructed by autopoiesis, this is no solution to the definitional problem since it assumes that delineating the boundaries between the system and the rest of society – the system's environment – is to be achieved by assigning certain social functions the character of systems. There is no objective reason for doing so, nor is there any objective way for any of these given systems to know what is legal, other than to assert that it is so. This can be seen in King's reasoning:

> Any act or utterance that codes social acts according to this binary code of lawful/unlawful may be regarded as part of the legal system, no matter where it was made and no matter who made it. The legal system in this sense is not confined, therefore, to the activities of formal legal institutions.
>
> (King, 1993, pp. 223–224)

On this basis, the question of what is legal is both relevant – since it entails the ascription of social meaning – and irrelevant – since the ascription of that meaning to law is presented as a foregone conclusion; the law always knows the meaning it ascribed to a social event was legal.

In the spirit of modesty which is at the heart of complexity theory, the concept of system is more malleable in complexity theory thinking. As will become clear from the following discussion, the idea that there are objectively identifiable systems is deeply problematic. This means, for example, that the 'legal system' in complexity theory thinking is not intended to indicate a discrete system *per se*. Instead, it should be taken as shorthand for a co-construction of different conclusions about social events and processes arising from the interactions between social assemblages that have (see again, DeLanda, 2006), for present explanatory purposes, been defined as legal. Similarly, those defining the legal system in this way, observers, are themselves as much a social assemblage as the subject of observation. The social assemblage is, again, intended only as shorthand for the collection of concepts, processes, objects and so on which go together to constitute a particular meaning in a given contingent time and place (for deeper elaboration see Philippopoulos-Mihalopoulos, 2015 pp. 47–49); it is a description of something which can be used *now* to productively engage in society (what is later referred to as a 'description strategy', see Cilliers, 2001, p. 141). Thus, to speak of complex systems is simply to ascribe descriptive parameters to a specified assemblage under observation for the purposes of analysis where it is thought that the assemblage displays the characteristics of a complex system (see Murray, Webb and Wheatley, in this volume). This does not mean that 'anything goes' (Cilliers, 1995, 1998, p. viii) in complexity theory thinking since any model which is patently nonsensical will not be engaged with. Rather, it entails a degree of pragmatism (Ansell and Geyer, 2017). A recognition that, since society is impossibly complex, and contingent, such that it defies modelling (Cilliers, 2000, p. 30, 2008, p. 46), to say anything useful at all we must set limits to our discussion, and seek out interaction with others to test and refine our limited descriptions.

Emergence and boundaries

Having outlined what I mean when I talk about complex systems, I can now briefly examine the key concepts within a complexity approach which are relevant to understanding how law's identity is dependent on a recognition of the concept of vulnerability. I begin with emergence and then consider the complexity understanding of boundaries.

Emergence is the essential first principle of complexity theory thinking, since without emergent behaviour, complex systems cannot exist. Emergence consists of two precepts: first, that *interaction between* the parts of the system – rather than the mere combination of the parts themselves – is what drives the creation of meaning within and between systems and thus the prospect of future interactions (Cilliers, 2010, pp. 6–7). Second, and following from this, the meaning, definition, scope or whatever other form of boundary one wishes to establish, of a social system, is dependent on that interaction (Richardson, 2004, p. 77;

Waldrop, 1994, pp. 63–66) and the context in which that interaction occurs (Cilliers, 2005, p. 263). That is, any boundary – a definition, an identity – is a product of emergent interaction between the assemblage under observation, and the observer, and is dependent on the context in which that interaction occurs (see DeLanda, 2006, pp. 10–11). Here one can see how the idea of the assemblage as shorthand for the way in which an observer's decision to establish specific analytical parameters has implications for the explanation they produce. This is because the construction of meaning, between observer and observed, is not one way; it is interactive, reflexive and 'determined relationally' (Cilliers, 2010, p. 6; de Villiers and Cilliers, 2010, p. 29). To understand this process in a little more detail it is necessary to consider the understanding of boundary in complexity theory thinking.

Boundary is a multifaceted concept (Webb, 2013). In complexity theory thinking, there are at least four understandings of boundary. The first, and most simplistic, understanding of boundary is as a dividing line. This construction demonstrates the importance of being able to differentiate oneself and one's descriptions from the environment to establish meaning and identity. The second understanding of boundary is intended to caution us against the overzealous use of the first. It entails the recognition that there is *no* boundary. That is, the recognition that society is irreducibly complex, and thus cannot be completely modelled, demonstrates that any boundary claim is merely temporary, a transient description to enable future interaction. On this view, any model is only ever a partial representation of the system, since any model purporting to describe society would have to be at least as complex as that which it seeks to describe (Cilliers, 2007, p. 161; Phillipopoulos-Mihalopoulos, 2010, p. 13). Nonetheless, having regard to the conclusions drawn about, first, the importance of being able to distinguish and, second, the impossibility of objectively distinguishing reveals the third meaning of boundary, as description strategy (Cilliers, 2001, p. 141). While accepting that creating complete models is an impossibility, it must also be acknowledged that to participate in meaningful interaction, an approximate understanding of society – *a* model – is needed.

The boundary as description strategy is a device which can be used to engage in productive interaction with society and reveal the fourth conception of boundary in complexity theory thinking: the boundary as interface. The description strategies of social assemblages, contingent descriptions of aspects of society, are employed to engage with other individuals to make sense of the world (Richardson, Cilliers and Lissack, 2001, pp. 8–9). Their subsequent form is a relational product of that interaction (see again Cilliers, 2010, p. 6; de Villiers and Cilliers, 2010, p. 29; see also: Cilliers, 1998, p. 4, 2001, p. 141; Philippopoulos-Mihalopoulos, 2015, p. 41 Richardson and Cilliers, 2001, p. 13; Webb, 2005, p. 237). By this process the interaction of models – description strategies – produces and reproduces emergence. It is the co-construction of the boundary that demonstrates why it is insufficient for law to merely assert its claim to be *the* legitimate site of legal decision-making. For complexity theory, law's identity is not constructed solely by law or legal processes; it is the product of emergent interaction, of the encounter between the self-understandings of assemblages, processes and concepts that claim the legitimacy afforded by identifying themselves as legal, and those assemblages compelled to engage with them; other social processes, and people.

Difference and exclusion

The ability to interact and reflexively reformulate accounts of boundaries brings us back to our first understanding of boundary as a device which establishes difference. One description strategy is not the same as another, it is contingent (Cilliers, 2005, p. 259), it is the product of nonlinearity (Cilliers, 2010, pp. 3–4), and it is valuable to other description strategies because of its unique perspective – because of its difference and capacity to be distinguished. The relative difference between description strategies is productively exploited to allow the observer to constantly revise their own imperfect understanding of the world. To flourish, therefore, complexity theory thinking reasons that individual description strategies require the existence of difference, established through the constant emergent renewal of the boundary. Productive interaction, that is, interaction which allows complex social assemblages to continually engage with the world, requires boundaries, and boundaries require productive interaction. In a very simplistic sense, we might observe that the adversarial legal system, with its deliberately divergent constructions of evidence (claimant/defendant), is a microcosm of the wider interactive, relational, and perpetual reconstruction of boundaries, because the interaction produces a new understanding; a verdict, a precedent, a judgment.

Although the concept of emergent interaction, and thus productive, constitutive, reflexive boundaries is superficially a positive one, it is predicated on assemblages possessing the capacity to access the interactive possibilities which the concept of emergence permits, and thus to establish difference, boundaries and identity. This recognition allows us to invert the concept of emergence to consider the systemic and human consequences of exclusion. If it is central to emergence that one is capable of interacting, of fuelling the relational experience which produces and reproduces meaning in society, then there must be correspondingly negative environmental, systemic and/or human consequences where this is not possible. Put simply, emergence is reliant on inclusion, and exclusion creates systemic, human and, by extension, environmental vulnerability (see further Neves, 2001, pp. 261, 263).

Neves (2001) observed that certain groups in society were much better placed than others to take advantage of the interactive possibilities presented by social systems, for example, because of their political, financial, or educational position. This meant that they could manipulate relationships, and their engagement with society, in ways simply not available to other individuals. Indicating that this arrangement was imbalanced, Neves referred to these individuals as being over-integrated and to those at the opposite end of the spectrum as being under-integrated (pp. 261–263). By *integration* both Neves and I mean the ability or inability to engage with systemic processes, rather than any pejorative meaning one might ascribe to 'social integration'. *Integration* in this context means that you have the facility to access a lawyer, an accountant, a political party. Over-integration would mean that, relative to your fellow citizens, you are more likely to secure a personally desirable outcome using the access to social interactions that your integration enables than if you were under-integrated. Clearly there are degrees of over- and under-integration (Neves, 2001, p. 262), and it is certainly relative. However, the point for complexity theory analysis is that access to the relational, interactive possibilities which allow one to establish difference from the environment, an identity, depends heavily on the degree of integration one can achieve. This is, in turn, partly dictated by the form which the structures for engaging in society take and the expectations they place on assemblages. Thus, if the law system is in principle premised on formal legal equality, but none-theless denies access to justice by placing financial barriers (e.g. court fees), or linguistic hurdles (e.g. jargon, the requirement to fill out complex forms), then this can have implications for the ability of individuals to access in practice the interactive – and, in the case of law, purportedly authoritative – processes available in theory.

In the light of this it can be seen that the principal risk to the individual of under-integration is exclusion from social interaction in a way that profoundly disadvantages you as an individual via the denial of access to the interactive possibilities which allow you to establish difference/identity (see also Philippopoulos-Mihalopoulos and Webb, 2015). Without the possibility of interaction, it is not possible to establish difference, and difference is an essential precursor to both further interaction and the maintenance of identity in complexity theory thinking. This may cause you to look for other solutions to the challenges you face than those offered by established frameworks. At the same time, those who occupy over-integrated, or at least sufficiently integrated frameworks are unable or unwilling to interact with you, because they are no longer able to differentiate you from the environment. Thus, exclusion is a profoundly dehumanising process. It has the effect of denying identity and thus access to those processes which might prevent the risks inherent in human vulnerability from being realised. This account of boundaries and their relationship to emergence has implications that are integral to understanding the challenge of vulnerability to establishing identity.

Vulnerable identities

By reflecting on how a traditional autopoietic understanding of vulnerability approaches change, where the realisation of uninitiated change is the essence of vulnerability, we can establish a useful counterpoint against which to understand the value of the model proposed by complexity theory. In so doing, it becomes possible to demonstrate the systemic risks of attempting to deny vulnerable individuals, such as refugees, the opportunity for unimpeded engagement with the legal system.

As already mentioned, the risk of change is at the heart of vulnerability. Autopoietic analysis stipulates that systems conceptualise change in their environment through their own internalised description of the environment via the processes of re-entry (Luhmann, 1992b, p. 411; Philippopoulos-Mihalopoulos, 2006, p. 226). Thus, change is always a product of how the system has, systemically, that is, self-referentially, understood its history and place in the environment and its difference from that environment. As Philippopoulos-Mihalopoulos has elaborated, at the core of functionally differentiated systems lies that which distinguishes them from the environment, 'identity is difference' (Philippopoulos-Mihalopoulos, 2010, p. 37; see also 2006, p. 226). In this way the system is to be defined *by itself* by what it is not, and accordingly it produces understandings of difference to maintain its identity:

> It produces the difference between the illusion of identification and the abyss of loss of identity. It also produces the difference between the system's continuous attempt to describe itself and a continuous interruption by its environmental exteriority which establishes a permanent dysfunction in the system . . . The system inclines to its form with its environment, clings onto it with a longing whose object is precisely the maintenance of this difference . . .
>
> (Philippopoulos-Mihalopoulos, 2010, p. 44)

One thus gets a sense of autopoietic identity as an inherently 'fragile, volatile, constructed thing' (Philippopoulos-Mihalopoulos and Webb, 2015, p. 457; see also Bankowski, 1996, p. 71). The fear that recognising fundamental shifts which the system is not able to conceptualise (or the possibility of such) presents autopoietic systems with the prospect that they will be unable to maintain their difference from the environment, dedifferentiation being tantamount to system death.

While the concept of identity in complexity theory is also built on the concept of difference (Cilliers, 2010, pp. 5–7, pp. 13–14), the earlier discussion of boundaries and emergence indicates that the notion of vulnerability and the value of change are to be embraced as creative forces. Whereas in autopoiesis the constant maintenance of difference via the perpetuation of self-referential functional differentiation is essential to the continuation of identity, in complexity theory identity is the product of interaction with other assemblages in the environment. Although the concept of identity is given a broad meaning, encompassing the 'myriad of influences that the self is exposed to every day (other people, the media, objects that it encounters, its own history, memories, perceptions, physical sensations)' it serves to demonstrate that the self is a product of its interactions (de Villiers and Cilliers, 2010, p. 27, note 30; see also Preiser and Cilliers, 2010, p. 267; Richardson, Cilliers and Lissack, 2001, p. 7). Identity exists relative to these structures, it 'has to form and operate within the structures and constraints provided by the environment, regardless of will, intellect and memory' (de Villiers and Cilliers, 2010, p. 33; also Cilliers, 2010, pp. 5–7). Furthermore, the self constitutes part of the environment of all other 'selves'; it is open to its environment such that 'it is impossible to point to some precise boundary where "we" stop and where the world begins' (de Villiers and Cilliers, 2010, p. 34). Not only is there no physical-conceptual boundary to self; there is also no temporal boundary to identity. It is subject to change over time, being the product of a set of prevailing interactions, influenced by our past; it is a 'network of traces' that forms a '(temporary) narrative' (de Villiers and Cilliers, 2010, p. 35; see also Cilliers, 2001, p. 146).

Nonetheless, while the boundary is not capable of final definition, it is always in a state of becoming; it *does* exist for the assemblage itself. Indeed, without the ability to conceptually disaggregate the self as an assemblage from its environment, one cannot speak of there being an assemblage at all – difference is 'a precondition for their [complex systems'] existence' (Cilliers, 2010, p. 5). Similarly, if one cannot differentiate one thing from another, it is not possible to give meaning to anything; 'meaning is the result of . . . distinctions, of the play of differences' (*ibid.*, p. 6). Whereas autopoiesis defines itself by its own internal self-construction of the other, for complexity theory, difference – identity – 'is determined relationally' (*ibid.*). An assemblage is not to be understood by reference to how it *sees itself* as being different but by an examination of, and interaction *with*, other systems.

It can therefore be said that in complexity theory identity is not an isolated concept, a function of the system's differentiation, but is a co-dependent, emergent product of the interaction of assemblages, via the interface of their boundary. Similarly, the concept of difference is not isolating or divisive; it acknowledges that we all have a unique experience of the world that informs our existence and that of others. However, this individuality of experience is only revealed through engagement. Difference is only discernible in the presence of others; it is a positive, relational consequence of interaction (Cilliers and Preiser, 2010, p. vii). If we return to the idea of relatively straightforward legal activity, the case hearing, we can see that a failure to make it through the doors of the court means that the experiences of those individuals who are palpably subject to the law will nonetheless remain largely unknown to it. From the perspective of the desire of the law system to maintain its legitimacy in the eyes of those people, this should be concerning for two reasons: first, because those experiences are denied to law, preventing the law system from refining its own expectations of how it should function in given circumstances, of refining its own description strategy. Second, the problems of those who cannot access law do not go away simply because access to law is denied. People with problems will look for solutions; they may seek alternative remedies via political action or, ironically, activity deemed unlawful by law. We thus understand that identity is fundamentally about being able to distinguish oneself from the environment, from other identities, and that the absence of an ability to distinguish is an intolerable problem. In complexity theory, identity permits interaction and reveals further affinities and differences between identities. This emergent process produces and is produced by the reflexive reformulation of identity in the face of interaction.

In what sense is this understanding of 'identity' vulnerable? To answer this question we need to consider the idea of vulnerability in a little more detail. If we start with a return to Fineman's definition given at opening of this chapter we see that vulnerability is the exposure of all things, especially humans and human systems, to the risk of change, especially a change that we are not equipped to resist (Fineman, 2008–09, p. 12). In recognising the urgent need to interact as being fundamental to the construction of identity, we can see that the loss of interaction – of integration, of inclusion in interactive processes – is necessarily detrimental. The risk of this loss is encapsulated in the concept of vulnerability because irresistible, often unlooked-for change breaks the interactive cycle I have been discussing. The loss of interactive opportunity excludes you, depriving you of the interaction that grants identity. It subsumes you into the background context of the environment.

This conceptual understanding of vulnerable identity must also be grounded in the material, especially in the bodily existence of humans, human institutions and the world which they inhabit. The necessity of developing the idea of vulnerability in this way springs, in part, out of the implicit and explicit connections between the discourse on vulnerability, bodies and feminist approaches to legal studies (for example Bottomley, 2002; Fitzgerald, 2010; Sherwood-Johnson, 2013), which has contributed to the exploration of vulnerability. However, that discourse has also made it clear that the concept of vulnerability is intuitively recognisable in all aspects of human social life. Thus, Fineman concludes that 'vulnerability is – and should be understood to be – universal and constant, inherent in the human condition' (Fineman, 2008–09, p. 1). The material aspect of vulnerable identity is therefore revealed in the recognition that, while one might mitigate some of the risks to which one is exposed through wealth and power, there is no getting away from the fact that your existence is a human one. The institutions on which we all rely – both public and private – consist, in part, of a physical infrastructure that is composed of mechanical and digital machinery and other humans (consider Philippopoulos-Mihalopoulos, 2015, pp. 41–42). What is more, though we might send satellites into space and put humans on the moon, every aspect of human existence currently depends on the continuation of life on Earth. The collapse of the systems which make our existence on Earth possible would demonstrate conclusively the universal nature of human vulnerability (see Fineman, 2008–09, pp. 8–10 and generally).

Having considered the material and conceptual aspects of vulnerable identity, we can now return to Neves' notion of under and over-integration to consider the possible consequences of exclusion arising from destabilising change. Neves would doubtless point to those less able to assert their legal rights and engage with political frameworks (Neves, 2001, p. 262) as being acutely unable to resist change. The inability to resist undesirable change can, in turn, make vulnerability move from an abstract feature of humanity to a burden on your existence. This can have consequences both for individuals and communities. Individually such change might mean you lack the financial resources to engage lawyers to assert your rights and thus render you unable to access the interactive processes of law systems. Where communities or particular groups are under-integrated, they may be unable to resist change imposed by the over-integrated with access to superior resources to command the attention of lawmakers, which might include something as simple as the right to vote or as contentious as capital to expend on lobbying, such as multinational organisations and political interest groups, which may argue for limitations on access to employment protection, housing or asylum.

This type of exclusion, and the concretisation of vulnerability which accompanies it, denies the productive, interactive, relational possibilities that complexity theory indicates are so important to the establishment of identity and the flourishing of ongoing emergent interaction. In this way, exclusion dehumanises the individual by removing the possibility of defining oneself in the context of a wider human society, preventing the formation of identity. A clear example of this can be found in the construction of refugees in societal discourse. Refugees are often homogenised and thus dehumanised; they are not thought of as individuals or even as individual bodies (Esses et al., 2013; Innes, 2010, p. 459; Khosravinik, 2010; Lewis, 2005, p. 7). The position of the refugee can be contrasted with those who are, systematically speaking, relatively well integrated into society, and who are often constructed as citizens with individual rights conceptually in terms of mental autonomy (for example, freedom of thought, conscience and religion); physically through their bodily integrity (rights against torture, unlawful imprisonment and to assembly); and procedurally (the right to a fair hearing). In this context, legal rights can be seen as a protection against vulnerability, a protection against certain types of change (in autopoiesis, a positive access route to social systems; see Verschraegen, 2002, pp. 264–268; Luhmann, 2008, p. 26). It should not, therefore, be surprising that those individuals, deprived of their individual, relationally differentiated identity through exclusion, seek riskier routes towards inclusion, towards the possibility of becoming individuals with a relationally differentiated identity again.

It is true that we might be concerned about how individuals seek to redress this balance because of the other negative consequences it produces, however, for complexity theory the answer is not to punitively contain these instances of circumvention as examples of counter-factual breaches of systemic expectations (see also Luhmann, 1992c, pp. 1426–1427). Indeed, reducing this consequence of individual vulnerability to a specific legal, often criminal wrong undermines *systemic* resilience and creates further vulnerability for systems by, *inter alia*, preventing deeper consideration of how other aspects of, for example, the law system may be seen as contributing to the sources of that counter-factual behaviour. Containing understanding in this limited way prevents 'more adventurous, deeper structural couplings between systems' between assemblages (Philippopoulos-Mihalopoulos and Webb, 2015, p. 457). Conversely, the internalisation of the concept of vulnerable identity, and the recognition that the concept attaches to all social assemblages, not just people, has the potential to enable new interactions (see also p. 456). Moreover, adopting a more flexible view of what constitutes an agent capable of engaging in social interaction opens new possibilities for non-traditional framings of social relationships for analytical purposes. While at an environmental and systemic level this does not 'solve' the challenges of what happens to individual assemblages when vulnerability bites, the very acceptance of the potential for systemic processes to produce situations which generate exclusion allows society and social interaction to internalise the concept of exclusion, and the nature of what it excludes, and so articulate a response to the risks of exclusion.

Excluding vulnerable bodies

Having recognised the conceptual possibility that vulnerable identity will produce negative consequences if both the prospect of vulnerability concretises, and the processes designed to respond to it prevent the establishment of difference, and thus identity, we can now explore the material consequences and opportunities of vulnerable identity. In particular, I propose to consider how the recognition of vulnerable identity in complexity theory thinking allows the production of stabilising forces and greater interactive possibilities. To do this, it is helpful to ground the discussion by briefly reflecting on how the body, as a key site for the concretisation of vulnerable identity, is conceived in complexity theory thinking. This, in turn, permits a consideration of the implications of vulnerable identity for assemblages that takes note of both the conceptual and material consequences which flow from it. Ultimately this shows the dangers of excluding assemblages, such as individual humans, from integration with systemic processes, and warns against any approach to the treatment of asylum applicants that seeks to make it harder for them to engage directly with legal processes.

In stark contrast to the exclusion of the physical and psychological existence of human bodies from systemic autopoietic contemplation (but see Philippopoulos-Mihalopoulos and Webb, 2015), those with connections to complexity theory thinking have expressly acknowledged the material and the psychological. Speaking of human bodies as the subject of punishment by law, for example, DeLanda writes,

> Like all social assemblages the material role in organizations is first and foremost played by human bodies. It is these bodies who are ultimately the target of punishment. But punitive causal interventions on the human body are only the most obvious form of enforcement of authority. Other enforcement techniques exist . . . a set of distinctive practices involved in monitoring and disciplining the subordinate members of, and the human bodies processed by, organisations.
>
> (2006, p. 72)

Such an understanding of bodies by organisations requires also that they be aware of the physical distribution of bodies in order to execute procedures on them and to know of their location in time in order to stipulate 'cycles and repetitions' of those processes (*ibid.*; also Philippopoulos-Mihalopoulos, 2010, p. 88). Space and time are also important to understand when a body is within and without the jurisdiction of the organisation (DeLanda, 2006, p. 73). In view of this, the material, psychological and temporal manifestation of the human body is to be viewed as both a site of interaction and an interactor with and within social assemblages. Ultimately these ways of acting on bodies help the organisation, for example, the law system, to maintain its legitimacy and authority to continue acting on those bodies. This is important because it allows the assemblage to exist alongside, in the context of, and as part of combinations of other assemblages 'as part of populations of other organizations with which they interact' (DeLanda, 2006, p. 75). The human body is thus fundamentally implicated in how organisations conceive of their own identity and how it is, in turn, perceived by other social systems. By extension, the material aspects of larger social assemblages are also integral to the operation of complex, emergent processes, in part, because of how they operate alongside human bodies. For example, the physical presence of bodies in courts come to be interpreted in the context of that space; they act according to the expectations demanded by the setting and are in the presence of other bodies there to carry out or witness the judicial process.

Just as the importance of the body is established in complexity theory–aligned thinking (DeLanda, 2006), the idea of embodiment is not a new concept in law (see, for example, Bottomley, 2002). It is therefore unsurprising that both complexity theory and legal studies have come to view the body, and the idea of the subject, as a constructive and disruptive force that presents new information for consideration and which must be engaged with (*ibid.*, p. 131). The creative/disruptive potential of the body lies in its contingency, the novelty which is established by each new interaction between the body and other assemblages, of the presence of 'continuous uncertainty and ambiguity' (Phillipopoulos-Mihalopoulos, 2015, p. 43). This recognition demands in both legal studies and complexity theory that the body be internalised by the system, and runs counter to theoretical approaches which call us to '[leave] our bodies behind', such as orthodox autopoiesis (Bottomley, 2002, pp. 130–131, see also 135–137). Without the internalisation of the body, especially the body as a site of vulnerability, we would struggle to see that the material is as integral to the perpetuation of emergence, and emergent identity for human systems, as the conceptual and procedural aspects of vulnerability (see also *ibid.*, pp. 140–146).

This understanding of embodiment as a positive force must also be coupled with the risks (and the possibility of growth) which vulnerability brings. As Fineman observes, vulnerability encapsulates both the positive aspects of embodiment – of potential, possibility and becoming – and the negative – of suppression and exclusion. The desire to maintain the positive and overcome the negative aspects of vulnerability are what 'make us reach out to others, form relationships, and build institutions' to engage in interaction (Fineman, 2012, p. 71). At the same time, if the physical presence of the body is denied – for example, by the extra-territorialisation of decision-making processes or the provision for appeals against asylum decisions to only be made out of country, then much of the weight offered by the body is lost.

In consequence, if the human body, and the material existence it evidences, is bound up with the understanding of law as a complex, productive, emergent assemblage, then actions which exclude the body from consideration, or which minimise the value ascribed to the experiences of the body at a systemic level, place the identity of the system in jeopardy. This is partly because the physical vulnerability of the body is a key aspect of the physical vulnerability of law as an assemblage. The identity of law is wholly reliant on being able to maintain its claim that it is *the* site to solve legal disputes. If the body is excluded as part of the more general invisibilisation of certain categories of person, then this creates further opportunities to call into question the validity of law's identity. In such circumstances, the body is forced to seek alternative ways of gaining inclusion – for example, by evading port authorities and not requesting asylum in an attempt to participate, to integrate – in the social processes of society, such as the economy. At the same time, to deny consideration of the body or to take measures which undermine it as an important site of emergence undermines the creative possibilities – the access to new, important information, new relational connections – which are of central importance to emergent processes.

How, then, will the exclusion of the vulnerable subject in the context of asylum increase the risk that the complex system's own material vulnerability will be engaged? For some indications of the risks to which a complex system would be exposed by not incorporating the vulnerable body we can consider the approach taken by autopoiesis. Autopoiesis seeks to conceptually invisibilise 'vulnerability and the possibility of dependency', assuming that such an act of cognitive denial and normative blindness has the effect of 'eliminat[ing] the experience of either in individual lives', but this is not the case (Fineman, 2012, p. 90). While such an action might have the effect of communicatively excluding the individual from participation in the functionally differentiated processes of society, and especially prevents engagement with legal processes, the body and its physical and psychological distress do not go away just because they are deliberately unobserved. Indeed, people *retain* their individual rights, but, because they struggle to engage with, or to be noticed by the system, they are prevented from realising those rights by the very system which gave them meaning in the first place (see further Philippopoulos-Mihalopoulos and Webb, 2015). This is especially problematic where the exclusion is made in pursuit of objectives which seem unreasonable.

In consequence procedures which show a preference, for example, for objective country information in place of subjective human experience actively minimises the role of humans. It is difficult to grapple with the disorderly nature of human experience, but the disinclination to engage in a reflexive process of considering that disorder exposes the material vulnerability of the system. While there are those who will abuse any system of immigration and asylum regulation, it is also empirically true that there are those who constitute examples of human suffering that do not easily meet the criteria of the 1951 Refugee Convention (Firth and Mauthe, 2013, pp. 500–501; see also Kelley, 2001). A reduction in the significance accorded to the marks on human bodies, and the damage to human minds, limits opportunities for creating meaningful, positively disruptive, substantive understandings that should enhance the richness of decisions and thus the quality of the reasons upon which they are based (see Baillot et al., 2014; Herlihy et al., 2010; Herlihy and Turner, 2015; Kagan, 2015; Sweeney, 2016). Furthermore, the patent existence of suffering bodies, coupled with a set of procedures which appears to either exclude them outright or to operate in such a way as to (inappropriately) exclude them in the final analysis, has implications for the legitimacy of the system. The perception and reality of procedural fairness is an essential component of legitimacy without which the legal system loses its authority to execute legal processes (DeLanda, 2006, p. 89). This is because, as I discussed earlier, in complexity theory thinking, human bodies and the actions practised on them are intimately bound up with the identity of social assemblages. Thus, the consequences of processes which enable the exclusion of vulnerable bodies have implications for how the identity of the law system emerges, because the meaning created by asylum applicants' bodies is bound up with the rest of their claim.

Where the legitimacy of a complex procedural assemblage is brought into question this can undermine the viability of the assemblage and damage its identity to the point that differentiation from other assemblages becomes impossible, for example, the differentiation of legal processes from politics. Just as systems conceived as autopoietic fear dedifferentiation, so, too, do those constructed as complex assemblages. While for complexity theory the motivation towards differentiation is not based on self-reference, but on emergent interaction, the cost of a failure of differentiation is still a loss of identity. The loss of legitimacy, on which organisations such as the legal system, and more specifically the network of adjudicative and administrative assemblages which constitute asylum processes depend, would deny law the exclusive jurisdiction to make pronouncements on these subjects. As legitimacy is the quintessence of legal identity, which differentiates it from mere political force; the loss of legitimacy is tantamount to a loss of identity.

To compound difficulties further, this loss risks a crisis (DeLanda, 2006, p. 90). While one might expect a relatively specialised aspect of the legal system, such as asylum, to have relatively isolated implications for the system as a whole, this is not necessarily the case. It is evident from popular discourses around the question of the United Kingdom's relationship with the European Union that immigration – in which asylum is inevitably, if inappropriately bound up – is a factor (Gietel-Basten, 2016). The constitutional changes wrought by *inter alia*, the 'Brexit' referendum and 2017 general election, while not necessarily amounting to a crisis, have evidently introduced a degree of uncertainty into the wider constitutional-legal assemblage for the time being.

Though it may seem counter-intuitive to suggest that the contribution towards instability of a popular concern with immigration could have been avoided if the vulnerable human bodies of refugees had been better incorporated into the thinking of the legal system and wider collection of social assemblages (especially, perhaps, the mass media), the earlier discussion demonstrates the negative consequences of not acknowledging such vulnerabilities. Similarly, the failure to acknowledge the vulnerabilities of refugees, to permit and engage in dehumanising discourses about them and to work to legally invisibilise them by, among other things, extra-territorialising decision-making processes, is as much a condemnation of their plight as it is of the concerns of those *citizens* who themselves are vulnerable because of their under-integration.

Observing vulnerability and exclusion

Until now I have mainly discussed the hypothetical risks to the body, system and environment of not internalising a concept of vulnerability – in particular, vulnerable identity – into legal-administrative justice processes. I have contended that one of the risks of the invisibilisation of vulnerability is that it undermines identity by challenging the capacity of an assemblage to differentiate itself from its environment. My central message has been that, when employing a complexity approach to analysis, the failure to relationally differentiate is a fundamentally bad thing because it compromises identity and, in law's case, the legitimacy of law systems to rule on matters which are purported to be within law's purview. What I have not discussed in any detail yet is that we can find examples in the case law, especially the case law concerning asylum applicants, of the law system taking measures to preserve its legitimacy in the face of the risk of dedifferentiation. By this I mean that we can see law processes acting to encourage inclusion and integration, and to limit the effects of attempts to promote the under-integration, invisibilisation, or complete exclusion of asylum applicants from law processes. In this way, asylum applicants are encouraged – at least in the qualified sense established in the two cases discussed in what follows – to engage with and articulate their problems to law rather than to seek alternative remedies to their problem. That is, the cases demonstrate two ways in which the law presents itself as an assemblage keen to engage in emergent interaction, to recognise vulnerability and to legitimately and convincingly assert that it is the proper site for the resolution of legal questions arising from asylum concerns.

The approach in two recent cases[3] supports the complexity perspective observation that the legitimacy of a legal system is bound up with how it internalises vulnerability. In *Detention Action* the High Court had concluded that the truncated nature of decision making under the so-called Detained Fast-Track application process was 'structurally unfair' such that it would lead to a 'serious procedural disadvantage' on the part of the asylum applicant (para. 60). In the Court of Appeal, Lord Dyson MR, agreeing with Nicol J (see variously paras. 19, 22, 24, 37, 38, 45), added that because the secretary of state for the Home Department was both the other party in the applicant's asylum claim, and in control of the decision to allocate a case to the fast-track process, they were 'able to gain a major litigation advantage by being able to decide that the appeal is suitable to be placed in [Fast-Track Review]' (para. 24, also paras.46–48). The courts' concern over the misuse of powers granted to the Home Office seeks to honour a commitment to legality and, more generally, to the rule of law by removing the exclusionary quality of the procedural arrangements. When these motivations are viewed from the perspective of complexity, it can be said that the judgment sought to rebalance the relative abilities of one party to an asylum appeal to engage with another: in Neves' language, to more adequately integrate both parties into the system, and to counteract the over-integration of the Home Office.

The individual vulnerability of any asylum applicant acting in good faith is obvious in this context. Their vulnerability exists regardless of the nature of the legal process to which they address the application. They have experienced, and continue to experience, substantial change in their circumstances that place them at the behest of others, and they are largely unable to influence their own destiny. While they can offer evidence in support of their application, how this will be received, and what other factors will be considered important is largely out of their hands. Nonetheless, they approach the law system with their own view of the world – a description strategy – and as part of this they give an account of how they have come to a point where they need to claim asylum. As one assemblage, they present themselves to another, the law. If legal processes are used to mini- mise the opportunity for their claim to be fully considered because it is con- structed with 'speed and efficiency' rather than 'justice and fairness' in mind (*Detention Action*, para.22) this does not eliminate the existence of the material, psychological and conceptual aspects of the claim from reality. Instead, it only eliminates – invisibilises – them from law's reality. This has consequences for a system which purports to be the only framework competent to process the claim.

The *Detention Action* case demonstrated the need for the legal system to assert both its legitimacy and its unwillingness to exclude vulnerable individuals in the face of political pressure. Had the decision concluded otherwise, this would have had the effect of, if not excluding, then marginalising the individuals subjected to Fast-Track procedures. As I have suggested, one consequence of excluding individuals from the legal system in this way is that they might feel compelled to seek resolutions to their real problems elsewhere. This would have under- mined the legitimacy of the legal system's claim to be authoritative in this field, and the wider claim of the legal system to legitimacy based on at least formally equal treatment of all before the law. With this in mind, the observation that it was possible for political pressure to be exerted on the legal-administrative structure responsible for designing the Fast-Track rules in the first place – the Tribunal Practice Committee – was especially troubling. Indeed, it raises con- cerns for the legitimacy of the process by which any amended rules, intended to take account of the judgment in *Detention Action*, are formulated (see Briddick, 2015, p. 324). These risks to legitimacy, and the possible systemic consequences, should be borne in mind when reflecting on any reforms having an impact on the ready accessibility of judicial and administrative remedies.

The *Public Law Project* case demonstrates even more starkly the compulsion of law to give substance to its claims to substantive legal equality for individuals in order to maintain its legitimacy, and thus, identity.[4] The secretary of state for jus- tice had sought to use Henry VIII powers contained in the Legal Aid, Sentencing and Punishment of Offenders Act 2012 to attach a residence test to the criteria to be met to qualify for legal aid. This would have prevented many non-residents of the United Kingdom from meeting the eligibility criteria to qualify for legal aid.[5] Lord Neuberger concluded that the residency test created by the secretary of state would have the effect of excluding individuals from access to legal aid by 'reduc[ing] the class of individuals who are entitled to receive those services . . .'.[6] In the language of complexity theory this would be achieved by invisibilising to law certain categories of individuals to control the degree of uncertainty posed by those individuals to systemic processes. While limiting the number of people able to access a service is not in itself necessarily nefarious, it was the attempt to base the question of eligibility on 'a personal characteristic or circumstance unrelated to the services' (*ibid.*) which made the provisions *ultra vires*.

The law here was faced with two tensions that had implications for its legitimacy. On one hand, it was recognised that legislation emanating from Parliament which authorised such a test would be legitimate because Parliament is *the* source of sovereign authority in the United Kingdom (implicit at para. 30). Provided that the courts were seen to honour the authority of Parliament, as any examination of the vires of executive action seeks to do, the legitimacy of the legal system could not be sensibly challenged in that respect.[7] On the other hand, the courts recognised that the exclusion of a category of vulnerable individuals from access to legal aid, and thus a degree of substantive equality before the law, *would* have implications for the legitimacy of the legal system. The conceptual and material difficulties at the root of the claims that would be made with the support of legal aid would, though they would be rendered invisible to law, not be factually eliminated just because legal aid was not available. Yet their bodily difficulties were concerned with rights – especially their rights to asylum and their human rights – such that, if law was to maintain its identity as *the* site of decision making for them, they could only reasonably be answered by legal processes. Similarly, their conceptual difficulties, namely the determination of their status as refugees or another category of migrant, was avowedly a legal question. The failure to address either the question of their legal rights as individuals, or to answer the law system's queries about their status both fairly and impartially, would present law with a challenge to its legitimacy.

The reason for this is that, in view of the complexity understanding of the need for interaction to establish differentiation, action which prevents interaction, and thus the relational co-production of both law and those approaching law, calls into question the appropriateness of law as *the* site for settling legal questions. If law denies, by legal constructs, that it is willing both in principle and in fact to deal with a purportedly legal issue – for example, because it effectively prevents the question being raised – then it denies its own identity as the right forum. Any decision on the part of the legal system which either makes it difficult for law to answer these questions, or which diverts patently legal questions elsewhere undermines the legitimacy of the legal system by challenging its claim to authority, and thus its identity. In consequence, the Supreme Court can be seen to have reasoned that, to maintain law's legitimacy, or at least to include in legal consideration as many instances of refugee material and conceptual concern, the courts were required to construe the way the Henry VIII powers were used as unlawful. They were thus able to maintain Parliament's sovereignty and law's legitimacy while enabling law to incorporate consideration of more cases and thus bolster its claim to be *the* legitimate site of resolution of questions arising from asylum applications.

Conclusion: vulnerability, identity, and emergent interaction

The physical, psychological and conceptual vulnerability of individuals is intimately bound up with the vulnerability of systems. Indeed, given the importance of relational co-production of difference proposed by complexity theory, the best approach to encourage interaction is to promote the interactive integration of individuals and other assemblages which compose the social environment. Thus, complexity theory indicates, far from seeking to invisibilise, marginalise or exclude the vulnerable individual as a potentially destabilising influence, the legal system must not merely confront, but embrace, interface with and integrate that vulnerability and apparent risk of destabilisation into its own processes. Why is this necessary? Put simply, vulnerability is a creative force for both individuals and for systems. The claimed instability posed by those who do not integrate neatly into established frameworks of understanding, in fact, represent an opportunity for creative interaction, for the expansion of law's competence to deal with legal issues and an increased resilience regarding challenges to its underlying legitimacy.

What is more, the universal nature of vulnerability – being a feature of assemblages at all scales – demonstrates the risks of exclusion of the emergent possibilities arising from interaction. I have discussed how the risks flowing from exclusion demonstrate how complex identity is bound up with the need to interact to establish difference from the environment. The inability to interact, of exclusion from communicative interaction, exposes the vulnerability of assemblages. At the same time, it is vulnerability that, as Fineman says, causes us to reach out, to form connections (Fineman, 2014, p. 22), that provokes the relational processes of emergence via interaction. In consequence, just as vulnerability has the potential to expose the risks of exclusion, it also works to counteract these and to promote emergent interaction. In this way, the pursuit of difference is turned from what might traditionally be considered a negative force, into a positive necessity of social existence that enables ongoing communication.

In the specific context of the asylum legal framework, the recognition of vulnerability should compel legislators and other actors to think differently about their participation in and contribution towards the character of that procedural assemblage. As I have shown, aspects of the process which seek to marginalise, minimise or entirely exclude individuals from systemic consideration are risky not just to the individual – for obvious, material and psychological reasons – but also to the system. This is because the identity of the law system, understood from a complexity perspective, is defined by its perceived legitimacy and the authoritative capacity to decide legal matters which flow from this. Actions which appear to undermine that legitimacy, which compel individuals to seek non-legal solutions to evidently legal problems, damage that identity by undermining that which differentiates the legal from its environment. This loss of difference is the loss of identity. It should always be remembered that the law system remains vulnerable to this loss wherever it is seen to enable the under-integration of assemblages, or where actions occur which dampen their integration, for example, by invisibilising or failing to incorporate the psychological, material or conceptual concerns of humans and other assemblages. If law forgets this, then it also forgets its own exposure, via the relationally constructed, emergent nature of its identity, to the effects of that under-integration and potential exclusion.

Notes

1 The author would like to thank Sara Fovargue, Jamie Murray, Andreas Philippopoulos-Mihalopoulos, Siobhan Weare and Steven Wheatley for their helpful comments on earlier versions of this chapter. Any errors remain my own.
2 *R (on the application of Detention Action) v First Tier Tribunal (Immigration and Asylum Chamber)* [2015] EWCA Civ 840; *R (on the application of Public Law Project) v Secretary of State for Justice* [2016] UKSC 39
3 Above, n.2
4 One might also consider, for example, *R* v *Lord Chancellor, ex parte Witham* [1998] QB 575
5 See Legal Aid Sentencing and Punishment of Offenders Act 2012, s.9(2)(b)
6 *R (on the application of Public Law Project) v Secretary of State for Justice* [2016] UKSC 39, para. 30.
7 However, consider the government's reasons for reforming judicial review (Mills, 2015).

Bibliography

Ansell, Christopher and Robert Geyer (2017), '"Pragmatic Complexity" a New Foundation for Moving Beyond "evidence based policy making"?' 38(2) *Policy Studies* 149.

Baillot, Helen, Sharon Cowan and Vanessa E Munro (2014), 'Reason to Disbelieve: Evaluating the Rape Claims of Women Seeking Asylum in the UK', 10(1) *International Journal of Law in Context* 105.

Bankowski, Zenon (1996), 'How Does It Feel to Be on Your Own? The Person in the Sight of Autopoiesis', in David Nelken (ed.) *Law as Communication* (Aldershot: Dartmouth Publishing), 63–80.

Bottomley, Anne (2002), 'The Many Appearances of the Body in Feminist Scholarship', in Andrew Bainham, Shelley Day Sclater and Martin Richards (eds.) *Body Lore and Laws* (Oxford: Hart), 127–148.

Briddick, Catherine (2015), 'Case Comment: *Detention Action v First-tier Tribunal (Immigration and Asylum Chamber)*', 29(3) *Journal of Immigration, Asylum and Nationality Law* 322.

Cilliers, Paul (1995), 'Postmodern Knowledge and Complexity (or Why Anything Does Not Go)', 14(3) *South African Journal of Philosophy* 124.

Cilliers, Paul (1998), *Complexity and Postmodernism: Understanding Complex Systems* (Abingdon and Oxon: Routledge).

Cilliers, Paul (2000), 'What Can We Learn from a Theory of Complexity?' 2(1) *Emergence* 23.

Cilliers, Paul (2001), 'Boundaries, Hierarchies and Networks in Complex Systems', 5(2) *International Journal of Innovation Management* 135.

Cilliers, Paul (2005), 'Complexity, Deconstruction and Relativism', 22 *Theory, Culture, Society* 255.

Cilliers, Paul (2007), 'Knowledge, Complexity and Understanding', in Paul Cilliers (eds.) *Thinking Complexity: Complexity and Philosophy Volume 1* (Mansfield, MA: ISCE Publishing), 159–164.

Cilliers, Paul (2008), 'Knowing Complex Systems: The Limits of Understanding', in Frédéric Darbellay, Moira Cockell, Jérôme Billotte and Francis Waldvogel (eds.) *A Vision of Transdiciplinarity, Laying the Foundations for a World Knowledge Dialogue* (Lausanne and Switzerland: EPFL Press), 43–50.

Cilliers, Paul (2010), 'Difference, Identity and Complexity', in Paul Cilliers and Rika Preiser (eds.) *Complexity, Difference and Identity (Issues in Business Ethics, Volume 26)* (London: Springer), 3–18.

Cilliers, Paul, Carlos Gershenson and Francis Heylighen (2007), 'Philosophy and Complexity', in Jan Bogg and Robert Geyer (eds.) *Complexity Science & Society* (Oxford: Radcliffe).

Cilliers, Paul and Rika Preiser (2010), 'Preface: Why Difference', in Paul Cilliers and Rika Preiser (eds.) *Complexity, Difference and Identity (Issues in Business Ethics, Volume 26)* (London: Springer), v–ix.

de Villiers, Tanya and Paul Cilliers (2010), 'The Complex "I": The Formation of Identity in Complex Systems', in Paul Cilliers and Rika Preiser (eds.) *Complexity, Difference and Identity (Issues in Business Ethics, Volume 26)* (London: Springer), 19–40.

DeLanda, Manuel (2006), *A New Philosophy of Society: Assemblage Theory and Social Complexity* (London: Continuum).

Esses, Victoria M., Stelian Medianu and Andrea S. Lawson (2013), 'Uncertainty, Threat, and the Role of the Media in Promoting the Dehumanization of Immigrants and Refugees', 69(3) *Journal of Social Issues* 518.

Fineman, Martha, A. (2008–2009), 'The Vulnerable Subject: Anchoring Equality in the Human Condition', 20 *Yale Journal of Law and Feminism* 1.

Fineman, Martha, A. (2012), ' "Elderly" as Vulnerable: Rethinking the Nature of Individual Responsibility and Societal Responsibility', 20 *The Elder Law Journal* 71.

Fineman, Martha, A. (2014), 'Equality, Autonomy, and the Vulnerable Subject in Law and Politics', in Martha A. Fineman and Anna Grear (eds.) *Vulnerability: Reflections on a New Ethical Foundation for Law and Politics* (London: Ashgate).

Firth, Georgina and Barbara Mauthe (2013), 'Refugee Law, Gender and the Concept of Personhood', 25(3) *International Journal of Refugee Law* 470.

FitzGerald, Sharron A. (2010), 'Biopolitics and the Regulations of Vulnerability: The Case of the Female Trafficked Migrant', 6(3) *International Journal of Law in Context* 277.

Fox O'Mohony, Lorna and James A. Sweeney (2010), 'The Exclusion of (Failed) Asylum Seekers from Housing and Home: Towards an Oppositional Discourse', 37(2) *Journal of Law and Society* 285.

Gietel-Basten, Stuart (2016), 'Why Brexit? The Toxic Mix of Immigration and Austerity', 42(4) *Population and Development Review* 673.

Herlihy, Jane, Kate Gleeson and Stuart Turner (2010), 'What Assumptions About Human Behaviour Underlie Asylum Judgments', 22(3) *International Journal of Refugee Law* 351.

Herlihy, Jane and Stuart Turner (2015), 'Untested Assumptions: Psychological Research and Credibility Assessment in Legal Decision-Making', 6 *European Journal of Psychotraumatology* 1.

Innes, Alexandria J. (2010), 'When the Threatened Become the Threat: The Construction of Asylum Seekers in British Media Narrative', 24 *International Relations* 456.

Kagan, Michael (2015), 'Believable Victims: Asylum Credibility and the Struggle for Objectivity', 16(1) *Georgetown Journal of International Affairs* 123.

Kelley, Ninette (2001), 'The Convention Refugee Definition and Gender-Based Persecution: A Decade's Progress', 13(4) *International Journal of Refugee Law* 559.

Khosravinik, Majid (2010), 'The Representation of Refugees, Asylum Seekers and Immigrants in British Newspapers: A Critical Discourse Analysis', 9(1) *Journal of Language and Politics* 1.

King, Michael (1993), 'The "Truth" About Autopoiesis', 20(2) *Journal of Law and Society* 218.

Lewis, Miranda (2005), *Asylum: Understanding Public Attitudes* (London: IPPR).

Luhmann, Niklas (1988), 'The Unity of The Legal System', in Gunther Teubner (ed.) *Autopoietic Law: A New Approach to Law and Society* (New York and Berlin: Walter de Gruyter Publishing), 12–35.

Luhmann, Niklas (1992a), 'The Coding of the Legal System', in Gunther Teubner and Alberto Febbrajo (eds.) *State, Law, and Economy as Autopoietic Systems: Regulation and Autonomy in a New Perspective* (Milan: Giuffrè), 145–185.

Luhmann, Niklas (1992b), 'Some Problems With <<Reflexive Law>>', in Gunther Teubner and Alberto Febbrajo (eds.) *State, Law, and Economy as Autopoietic Systems: Regulation and Autonomy in a New Perspective* (European Yearbook in the Sociology of Law, Double Issue) (Milan: Giuffrè), 389–415.

Luhmann, Niklas (1992c), 'Operational Closure and Structural Coupling: The Differentiation of the Legal System', 13 *Cardozo Law Review* 1419.

Luhmann, Niklas (2008), 'Are There Still Indispensable Norms in Our Society?' 14(1) *Soziale Systeme* 18.

Matthews, Julian and Andy R. Brown (2012), 'Negatively Shaping the Asylum Agenda? The Representational Strategy of a Tabloid News Campaign', 13 *Journalism* 802.

Mills, Alex (2015), 'Reform of Judicial Review in the Criminal Justice and Courts Act 2015: Promoting Efficiency or Weakening the Rule of Law?' *Public Law* 583.

Neves, Marcelo (2001), 'From the Autopoiesis to the Allopoiesis of Law', 28(2) *Journal of Law and Society* 242.

Pearce, Julia, M. and Janet E. Stockdale (2009), 'UK Responses to the Asylum Issue: A Comparison of Lay and Expert Views', 19 *Journal of Community and Applied Social Psychology* 142.

Philippopoulos-Mihalopoulos, Andreas (2006), 'Dealing (with) Paradoxes: On Law, Justice and Cheating', in Michael King and Chris Thornhill (eds.) *Luhmann on Law and Politics, Critical Appraisals and Applications* (Portland, Oregon: Hart Publishing), 217–234.

Philippopoulos-Mihalopoulos, Andreas (2010), *Niklas Luhmann – Law, Justice and Society* (Abingdon: Routledge).

Philippopoulos-Mihalopoulos, Andreas (2015), *Spatial Justice* (Oxford: Routledge).

Philippopoulos-Mihalopoulos, Andreas and Thomas E. Webb (2015), 'Vulnerable Bodies, Vulnerable Systems', 11(4) *International Journal of Law in Context* 444.

Philo, Greg, Emma Briant and Pauline Donald (2013), 'The Role of the Press in the War on Asylum', 55 *Race & Class* 28.

Preiser, Rika and Paul Cilliers (2010), 'Unpacking the Ethics of Complexity: Concluding Reflections', in Paul Cilliers and Rika Preiser (eds.) *Complexity, Difference and Identity (Issues in Business Ethics, Volume 26)* (London: Springer), 265–288.

Richardson, Kurt A. (2004), 'Systems Theory and Complexity: Part 1', 6(3) *Emergence: Complexity and Organisation* 75.

Richardson, Kurt A. and Paul Cilliers (2001), 'What Is Complexity Science? A View from Different Directions', 3(1) *Emergence* 5.

Richardson, Kurt A., Paul Cilliers and Michael Lissack (2001), 'Complexity Science: A "Gray" Science for the 'Stuff in Between', 3(2) *Emergence* 6.

Sherwood-Johnson, Fiona (2013), 'Constructions of 'vulnerability' in Comparative Perspective: Scottish Protection Policies and the Trouble with 'adults at risk', 28(7) *Disability & Society* 908.

Sweeney, James A. (2016), 'A "credible" Response to Persons Fleeing Armed Conflict', in Matthew Happold and Maria Pichou (eds.) *The Protection of Persons Fleeing Armed Conflict and Other Situations of Violence* (Brussels: Larcier), 81–103.

Teubner, Gunther (1993), *Law as an Autopoietic System* (Oxford and Cambridge: Wiley-Blackwell).

Verschraegen, Gert (2002), 'Human Rights and Modern Society: A Sociological Analysis from the Perspective of Systems Theory', 29(2) *Journal of Law and Society* 258.

Waldrop, Mitchell M. (1994), *Complexity: The Emerging Science at the Edge of Order and Chaos* (London: Penguin Books).

Webb, Julian (2005), 'Law, Ethics, and Complexity: Complexity Theory and the Normative Reconstruction of Law', 52 *Cleveland State Law Review* 227.

Webb, Thomas E. (2013), 'Exploring System Boundaries', 24(2) *Law and Critique* 131.

Section III

Complexity beyond the state: human rights and international law

5 Explaining change in the United Nations system

The curious status of Security Council Resolution 80 (1950)

Steven Wheatley

This chapter looks to make sense of the paradox of change in the United Nations (UN) system, whereby an alteration in the behaviours of the member states can modify the rules of the organisation that bind the same members: rule-breaking behaviour thus becomes rule-making behaviour. By looking to complexity theory, which is fundamentally a theory of change, it shows how we can think of the UN 'system' as the emergent property of the actions and interactions of the member states, evolving as they respond to new information about events in the outside world or the unexpected actions of another state.

The work takes as its case study the amendment of the rules for voting in the Security Council that followed the 'empty chair' policy assumed by the Union of Soviet Socialist Republics (USSR) in 1950 in response to the refusal of the organisation to recognise Beijing as the legitimate government of China. The literal text of the UN Charter makes clear the positive support of all five permanent members (the 'P5') is required for the adoption of a substantive resolution, yet the International Court of Justice later accepted that the practice of the Security Council, whereby the absence of a permanent member did not prevent the passing of a resolution, had changed the meaning of the relevant provision, so all that was needed was the absence of a veto.

Whilst an evolution in the plain meaning of Charter provisions can be explained by the role of subsequent practice in the interpretation of treaties, there remains the problem of the status of the first resolution adopted under any 'new' procedure. Security Council Resolution 80 (1950) was the first clearly substantive resolution to be adopted whilst a P5 member was away from the Council, calling on the governments of India and Pakistan to make immediate arrangements for the demilitarisation of Jammu and Kashmir. The resolution was not adopted in accordance with the old (literal) rule, requiring the positive support of the 'P5', but nor could there be a new pattern of behaviour, meaning only the absence of a veto was needed. The logical conclusion must be that Resolution 80 (1950) was not validly adopted under either the old or the new rules, yet the UN official records report it was '[a]dopted . . . by 8 votes to none, with 2 abstentions'. A footnote reads, 'One member (Union of Soviet Socialist Republics) was absent'.[1]

The contention here is that, by thinking of the United Nations organisation as a complex system of regulatory authority, we can make sense of this evolution in the procedural rules of the Security Council and explain how an innovation in practice, like Resolution 80 (1950), can establish a new rule of behaviour. The key is that change is observed within a temporal frame, which also changes over time.

The chapter begins by explaining the central ideas behind complexity theory, focusing on emergence, the notion of the whole being greater than the sum of the component parts. Complexity highlights the interdependence of the system and its component agents, emphasising the importance of contingency in the evolution of complex systems and the power of events in bringing about change. The work demonstrates the ways these insights can help us make sense of the UN system, specifically the alterations in the rules of the organisation following adaptions in the behaviours of member states. The analysis then turns to the problem of innovations in practice, explaining how the problematic status of Resolution 80 (1950) can be resolved by foregrounding the factor of time.

Complexity theory

Complexity theory developed in the 1980s as a further challenge to the Newtonian model of a Clockwork Universe that could be taken apart and subjected to analysis, with the presumption being that all systems, even highly complicated systems, were 'the sum of their parts', and the future shape and form of any system could, in principle, be predicted – think of the mechanical models of the solar system. The properties of complex systems were seen to be the result of the behaviours of the individual components *and* their interactions with each other *and* their interactions with the external environment, with the consequence that 'the whole was more than the sum of its parts'. Complexity thinking is firmly established in the natural sciences and has been used, for example, to explain the workings of insect colonies, the way the immune system functions, and the relationship between the mind and the brain. The insights from complexity have also been applied in the social sciences, including analyses of the global financial system, the World Wide Web and the actions of human populations, from the phenomenon of the standing ovation in the concert hall to the organisation of nation-state societies.

Given the characterisation of law in general, and international law in particular, as a 'system', there seems an obvious argument for looking to complexity theory as a methodology and there is an emergent body of scholarship that applies the insights from complexity to domestic law systems (Webb, 2005; Jones, 2008; Ruhl, 2008; Webb, 2014) and international law (McGoldrick, 2004; Carline and Pearson, 2007; Chinen, 2014; Wheatley, 2016, 2018; Morin et al., 2017; and Prior, 2017). Complexity gives us a new way of thinking about the international law system and a new language to describe the ways it functions by looking to the insights developed in the hard sciences of physics, chemistry and biology on the workings of complex systems.

Whilst many scholars use the word *complexity*, there is no agreed-on definition of the term. Melanie Mitchell reports that when she asked an expert panel, 'How do you define complexity?' everyone laughed because the question was so straightforward, but then each member proceeded to give a different definition (Mitchell, 2009, p. 94). When applied to the study of law, complexity can have one of four possible meanings. The first equates complexity with complicatedness, the notion that the law system is simply too complicated, or complex, for any mortal lawyer to understand. Second, and related to complicatedness, is the idea of computational complexity, which draws on the mathematical theory of complexity outlined by computer scientists to develop computational models of law systems. Third, there is the general, or postmodern, approach that regards attempts to produce general laws of complexity as a negation of the central insight that some systems cannot be modelled perfectly because they are complex systems. Finally, there is the methodology associated with the early work of the Santa Fe Institute that sees emergence as the distinguishing feature of complexity. John Holland, who has written widely on the subject, draws a clear distinction between computational and other forms of complexity on the basis that computational complexity is not concerned with emergence (he does not deal with postmodern complexity; Holland, 2014, p. 4).

This work shares this understanding that emergence is the defining feature of complexity that distinguishes complex systems (properly so called) from other systems. The concept is not new, with an early example being provided by John Stuart Mill's observation that the properties of water (H_2O) are different from the properties of its component elements, hydrogen and oxygen (Mill, 1858, p. 255).

Writers on complexity use the term emergence to describe phenomena that result from the actions and interactions of component elements, without the need for a controlling power or guiding hand (Waldrop, 1992, p. 11). Emergence is often referred to in terms of 'the whole being more than the sum of its parts', but in a complex system the whole also determines the behaviour of the parts (Küppers, 1992, p. 243). The notion the whole influences the parts is called downward causation, defined by Achim Stephan as the 'causal influence exerted by the system itself (or by its structure) on the behaviour of the system's parts' (Stephan, 2002, p. 89). A complex system is, then, the result of the interactions of its component agents and their interactions with the outside world; the system then influences the behaviours of the same component agents that brought it into existence in the first place.

The emergence and evolution of complex systems

A key insight from complexity is that complex systems emerge without the need for a controlling power or guiding hand and to make sense of these types of system, we must be attentive to the actions of component agents, their interactions with each other, and their interactions with the outside world. The point applies equally to colonies of insects, the human populations of cities, and the international law system. Complex systems evolve as component agents alter their behaviours (what theorists call synchronic emergence), but they change from one state to another (the idea of diachronic emergence, seen also in the evolution of species). We explain the evolution of complex systems by describing the system at one instance and comparing that description with another at a different moment and then explaining what change has occurred and how that change occurred (Byrne, 1998, p. 24).

Whilst there is no hard science of complexity or established philosophy, we can draw several lessons from the available literature. Complexity tells us, for example, that everything is, ultimately, in some way, connected to everything else and to study a complex system, we must separate the system from its environment. This is no easy task, because, in a complex system, component agents interact directly with actors and elements in the outside world, and, moreover, the system is dependent on these interactions to provide the necessary energy to evolve. Complex systems transform their structures as their component agents alter their behaviours in response to information that provokes a change that cascades through the system.

Where significant reshaping occurs, the system can take on a completely different form, but this is context-dependent and highly contingent, and there is no guarantee the system would evolve in the same way if we could rewind the clock. This is noteworthy because in classical physics most phenomena are time-reversible, the mathematics of any physical process makes sense in either temporal direction (Dainton, 2001, p. 439). Ilya Prigogine won the Nobel Prize for Chemistry for his work which showed that far from equilibrium systems (a type of complex system; Yin and Herfel, 2011, pp. 390–391) follow the second law of thermodynamics, which establishes that some processes only make sense in one direction. We know that omelettes cannot be made in eggs, that the hot cup of coffee always cools, and that far from equilibrium systems can only move forward in time. Time irreversibility in far from equilibrium systems (the standard example is the Rayleigh–Bénard convection) results from the fact that significant changes ('bifurcations') are a combination of knowable ('deterministic') elements and unknowable ('probabilistic') elements. There is, then, an arrow of time in complex systems, which are in a constant state of evolution from the past, through the present, into the future. This time irreversibility is a consequence of the fact significant changes ('bifurcations' in the language of complexity) are the result of a combination of knowable factors and unknowable ingredients. At a moment of bifurcation, it is not possible to tell what direction a system will move – it all depends on a mixture of necessity and chance, as the system is seen to take one path 'over a number of other equally possible paths' (Prigogine and Stengers, 1985, p. 176).

Complexity as a methodology in international law

Whilst complexity is often used to explain systems with large numbers of component agents (a rainforest ecosystem would be the standard example), it can also be applied to systems with relatively few actors, where 'the agents are complex and the communication language is complex' (Sawyer, 2005, p. 176) – like the international law system.

Drawing on the insights developed by our colleagues in the physical sciences, we can think about international law as a complex, self-organising system that emerges from the actions and interactions of states (along with certain non-state actors), without the requirement for a global sovereign power. Complexity reminds us there is, as any undergraduate student of international law knows, no central controller, guiding hand or sovereign power in international law, which can be thought of as a complex system that emerges from the actions and interactions of states and then constrains the future behaviours of the very actors that brought it into existence. Christian Tomuschat expresses the point this way: 'international law can be perfectly conceived as an autonomous régime, created by the same States that are at the same time its addressees' (Tomuschat, 1993, p. 235).

Change in the international law system can be a consequence of a major conflagration – the adoption of the UN Charter after World War II, for example, but it can also happen in response to seemingly minor occurrences, such as the development of a body of space law following the launch of Sputnik 1 in 1957. This is explained by the nonlinearity of complex systems, whereby small changes in the behaviours of component agents or relatively minor events can have a significant impact on the structure of the system (also called the *Butterfly effect*). Nonlinearity is a consequence of two factors: first, the unpredictability of the reactions of component agents to the same piece of information and, second, the fact the system can provide positive feedback to the component agents, encouraging further change in the same direction.

The international law system evolves as states respond to new information, such as a change in the behaviours of other states, or the actions of a non-state actor, like the International Court of Justice, or to events in the external environment, including developments in science and technology. The following section builds on these basic understandings to explain change in the UN system.

The UN system

The United Nations was established by the original signatories, with the preamble to the Charter proclaiming, 'We the peoples of the United Nations . . . do hereby establish an international organisation to be known as the United Nations.' By declaring its existence and then acting as if the United Nations had regulatory power over them, the signatories established the institutional fact of a world organisation with regulatory power. In the more familiar words of the International Court of Justice: 'fifty States, representing the vast majority of the members of the international community, had the power, in conformity with international law, to bring into being an entity possessing objective international personality'.[2]

When the Charter came into force on 24 October 1945, there was no UN system. The notion of a system implies something with multiple elements and, in the case of a complex system, an entity which is adaptive and capable of changing over time. Neither description applies to the text of the constituent instrument of an international organisation. It was only when the UN bodies started to function and member states responded to UN bodies that the UN system emerged, and after this moment, member states were no longer in control. As Nigel White points out, 'although States may control the creation of [an international organisation], once created it takes on a life of its own.' The fact of majority rule means the organisation is no longer under the control of the member states: 'Indeed the reverse is true' (White, 2005, p. 20).

In *Reparations for injuries suffered in the service of the United Nations*, the International Court of Justice confirmed that the United Nations enjoyed an objective international personality distinct from that of its member states but that it was not 'a state, still less '"a super-State" whatever that might mean.'[3] The United Nations does not, then, enjoy sovereign authority in the same way as its member states, and, like any other international organisation, it can only act within the powers given to it by the members in the form of its constituent instrument, in this case the UN Charter. Increasingly, scholars refer to this exercise of regulatory authority in terms of 'UN law' (White, 2002, p. 18), and Oscar Schachter has written that this UN law emerges from the 'complex patterns' of the actions of the various specialised bodies and agencies (Schachter, 1988, p. 16).

The regulatory powers of the United Nations depend on an express grant of authority in the Charter (the doctrine of attribution) and the recognition of powers regarded as essential to the performance of its duties (the implied powers doctrine). The requirement for implied powers, follows, as Jan Klabbers observes, the fact that, 'while the notion of attribution may be a nice point of departure[,] organisations are usually held to be dynamic and living creatures, in constant development, and it is accepted that their founding fathers can never completely envisage the future' (Klabbers, 2009, p. 58).

Much of the literature on international organisations reflects an attempt to make sense of the paradox of a regulatory system emerging from the actions of the subjects the system seeks to control. The original delegation of authority is seen to be supplemented by the doctrine of implied powers but with unsatisfactory results, given the unclear boundaries. Arguments are then made that the powers of the organisation should be limited to those expressly included in the constituent instrument or that can be attached to an express power, to the necessary inherent powers of the organisation, to those required for it to carry out its functions or that constitutional limits should be placed on the exercise of regulatory authority.

The arguments reflect the tension between the apparent desire of the organisation for more power and the concerns of member states about regulatory overreach. But, of course, an international organisation is not a separate entity; it is composed of the very agents it seeks to regulate. In Klabbers's terms, international organisations, like the United Nations, 'are, at one and the same time, independent of their members [. . .] and fundamentally dependent on them'

(Klabbers, 2009, p. 36). The organisation 'may aspire desperately to gain independence from the member states and impose its will on those member states, yet at one and the same time the organization can only act to the extent the member states allow this' (Klabbers, 2004, p. 43).

Establishing the rules of the organisation

The basic point of reference for establishing the regulatory authority of an international organisation is its constituent instrument, and the UN organisation can only act within the express and implied powers allocated to it under the Charter, which is 'a multilateral treaty, albeit a treaty having certain special characteristics'.[4]

Article 31(1) of the Vienna Convention on the Law of Treaties outlines the general rule: 'A treaty shall be interpreted in good faith in accordance with the ordinary meaning to be given to the terms of the treaty in their context and in the light of its object and purpose.' Article 31(3)(b) further directs us to take into account '[a]ny subsequent practice in the application of the treaty which establishes the agreement of the parties regarding its interpretation'. This allows those interpreting the Charter to make sense of its provisions by looking to the behaviours of the member states. Richard Gardiner explains the point this way: 'Words are given meaning by action' (Gardiner, 2015, p. 253). The justification for this is explained by the International Law Commission in terms of subsequent practice providing 'objective evidence of the understanding of the parties as to the meaning of the treaty'.[5] The idea is sometimes referred to as auto-interpretation, capturing the fact that 'the interpretive agent in this context is, so to speak, both judge and party' (Provost, 2015, p. 293).

Subsequent practice has allowed the United Nations to expand its regulatory authority in significant ways without formal amendment of the Charter, from the legislative activity of the UN Security Council to the periodic review of member states' human rights performance by the UN Human Rights Council. This has been notable for two reasons. First, the relevant practice has often been the practice of UN bodies, such as the Security Council and General Assembly and not the member states (although the representatives on UN bodies are the member states). Second, looking to the activities of UN bodies admits the problem of majority rule, as none require unanimity in their decision-making procedures,[6] and not all member states are represented on all UN bodies all the time (only 15 of the 193 member states sit on the Security Council, and five of those are permanent members). Nonetheless, there is widespread recognition that a practice generally accepted by UN member states can lead to a new understanding of the Charter (Amerasinghe, 2005, p. 52; Cot, 2011, para, 60), meaning the regulatory role of the organisation can evolve with changes in the behaviours of its members.[7]

We can only make sense of the regulatory authority of the United Nations, and any evolution in that regulatory authority, by looking at the behaviours of the member states and the practices of the primary and secondary organs of the organisation. Subsequent practice in the interpretation of the UN Charter can include, then, the practice of the member states, the practice of the organisation or some combination of the practice of the member states and the organs of the United Nations. The International Law Commission's special rapporteur on subsequent practice, Georg Nolte, concludes that it is the 'general practice of the organisation' that seems to carry more weight as a means of interpretation than an 'established practice' of a particular organ of the organisation (Nolte, 2015, para. 79). The implication is that our understanding of the constituent instruments of the United Nations is an emergent property of the actions and reactions of member states and the actions and reactions of the organs of the organisation.

The UN system is, like all complex systems, remodelled with alterations in the behaviours of its component agents. The multitudinous possibilities of action and interaction mean we cannot be sure how it will change, but that does not mean 'anything goes'. In *Reservations to the Convention on Genocide*, Judge Alejandro Alvarez argued the evolution of treaty systems

> can be compared to ships which leave the yards in which they have been built, and sail away independently, no longer attached to the dockyard. These conventions must be interpreted without regard to the past, and only with regard to the future.[8]

Alvarez was right to note a treaty can change over time, but wrong to conclude the past is irrelevant. Whilst the meaning of the UN Charter evolves with changes in the behaviours of member states, those alterations build on past actions and interactions and the previous understandings of Charter provisions. History matters in the development of complex regulatory systems, like the United Nations. In the words of Judge Hersh Lauterpacht, when trying to make sense of the powers of the United Nations, '[we] must take into account not only the formal letter of the original instrument, but also its operation in actual practice and in the light of the revealed tendencies in the life of the Organization'.[9]

The power of an event: the Soviet Union 'empty chair' policy

Complex systems evolve with changes in the conduct of component agents as they respond to new information. Once a system has adopted a set of responses to known problems, there is no reason for it to change the way it responds, unless some event or other influence causes the system to change course. This is the idea of path-dependence, a term invented by the economic historian Paul David to describe situations in which present conditions can only be explained by contingent decisions taken in the past. History, David explains, simply describes a scenario in which 'one damn thing follows another' (David, 1985, p. 332). His standard example is the QWERTY keyboard, designed by Remington in the 1880s and still used on my iPad today. We continue to use the QWERTY layout for typing because we have used it for over a hundred years (nothing more). Path-dependence explains the importance of history to international law, as key decisions, at key moments, establish the rules and institutional structures in which the international lawyers must operate.

Path-dependence tells us that, in the absence of a reason to change, the component agents in a system will not change their behaviours (Djelic and Quack, 2007, p. 167). Change in a complex system is initiated by some event that causes component agents to act in a different way. The notion of an event here refers to information coming from within the system (the actions of other component agents) or from the external environment. The UN system evolves then as the behaviours of the member states change as they respond to new information and we will not see change unless there is some event that provokes an alteration in the behaviours of component agents, either in response to some new piece of information about the actions of other component agents or in an event in the outside world. The following sections examine the change that occurred in the UN system around the rules for voting in the Security Council, demonstrating how this came about in response to the 'empty chair' policy adopted by the Soviet Union in 1950.

Voting procedures in the Security Council

Under Article 24(1) of the Charter, UN member states confer primary responsibility for the maintenance of international peace and security on the Security Council and decisions adopted by the Council under Chapter VII of the Charter are binding for member states. Article 27(3) establishes that decisions of the Security Council 'shall be made by an affirmative vote of nine members including the *concurring* votes of the permanent members' (emphasis added). At the time of the adoption of the Charter, the accepted position was that the requirement for the 'concurring votes' of the permanent members 'could hardly be interpreted in any other way than that, in the absence of affirmative votes cast by all the five permanent members, a decision on a non-procedural matter could not be taken by the Security Council' (Liang, 1950, p. 695). In other words, a binding resolution required the positive support of the P5: China, France, the USSR Republics, the United Kingdom and the United States.

In terms of practice, it was quickly established that voluntary abstention, a situation where a permanent member was present but not voting, did not prevent the adoption of a substantive resolution. The question remained as to the status of a resolution voted on when a permanent member was absent from the Security Council. In 1950, the Soviet Union assumed an 'empty chair' policy, boycotting Security Council meetings, in protest at the UN's refusal to recognise the People's Republic of China as the official representatives of China (Stueck, 2008; until 1971, the Chinese seat at the United Nations was held by Taiwan, the 'Republic of China'). One consequence was that the USSR was not in the Security Council in June and July 1950 to veto the resolutions that determined that the attack by North Korea on South Korea constituted a breach of the peace, called on UN member states to furnish assistance to South Korea, and for those members providing military forces to make such forces available to a unified command under the United States. Yuen-Li Liang, then director of the Division for the Development and Codification of International Law in the UN Secretariat, notes that the legal validity of these resolutions was contested by several states because of the Soviet absence (Liang, 1950, p. 703).

The rules for voting in the Security Council eventually came before the International Court of Justice in 1971 in *Legal Consequences for States of the Continued Presence of South Africa in Namibia* when South Africa challenged the validity of the Security Council resolution requesting an advisory opinion on the ground that two of the five permanent members had abstained from the vote. The court observed that the members of the Council, in particular, its permanent members, had 'consistently and uniformly interpreted the practice of voluntary abstention by a permanent member as not constituting a bar to the adoption of resolutions'.[10] The practice of states, the court determined, had replaced the ordinary meaning of the word *concurring*, that is to be in positive agreement with. The judgment did not conclude that practice alone could revise the meaning of Charter provisions, with the court observing the relevant procedure had been 'generally accepted by Members of the United Nations and evidences a general practice of that Organization'.[11] There are two requirements, then, to establish a revised understanding of the Charter by way of subsequent practice: first, we must be able to identify a consistent and uniform practice; second, we have to show the practice has been accepted by UN member states. Whilst *Legal Consequences* is concerned with abstention, the position of most international lawyers is that the reasoning also applies to absence and that a P5 member must veto, that is *vote against*, a resolution to prevent it from being adopted (Fassbender, 1998, p. 178), a position that conflicts with a literal reading of the UN Charter, which requires a *vote for* the resolution by each of the P5.

The paradox of change in the UN system

We have seen that change in the rules for voting in the Security Council followed the situation where not following the rules led to a new set of rules, that by not requiring the positive ('concurring') vote of the Soviet Union, the meaning of the phrase 'concurring votes of the permanent members' changed to require the positive exercise of the veto (the reverse of the literal meaning). This happened because UN member states accepted the validity of a series of Security Council resolutions in which the concurring vote of all permanent members was not required, but this appears paradoxical, as each step from one set of valid rules for voting to another contradictory set would involve a rule-breaking act, and only where sufficient steps of illegality (or invalidity) had been taken would the new position have been reached.

It is, in some ways, reminiscent of Zeno's paradox of Achilles and the tortoise, in which Achilles never catches the tortoise because the animal always holds a lead, however small. The story goes like this. At the start of a race, Achilles, who runs ten times faster than the tortoise, gives the animal a 100-metre head start. When Achilles has run the 100-metre, he finds that the tortoise has run

10 metres; Achilles then runs the further 10 metres but finds that the tortoise has run another 1 metre; Achilles runs the 1 metre, but the tortoise has run 10 centimetres; Achilles runs the 10 centimetres, but the tortoise has run 1 centimetre; and so on (Black, 1951, p. 91). In case of the United Nations, it appears paradoxical that change from one rule under the Charter scheme to a different rule can come about through a series of rule-breaking steps.

The function of a paradox is to get us to ask the right questions or to be clear about the problems we face. In the case of Achilles and the tortoise, the problem is to explain how a finite space (the distance Achilles and the tortoise) can be conceptually divided into an infinite number of smaller spaces. In the case of change in the UN system, the problem is to explain how rule-breaking behaviour can become rule-making behaviour. The problem is often examined in terms of the principle *ex injuria jus non oritur* (a right does not arise from wrongdoing). If, as a general principle of international law, as Dionisio Anzilotti argued in his dissenting opinion in *Legal Status of Eastern Greenland*, 'an unlawful act cannot serve as the basis of an action at law',[12] it would not be possible to change the rules of an international organisation by way of subsequent practice, yet we know this is possible, because it has been recognised by the International Court of Justice.

The key to making sense of transformations in the UN system from one position ('Do X') to a contradictory position ('Do non-X') is the factor of time. The idea we should make sense of change by recognising the function of time is not new, 'now', as Aristotle observed, is the difference between 'before and after' (Aristotle, 1967, p. 13). It is only because of time, we can speak of change, showing how, for example, the traffic light can be red *and* green and how the voting procedures in the UN Security Council can require the positive support of the five permanent members *and* the absence of a positive veto but only at different moments in time.

The intertemporal law of treaties

Law scholars have not traditionally concerned themselves with the subject of time (see generally French, 2001). Those legal philosophers who have written on the subject generally conclude that law systems exist in time and that the function of time is to help us distinguish between the past, the present and the future. Martti Koskenniemi has, for example, written that international law 'looks both backwards and forwards in time' (Koskenniemi, 2012, p. 23), whilst Phillip Allott describes international law as 'a bridge between the social past and the social future through the social present' (Allott, 1999, p. 1).

The classic statement on the subject of time in international law is provided by Judge Max Huber in the *Island of Palmas* case (1928). The United States' claim to the island was primarily based on its discovery by Spain in the 16th century and the subsequent transfer of Spanish title to the United States in the Treaty of Paris 1898. The Dutch claim was based on effective occupation from the 18th century onwards. On the basis of the unchallenged acts of the peaceful display of sovereignty from 1700 to 1906, Huber concluded that the Netherlands had acquired sovereignty to the Island of Palmas, which passed to Indonesia upon independence in 1949.

Both parties to the *Island of Palmas* arbitration agreed on the timeline of events and accepted that 'a juridical fact must be appreciated in the light of the law contemporary with it, and not of the law in force at the time when a dispute in regard to it arises or falls to be settled'.[13] This is the first branch of the doctrine of intertemporal law; it establishes that the legality or validity of an act is to be judged in accordance with the norms of the momentary international law system in force at the time the act takes place. This is recognised as a general principle of international law (Elias, 1980, p. 285), and the principle of non-retroactivity is applicable to the law of treaties (Corten and Klein, 2012, p. 719).

Huber went on to conclude that the principle of intertemporal law also required that 'the existence of the right, in other words its continued manifestation, shall follow the conditions required by the evolution of law'.[14] Whether or not Spain had acquired title to the island in the 16th century, by the 18th century, international law had changed, establishing a principle of effective occupation and a state that did not effectively occupy a territory it had discovered risked losing it to another state. This is the second branch of the doctrine. There are two aspects to the second branch: it confirms the possibility that international law can change over time (a point often overlooked) and makes clear that changes in the international law system can have consequences for the rights and duties of states.

Max Huber's doctrine of intertemporal law reflects both the A- and B-series conceptions of time familiar to philosophers of time and first outlined by the metaphysician John McTaggart (McTaggart, 1908, p. 458). A-series conceptions are concerned with dynamic change over time, the idea that something will be future, is present, and was past – think of the way that the ticking of a clock measures the coming and going of things in terms of past, present, future. B-series conceptions are concerned with the allocation of fixed and unchanging relations between events that occurred at different times (or the same time) – the kind of information we get by looking at a calendar (Bardon, 2013, p. 81).

The first branch of the doctrine of intertemporal law accords with the static B-series: events are aligned (and fixed) in date order to determine the valid laws of the international law system applicable at a fixed moment in time. The law (and legal position of actors) is fixed and unchangeable: if the act was lawful at that moment in time, it does not become unlawful because of a subsequent change in the law; likewise, if it were unlawful, it does not become lawful because of a change in the law. The second branch accords with the dynamic A-series, affirming the ephemeral nature of international law, which is always evolving into something else; it is concerned with the possibility of change over time. To make sense of international law, we must, in the words of the *Institute de Droit International*, appreciate 'the dual requirement of development and stability' in the system.[15]

Innovations in regulatory practice

The rules on the interpretation of treaties allow for an evolution in our meaning of treaty terms where the subsequent practice of states parties reflects a revised understanding of their obligations. In the case of the UN system, our initial understanding was that the positive support of all the permanent members was required for the adoption of a substantive resolution of the Security Council, but over time, by way of subsequent practice, we came to understand that a resolution could pass where a P5 member abstained or was absent, provided the other procedural requirements were met, for example, 9 positive votes out of 15, after the 1965 reforms.[16]

Making sense of change in the rules of the United Nations depends on both a fixed and a dynamic conception of time. We know the rules on voting in the Security Council changed because our knowledge of the applicable rules altered from one position to another (contradictory) position, reflecting both the fixed conception of time (we know what the rules were on different days) and the dynamic conception of time (we know the rules changed). We can model the evolution of a complex system like the United Nations by taking snapshots of the system at different moments in time (think of the way different editions of the international law textbooks explain the evolution of the UN system) – and then developing a story to explain what change has occurred and how (and why) that change occurred.

The curious status of Resolution 80 (1950)

International law recognises the possibility the meaning of the provisions of the constituent instrument of an international organisation can change over time because of a change in the behaviours of the member states. The emergence of a new pattern of behaviour amongst UN members can, then, result in a change in the rules of the UN organisation – in this case, the provisions on the rules concerning voting procedures in the UN Security Council. But what about the status of the first resolution adopted under the new procedural requirements?

Security Council Resolution 80 (1950), on 'The India-Pakistan Question', adopted at the 470th meeting on 14 March 1950, called on the governments of India and Pakistan to make immediate arrangements for the demilitarisation of Jammu and Kashmir, whose status had become a point of potential conflict following the partition of British India. Was the resolution binding? Were India and Pakistan in breach of their obligations under the Charter when they failed to comply with the demands of the Security Council?[17]

Security Council Resolution 80 (1950) was the first clearly substantive resolution to be adopted whilst a permanent member was absent from the council (Liang, 1950, p. 702). It was not passed in accordance with the (old) literal understanding of Article 27(3) of the UN Charter, requiring the positive support of the P5, nor could there be a new understanding resulting from a consistent and uniform practice, requiring only the absence of a veto. The logical conclusion must be that Resolution 80 (1950) was not validly adopted under either the old procedure or the new, leaving it in a legal no man's land. Yet, the UN official records report the Resolution as being adopted by 8 votes to 0, with two abstentions. A footnote simply observes that '[o]ne member (Union of Soviet Socialist Republics) was absent.'

The rules on voting in the Security Council depend on our reading of Article 27(3) of the Charter, which provides that decisions 'shall be made by an affirmative vote of nine members including the concurring votes of the permanent members'. To explain a change in the voting procedures without formal amendment of the Charter, we must explain how UN law changed from one position to another (contradictory) position through the emergence of a new 'consistent and uniform practice' that removed the requirement for the positive support of each of the P5. The problem is that one resolution cannot establish a 'consistent and uniform practice' – hence the difficulty with the status of Security Council Resolution 80 (1950).

For the UN system to change because of a change in the behaviours of the member states, there must be a change from one position of legal validity to another position of legal validity, and this change must pass through at least one step of legal invalidity – in this case Security Council Resolution 80 (1950), which cannot be valid under either the old or new procedures. Yet the resolution was regarded as valid by the UN system. How is it possible, then, that an innovation in the practice of member states can be valid on the basis that it reflects a new pattern of behaviour?

The key to understanding change in complex systems is, as we have noted, the factor of time. To make sense of any evolution in the meaning of Charter provisions, we must recognise that international law is, at one and the same time, both a fixed and a dynamic system. The first branch of Max Huber's doctrine of intertemporal law establishes that the validity of an act is to be judged in accordance with the rules in place at a fixed moment in time; the second branch confirms the dynamic and ephemeral nature of international law, which is always evolving into something else. To explain change in the provisions of the UN Charter, we must explain how meaning changed from one position to another, in this case, from a literal interpretation, whereby the positive support of the P5 was required, to a position where abstention or absence did not constitute a veto.

Our understanding of the Charter scheme evolved because of a change in the behaviours of states in the UN Security Council, especially the P5, demonstrated a different understanding of the voting procedures from that we had expected from reading Article 27 of the Charter and that different interpretation was accepted by the wider body of UN membership.

We can make sense of the change in the pattern of behaviours of member states on the procedural rules for the adoption of substantive UN Security Council resolutions by aligning the evidence along the x-axis of time. Consider the following sequence (which develops over time), where '0' represents evidence that supports the old rule (i.e. the adoption of a resolution requires the positive support of the P5) and '1' evidence that supports the new (adoption of resolution requires absence of a veto by one of the P5): 0, 0, 0, 0, 1, 1, 1, 1, 1. For us to conclude there is a new rule, there must be a clear pattern of 1s (as here). To see whether there is a clear pattern, we place brackets around the digits that (in our opinion) contain the pattern. Here, we can see that the pattern changes from the old rule ({0, 0, 0, 0,} 1, 1, 1, 1, 1 . . .) to the new (0, 0, 0, 0, {1, 1, 1, 1, 1 . . .}).

Because the pattern changes over time, these brackets establish our temporal frame of reference, containing the first and latest digits in the pattern. The pattern is seen to emerge at the *first* point within our temporal frame, immediately after the opening bracket; it concludes (and the temporal frame closes) after the *latest* piece of evidence for the simple reason that we cannot consider evidence from the future, only that available today and from the past.

In the process of transition from one set of valid law norms to another (contradictory) set, there will often be some uncertainty about the outcome of the process, whether a new rule is emerging or will emerge. But law systems cannot take a position of 'not knowing'. The reason for this is straightforward. Law, as Niklas Luhmann has shown, is a system of communications identified through its use of the binary code 'legal/illegal' (Luhmann, 2004, p. 58). Law involves the application of law norms to facts, with the outcomes expressed in terms of 'legal' or 'illegal' behaviour (ibid., p. 93). When called on to decide on the status of an act, a law-applier, including a court or tribunal, must decide whether the behaviour was 'legal' (in accordance with valid law norms) or 'illegal' (not in accordance), and to do this the law-applier must always be able to identify the valid law norms in place on any given day. This is expressed in terms of the general principle of legal certainty or the rule of law. In this case, the law-applier would have to decide whether a Security Council resolution was validly adopted or not validly adopted in accordance with the rules of the UN system – not knowing is not an option.

Where it is clear a law system has undergone a process of transition, the court or tribunal must confirm the fact of transition at the earliest moment that it was clear the new rule had emerged, at the opening of the temporal frame that contains evidence for the new rule. The alternative would be to insist on a repeated pattern of behaviour for the existence of the new rule, but that would leave all instances of practice before the tipping point in legal limbo, as they would (necessarily) be inconsistent with the old rule, but their legality and validity could not be confirmed by the new requirement. If we accept Security Council Resolution 80 (1950) was validly adopted, we must accept that this innovation in the practice of UN member states established a new rule of behaviour concerning voting procedures in the UN Security Council; the only other possibility would be to regard Resolution 80 (1950), along with subsequent resolutions adopted under the 'new' procedure, as invalid, until a sufficient number of invalid resolutions had been adopted to confirm the existence of a new general practice within the UN organisation.

But how can we explain this? A single resolution cannot amount to a 'consistent and uniform' practice as required by the International Court of Justice.[18] A single event cannot establish a new pattern of behaviour (i.e. '1̲, . . . '). If, however, we add four more validly adopted Security Council resolutions in which one of five permanent members abstained, there would be a clear pattern ('1̲, 1, 1, 1, 1 . . . '), which includes Resolution 80 that we decided (at the time) did not (by itself) constitute a new pattern of practice ('1̲, . . . ').

Had we been asked about the status of Security Council Resolution 80 (1950), on 'The India-Pakistan Question', on the date of its adoption, we would have answered that this was an invalid resolution and there was no international law obligation on India and Pakistan to demilitarise Jammu and Kashmir. This must be the case, because we could not be certain that a new pattern of practice would emerge – the reaction of UN member states might have provoked a return to the status quo ante {0, 0, 0, 0, $\underline{1}$, 0, 0, 0. . .}. But with the passing of time and adoption of more Security Council resolutions under the new procedure, we saw a new pattern of practice emerging {0, 0, 0, 0, $\underline{1}$, 1, 1, 1, 1. . .}.

With the passing of more resolutions, we came to understand that Resolution 80 (1950) had – as a legal fact – been adopted by an appropriate procedure of the UN Security Council, but we could only see this by reconciling the fixed and dynamic nature of the international law system. Whereas we used to think that the new behaviour was a deviation from the old procedures ({0, 0, 0, $\underline{1}$, . . .}) and therefore invalid, we can now see it represented the emergence of a new rule (0, 0, 0, {$\underline{1}$, 1, 1, 1, 1. . .}), and was, consequently, validly adopted.

Resolution 80 (1950) was an innovation in practice when it was adopted on 14 March 1950, reflecting a new understanding of the requirements for voting in the UN Security Council. Looking back in time, we can see it was the start of a consistent new pattern of behaviour which emerged from that date, allowing us to conclude that Resolution 80 (1950) was validly adopted and therefore created an obligation under the UN Charter for India and Pakistan to demilitarise the territory of Jammu and Kashmir. But we could only reach this conclusion by taking into account the factor of time when looking to make sense of the evolution of complex regulatory systems like the United Nations.

Conclusion

The objective of this chapter was to make sense of the transformations in the UN system that result from changes in the behaviours of the member states. Our case study was the alteration in the voting procedure of the UN Security Council from a position whereby the positive support of the P5 was required for the adoption of a binding resolution, to a situation in which absence or abstention did not count as a veto, with Security Council Resolution 80 (1950) being the first resolution adopted under the new procedure. The work looked to complexity theory to make sense of the issue, understanding the UN system as the emergent property of the actions and interactions of the member states, with the organisation then regulating the very same component agents that brought it into existence. The analysis showed how the UN system evolved with changes in the behaviours of its component agents and how the alteration in the voting procedures in the Security Council occurred in response to the 'empty chair' policy adopted by the USSR and determination of other member states to carry on with business as usual. By isolating the event that triggered the evolution in UN law, we developed an account about how and why the rules for voting in the Council changed over time.

An alteration in the meaning of Charter provisions because of member states acting differently is explained in international law doctrine by the role of subsequent practice in the interpretation of treaties, allowing a new understanding of the constituent instrument to emerge a consequence of a change in behaviour. Even if we accept (as we must) the paradox of rule-breaking behaviour becoming rule-making behaviour (a possibility recognised by the International Court of Justice and Vienna Convention on the Law of Treaties), there remains the problematic status of the first iteration of any new practice. The first act of deviant behaviour cannot be part of a new pattern of activity, leaving it, it would seem, in legal limbo, inconsistent with the old rule but without there being a new pattern of behaviour confirming the emergence of a new rule.

We cannot make sense of the problem of change by a retrospective change in the status of Security Council Resolution 80 (1950). Time travel is not possible in international law, but time is central to understanding and resolving this problem. We make sense of change over time in complex law systems by comparing one (earlier) description of the system with another (later) description and then developing a narrative to explain what has changed and how it changed. By isolating the event that triggered the change, we construct an account about how and why the UN regulatory system changed. Here, the event that initiated the change in the voting procedures was the absence from the UN Security Council of the Soviet delegate in protest of the refusal of the United Nations to recognise the Beijing government as the representative of China.

On 14 March 1950, we would have concluded that Security Council Resolution 80 (1950) was an invalid resolution, but, looking back, we can see it was part of a new pattern of behaviour in the Security Council which emerged from that date, and that Resolution 80 was in fact validly adopted. In other words, our knowledge of the content of the rules of the UN organisation on that day changed over time as more evidence of the behaviours of member states became available with the adoption of further resolutions without the support of the P5. We knew (and know) what the valid laws were on Tuesday 14 March 1950 by looking at the available evidence (the first branch of the doctrine of intertemporal law), but our understanding and knowledge of the rules in force on that date changed over time as we widened our temporal frame (the second branch).

Looking to complexity theory emphasised the dynamic nature of the international law system, which is always in a constant state of evolution, always in a condition, in the words of the Nobel laureate Ilya Prigogine, of 'being' and 'becoming' (Prigogine, 1980, p. 13). By looking to complexity, and by understanding the United Nations as an emergent regulatory system that evolves with changes in the behaviours of the member states, we were able to make sense of change over time in the UN system. The analysis demonstrated that whilst we can always tell the content of UN law on any given day, by looking at the available evidence (the first branch of the doctrine of intertemporal law), our understanding and knowledge of the valid rules can (and often will) change over time as we broaden our temporal frame (the second branch). In other words, we cannot make sense of a complex regulatory system, like the United Nations, without taking into account the factor of time and the possibility of change over time.

Notes

1 Before 1965, 7 positive votes out of 11 were required for the adoption of a substantive resolution.
2 *Reparation for Injuries Suffered in the Service of the United Nations*, ICJ Rep. 1949, p. 174, p. 185.
3 Ibid., p. 179.
4 *Certain Expenses of the United Nations*, ICJ Rep. 1962, p. 151, p. 157.
5 Yearbook of the International Law Commission (1966), vol. II, p. 222.
6 Recall that article 31(3)(b) allows recourse to subsequent practice that 'establishes the *agreement* of the parties regarding its interpretation'.
7 *Legality of the Use by a State of Nuclear Weapons in Armed Conflict*, Advisory Opinion, ICJ Rep. 1996, p. 66, p. 75.
8 *Reservations to the Convention on Genocide*, ICJ Rep. 1951, Dissenting Opinion of M. Alvarez, p. 49, p. 53.
9 Separate Opinion of Judge Lauterpacht, *Voting Procedure on Questions Relating to Reports and Petitions Concerning the Territory of South-West Africa*, Advisory Opinion, I.C.J. Reports 1955, p. 67, p. 90, p. 106.
10 *Legal Consequences for States of the Continued Presence of South Africa in Namibia* (South-West Africa) *notwithstanding Security Council Resolution 276 (1970)*, ICJ Rep. 1971, p. 16, para. 22
11 Ibid., para. 28.
12 Dissenting Opinion of M. Anzilotti, in *Legal Status of Eastern Greenland*, PCIJ Ser. A/B 53, p. 76, at p. 95.
13 *Island of Palmas* (Netherlands v USA) (1928) 2 RIAA 829, 845.
14 Id.
15 Preamble, *Institut de Droit International*, Resolution on 'The Intertemporal Problem in Public International Law', Session of Wiesbaden (1975). <http://www.idi-iil.org/en/sessions/wiesbaden-1975/>, accessed 10 May 2018.
16 Before 1965, 7 positive votes out of 11 were required for the adoption of a substantive resolution.
17 The Security Council noted the lack of progress on the issue in Resolution 91 (1951), 'The India-Pakistan Question', adopted 30 March 1951, adopted by 8 votes to 0, with 3 abstentions (including the Union of Soviet Socialist Republics).
18 *Legal Consequences for States of the Continued Presence of South Africa in Namibia* (South-West Africa) *notwithstanding Security Council Resolution 276 (1970)*, ICJ Rep. 1971, p. 16, para. 22

Bibliography

Allott, Philip (1999), 'International Law and the Idea of History', 1 *Journal of the History of International Law* 1.
Amerasinghe, C. F. (2005), *Principles of the Institutional Law of International Organizations* (Cambridge: Cambridge University Press, 2nd ed.).
Arato, Julian (2013), 'Treaty Interpretation and Constitutional Transformation: Informal Change in International Organizations', 38 *Yale Journal of International Law* 289.
Aristotle. (1967), 'Time', from *Physics, Book IV*, reprinted in Gale, Richard M. (ed.) *The Philosophy of Time: A Collection of Essays* (New York: Anchor Books), 9.
Bardon, Adrian (2013), *A Brief History of the Philosophy of Time* (Oxford: Oxford University Press).
Black, Max (1951), 'Achilles and the Tortoise', 11(5) *Analysis* 91.

Byrne, David (1998), *Complexity Theory and the Social Sciences* (London: Routledge).

Carline, Anna and Zoe Pearson (2007), 'Complexity and Queer Theory Approaches to International Law and Feminist Politics: Perspectives on Trafficking', 19 *Canadian Journal of Women and the Law* 73.

Chinen, Mark (2014), 'Complexity Theory and the Horizontal and Vertical Dimensions of State Responsibility', 25 *European Journal of International Law* 703.

Corten, Olivier and Pierre Klein (2012), *The Vienna Conventions on the Law of Treaties: A Commentary. Vol. 2* (Oxford: Oxford University Press).

Cot, Jean-Pierre (2011), 'United Nations Charter', *Max Planck Encyclopedia of Public International Law*. <http://opil.ouplaw.com/home/EPIL>, accessed 10 May 2018.

Dainton, Barry (2001), *Time and Space* (Chesham: Acumen).

David, Paul A. (1985), 'Clio and the Economics of QWERTY', 75(2) *The American Economic Review* 332.

Djelic, Marie-Laure and Sigrid Quack (2007), 'Overcoming Path Dependency: Path Generation in Open Systems', 36 *Theory and Society* 161.

Elias, T. O. (1980), 'The Doctrine of Intertemporal Law', 74 *American Journal of International Law* 285.

Fassbender, Bardo (1998), *UN Security Council Reform and the Right of Veto: A Constitutional Perspective* (The Hague: Kluwer Law International).

Frank, Adam (2011), *About Time* (Oxford: Oneworld).

French, Rebecca R. (2001), 'Time in the Law', 72 *University of Colorado Law Review* 663.

Gardiner, Richard (2015), *Treaty Interpretation* (Oxford: Oxford University Press, 2nd ed.).

Holland, John H. (2014), *Complexity: A Very Short Introduction* (Oxford: Oxford University Press).

Jones, Gregory Todd (2008), 'Dynamical Jurisprudence: Law as a Complex System', 24 *Georgia State University Law Review* 873.

Klabbers, Jan (2004), 'Constitutionalism Lite', 1 *International Organizations Law Review* 31.

Klabbers, Jan (2009), *An Introduction to International Institutional Law* (Cambridge: Cambridge University Press, 2nd ed.).

Koskenniemi, Martti (2012), 'Law, Teleology and International Relations: An Essay in Counterdisciplinarity', 26 *International Relations* 3.

Küppers, Bernd-Olaf (1992), 'Understanding Complexity', in Ansgar Beckermann, Hans Flohr and Jaegwon Kim (eds.) *Emergence or Reduction?* (Berlin: Walter de Gruyter), 241.

Liang, Yuen-Li (1950), 'Abstention and Absence of a Permanent Member in Relation to the Voting Procedure in the Security Council', 44 *American Journal of International Law* 694.

Luhmann, Niklas (2004), *Law as a Social System*, translated by Klaus Ziegert (Oxford: Oxford University Press).

McGoldrick, Dominic (2004), *From '9–11' to the 'Iraq War 2003': International Law in an Age of Complexity* (Oxford: Hart Publishing).

McTaggart, J. Ellis (1908), 'The Unreality of Time', 17 *Mind* 457.

Mill, John Stuart (1858), *A System of Logic, Ratiocinative and Inductive* (New York: Harper and Brothers).

Mitchell, Melanie (2009), *Complexity: A Guided Tour* (New York: Oxford University Press).

Morin, Jean Frédéric et al. (2017), 'The Trade Regime as a Complex Adaptive System: Exploration and Exploitation of Environmental Norms in Trade Agreements', 20 *Journal of International Economic Law* 365.

Nolte, Georg (2015), 'Third Report on Subsequent Agreements and Subsequent Practice in Relation to Treaty Interpretation', UN Doc. A/CN.4/683, 7 April 2015.

Prigogine, Ilya (1980), *From Being to Becoming: Time and Complexity in the Physical Sciences* (San Francisco: W. H. Freeman).

Prigogine, Ilya and Isabelle Stengers (1985), *Order Out of Chaos: Man's New Dialogue With Nature* (London: Flamingo).

Prior, Tahnee Lisa (2017), 'Engaging Complexity: Legalizing International Arctic Environmental Governance', in Holly Cullen, Joanna Harrington and Catherine Renshaw (eds.) *Experts, Networks, and International Law* (Cambridge: Cambridge University Press).

Provost, René (2015), 'Interpretation in International Law as a Transcultural Project', in Andrea Bianchi, Daniel Peat and Matthew Windsor (eds.) *Interpretation in International Law* (Oxford: Oxford University Press), 290.

Ruhl, J. B. (2008), 'Law's Complexity: A Primer', 24 *Georgia State University Law Review* 885.

Sawyer, Keith R. (2005), *Social Emergence: Societies as Complex Systems* (Cambridge: Cambridge University Press).

Schachter, Oscar (1988), 'United Nations Law', 88 *American Journal of International Law* 1.

Stephan, Achim (2002), 'Emergentism, Irreducibility, and Downward Causation', 65 *Grazer Philosophische Studien* 77.

Stueck, William (2008), 'The United Nations, the Security Council, and the Korean War', in Vaughan Lowe, Adam Roberts, Jennifer Welsh and Dominik Zaum (eds.) *The United Nations Security Council and War: The Evolution of Thought and Practice Since 1945* (Oxford: Oxford University Press), 265.

Tomuschat, Christian (1993), 'Obligations Arising for States Without or Against their Will', 241 *Recueil des Cours* 195.Waldrop, M. Mitchell (1992), *Complexity: The Emerging Science at the Edge of Order and Chaos* (London: Penguin Books).

Webb, Julian (2005), 'Law, Ethics, and Complexity: Complexity Theory & The Normative Reconstruction of Law', 52 *Cleveland State Law Review* 227.

Webb, Thomas E. (2014), 'Tracing an Outline of Legal Complexity', 27 *Ratio Juris* 477.

Wheatley, Steven (2016), 'The Emergence of New States in International Law: The Insights from Complexity Theory', 15(3) *Chinese Journal of International Law* 579.

Wheatley, Steven (2018), *The Idea of International Human Rights Law* (Oxford: Oxford University Press).

White, Nigel D. (2002), *The United Nations System: Toward International Justice* (London: Lynne Rienner Publishers).

White, Nigel D. (2005), *The Law of International Organisations* (Manchester: Manchester University Press, 2nd ed.).

Yin, Gao and Herfel, William (2011), 'Constructing Post-Classical Ecosystems Ecology', in Hooker, Cliff (ed.), *Philosophy of Complex Systems* (Amsterdam; London: North Holland) 389.

6   The 'Consensus Approach' of the European Court of Human Rights as a rational response to complexity

Dimitrios Tsarapatsanis

Introduction

The present chapter uses complexity theory to argue that the so-called 'consensus approach' of the European Court of Human Rights (henceforth 'the Court' or 'ECtHR') can be a rational response to the cognitively demanding task of interpreting and applying the European Convention of Human Rights (henceforth 'the Convention' or 'ECHR') to member states of the Council of Europe. The chapter begins by setting the stage in two ways. First, drawing on recent literature on the subject, I provide a succinct sketch of a number of complexity theory concepts and argue that they can be relevant to the study of the ECHR. Second, I briefly present the consensus approach and some of the criticisms that have been addressed against it, with specific reference to the moral reading of the Convention. The moral reading of the ECHR, associated with Ronald Dworkin's legal interpretivism and defended by leading commentators such as George Letsas (Letsas, 2007), is one of the most forceful sources of criticism of the consensus approach. It is also an independently plausible and sophisticated theory of interpretation of the ECHR. Thus, using complexity theory to show that, despite initial appearances, the moral reading of the Convention could be compatible with the consensus approach is an interesting result in itself.

I aim to do this in the main body of the chapter by first outlining the decision problem that the Court faces in a number of hard cases. These involve human rights review of state measures emerging through complex patterns of institutional interaction at the domestic level. I then provide an outline of a number of constraints that limit the epistemic capacity of the ECtHR and can spawn uncertainty about the correct outcome. The difficulty specifically stems from the combination of complexity with limited epistemic resources. Next, supposing for the purposes of my argument that the moral reading of the Convention is the correct theory of interpretation of the ECHR, I claim that, under circumstances of uncertainty, consensus can best be understood as a reasoning strategy, not as a criterion of truth about ECHR rights. It is thus not necessarily incompatible with a moral reading of the Convention. Last, I suggest that, *qua* reasoning strategy, the consensus approach could perhaps be best understood and assessed as a collective intelligence device. Throughout, the chapter is exploratory rather than conclusive. Sketching a possibility is a long way from defending it against all, or even the most important, objections. My main goal, rather, is to make conceptual space for further and more detailed future work along the lines suggested here.

Setting the stage (a): complexity theory, domestic legal systems and the ECHR

The theory of 'complex adaptive systems' or simply 'complexity theory' roughly designates a family of approaches that initially emerged in natural sciences, such as ecology and neuroscience, to explain ways in which patterned order could emerge from the unplanned interactions of a number of heterogeneous agents or elements, be they individual neurons vis-à-vis the brain or colonies of insects (Wheatley, 2016, p. 581 and 587–588; Page, 2010; Ruhl, 2008). One of the most important guiding ideas behind complexity approaches is the recognition that the state of such systems is not reducible to that of their constitutive elements or agents, the system being 'larger than the sum of its parts' (Wheatley, 2016, p. 587). Because of the success of complexity approaches in accounting for the function of such natural systems, the approaches were subsequently used to shed light on social systems presenting similar attributes, such as economies, domestic political systems or systems of states (Page, 2010; Wheatley, 2016). Importantly, there is also now an emerging literature that applies complexity theory to domestic legal systems and to international law (Wheatley, 2016, p. 580).

Whilst there is no canonical definition of complexity (Wheatley, 2016, p. 589), there is broad agreement that a given system, be it natural or social, can be considered *complex* as opposed to merely complicated (Page, 2010, p. 7) when (at least some of) the following features are present (Page, 2010; Wheatley, 2016). First, the system involves the interaction of a multiplicity of heterogeneous elements or agents. Second, the system is open to an environment that exists outside of it. Third, agents can adapt their behaviour on the basis of feedback they receive from other agents or from the environment of the system. Fourth, because of heterogeneity and adaptation, the system *qua* system can have properties which are 'emergent' in the specific sense that they supervene on but are not reducible to any of the properties of the individual agents composing the system. Fifth, systems may achieve various states of 'stable disequilibrium' and are to this extent 'self-organising'. Importantly, this means that the structure of complex systems can sometimes achieve a level of spontaneous stability, that is stability which is not the result of a 'central controller' but, rather, the patterned result of the numerous interactions between its constitutive agents. Equally important, such stable states are temporary insofar as they are continuously challenged from agents under pressures by the system's environment. Sixth, emergent properties, as well as temporary states, of stable disequilibrium of a system may change in 'non-linear ways', that is in ways which are not the direct consequence of the behaviour, intended or otherwise, of any individual agent. As a result, individual behaviours appearing insignificant or innocuous may have a wide systemic impact in ways difficult to predict in advance. Seventh, some changes of the emergent properties of the system become practically irreversible because of path-dependence mechanisms that steer and 'lock in' the system towards one particular direction.

Even this brief glance at some generic concepts from complexity theory suffices to indicate their potential fruitfulness when it comes to understanding the relationship of the ECHR with domestic legal systems. For the purposes of the present chapter, in particular, two aspects are essential. The first is to do with the fact that domestic legal systems are complex in the sense roughly specified earlier. To begin with, they are the product of the interaction of a multiplicity of heterogeneous agents who can adapt their behaviour to that of other agents within the system. This is especially the case insofar as distributions of power allow some agents to block or alter the decision made by other agents. As Adrian Vermeule puts it, under these conditions law is the product of the concerted action of aggregates of individuals (which compose institutions), as well as of nested aggregates of aggregates (relationships between institutions themselves; Vermeule, 2012, chapter 1). The solution provided to some issue under domestic law can thus be an emergent property of the system in the sense that it is not necessarily reducible to the action or intention of any one agent within the system. Moreover, legal solutions to issues are frequently merely the outcome of temporary stable disequilibria, apt to change under a different configuration of interactions on the part of the agents composing the system. Last, changes to such solutions are prone to phenomena of path-dependence in the sense that it may be much costlier or practically impossible to revert the system back to its prior state once changes have taken place.

The second aspect is to do with the fact that the ECHR *itself* can be understood as a system prone to complexity effects. Thus, and in a non-exhaustive manner, domestic agents may adapt their behaviour to accommodate decisions of the ECtHR in potentially disruptive ways through, for example, strategic interaction, which may result in the system having emergent properties not intended by any one agent. Likewise, the stability of the system may be the result of a temporary, stable but dynamic disequilibrium resulting from the interactions between the Court and nested aggregates of domestic agents or even between individual judges composing the Court.

An important premise of this chapter is that both aspects, that is those pertaining to the complexity of domestic legal systems and those to do with the ECHR itself as a complex system, can affect the review of state practices with regard to their compatibility with the Convention in a number of important ways. As already indicated, it is not my ambition to analyse or even outline all these ways. Instead, I draw out some of the consequences of complexity with regard to the more specific issue of the so-called consensus approach used by the ECtHR, perceived as a reasoning strategy.

Setting the stage (b): the moral reading of the ECHR, the consensus approach and its criticisms

With these preliminary points concerning complexity theory in place, I now move on to outline the approach of interpretation of the ECHR that I shall assume throughout the chapter. Following Letsas (Letsas, 2013, pp. 122–141) and Dworkin (Dworkin, 1996), I shall call it 'moral reading' of the Convention.

The approach draws on Dworkin's legal interpretivism to claim that objective moral considerations about the point of abstractly formulated ECHR rights necessarily figure among the truth conditions of propositions about the content of those rights. In his book-length defence of the moral reading of the Convention (Letsas, 2007), Letsas suggested that the abstract moral language of the ECHR lends itself quite naturally to such a rendering. The main idea is that applying the Convention's abstractly formulated rights to particular cases necessitates specification of their content through an interpretation of the moral values underpinning and justifying these rights. Furthermore, and as a matter of substantive political morality, Letsas favours a liberal egalitarian theory of Convention rights, which is robustly anti-perfectionist and anti-majoritarian (Letsas, 2007, chapter 5). Under such a theory, the purpose of ECHR rights is to shield individuals from the hostile preferences of majorities in a wide variety of situations. Within this picture, the Court successfully discharges its role by acting as an international guardian of equal individual liberty whenever the judicial institutions of contracting states have failed themselves to accomplish this essential task.

Letsas plausibly maintains that the Court's interpretive practice provides sufficient evidence of endorsement of the moral reading (Letsas, 2007, chapter 3). Indeed, through its 'autonomous concepts' and 'living instrument' approaches, the Court has opted for a purposive interpretation of the Convention, relatively detached from contracting states' understandings of ECHR rights. Especially in recent years, Letsas convincingly contends (Letsas, 2013, pp. 115–122), the Court's practice seems to aim at discovering the objective moral truth about the content of ECHR rights. Importantly, Letsas argues, such an approach depends on substantive moral considerations and not on member states' shared understandings. Thus, on the moral reading, the sheer fact that a majority of states happens to share a moral view does *not* make that view true. Accordingly, the goal is to establish the *objective* content of Convention rights, which is not reducible to the content that contracting states merely believe these rights have (Letsas, 2004).

One central point of contention addressed by the present chapter is whether the moral reading of the ECHR is consistent with the so-called consensus approach adopted by the Court. In order to tackle it, we must have some idea of what the approach entails (Dzehtsiarou, 2015). At a first take, the consensus approach consists in interpreting and applying Convention rights according to a rough requirement of identification of shared understandings and practices across contracting states. These shared understandings and practices serve as a standard whereby to evaluate the performance of individual states on the human rights issue adjudicated by the Court. The consensus inquiry typically consists in a comparative examination of national, European or international law and practice. Whilst there are different ways to understand the approach's function in the Court's reasoning, at a minimum it grounds a 'rebuttable presumption', if not always a conclusive reason, in favour of a given outcome. Moreover, the Court typically links consensus to the 'dynamic interpretation' of the ECHR, whereby new interpretations of Convention rights are provided in order to treat novel kinds of human rights issues. When it comes to deciding on these new interpretations, the Court considers shared states' understandings and practices as a particularly important factor.

A non-exhaustive review of the case law reveals a number of features of the Court's consensus inquiry that are of particular interest for the purposes of the present chapter. First, the Court formulates the issue on which it bases its comparative law inquiry in a way specifically tailored to the outcome of the case and to the level of abstraction appropriate to the case's particular facts.[1] Second, by doing this, the Court typically does not delve into the reasoning process that led to the particular political decisions made by the contracting states but merely compares the end results of these decision-making processes, to wit, the decisions themselves.[2] Third, and in view of the preceding, the Court does not provide any deep analysis of the moral point or purpose of the human right involved, nor does it engage in direct moral reasoning. Instead, it focuses on the facts of the case and to the results of the comparative law inquiry, deferring to common understandings of contracting states in deciding the issue at hand. Fourth, the Court's approach consists in loosely aggregating contracting states' solutions with respect to the identified issue.[3] Fifth, the choice made by the majority of contracting states is seen as providing a particularly weighty but not necessarily conclusive reason in favour of deciding the issue at hand in the same way (Dzehtsiarou, 2015, pp. 24–30). Sixth, the Court's comparative law inquiry is seldom systematic or comprehensive (Dzehtsiarou, 2015). Seventh, the Court may consider that the existence of state consensus with respect to some issue is a factor that narrows the margin of appreciation of the respondent state,[4] but it may also hold that the absence of consensus widens the respondent state's margin of appreciation and lowers scrutiny[5] or, alternatively, that the existence of consensus in favour of a particular state measure furnishes a rebuttable presumption that the measure is not in violation of the Convention.[6]

Commentators' reactions to uses of the consensus approach have been mixed. While the approach has generally been considered as a legitimacy-enhancing mechanism (Dzehtsiarou, 2015, pp. 143–176), it has also been criticised on a number of different grounds. We can usefully group such criticisms into two general categories. The first revolves around the perceived indeterminacy and lack of precision of the appropriate doctrinal test (Helfer, 1993; Ambrus, 2009), which, critics argue, often fails to provide clear guidance to states and is sometimes even characterised as 'random' (Ambrus, 2009, p. 354) or else applied in an imprecise and inconsistent fashion. Critics have also complained that the Court frequently fails to define in a clear and consistent way whose consensus should be taken into account and how, as well as the correct level of abstraction at which it should be formulated (Dzehtsiarou, 2015, pp. 14–23). As a result, the manner in which the Court uses the consensus approach is often in tension with the rule of law values of legal certainty, predictability and equality before the law. Importantly, critics also propose specific ways of reconstructing the approach so as to better promote these values (Ambrus, 2009, pp. 362–370). We can thus label this first kind of criticism 'ameliorative'. Scholars engaging in it generally agree that the consensus approach is intrinsically valuable, suggesting ways in which its application by the Court could be normatively enhanced, once rid of inconsistencies and ambiguities.

In this chapter, I shall not take issue with such ameliorative criticism. Instead, I focus on a second, more radical kind of reproach. Authors who subscribe to this kind of argument maintain that the approach is fundamentally at odds with moral requirements stemming from the very idea of human rights protection. Accordingly, they urge that it be abandoned in favour of direct moral reasoning by the Court. Two mutually reinforcing claims are advanced. The first, weaker, claim is to the effect that frequently the consensus approach appears to be superfluous. The indeterminacy of the consensus test, critics argue, shows that fleeting mention of common understandings in ECtHR judgments merely bolsters conclusions already arrived at by recourse to substantive moral reasoning at a prior stage (Letsas, 2013, pp. 108–115). As Eyal Benvenisti puts it, '[consensus] is but a convenient subterfuge for implementing the court's hidden principled decisions' (Benvenisti, 1999, p. 852). On its face, this claim is compatible with the ameliorative view, at least if it turned out that it is possible to formulate the consensus test with a degree of precision sufficient to provide a well-structured decision procedure. However, the superfluity criticism serves as a prelude to a second, much stronger, claim, to the effect that, even if it were possible to arrive at a precise formulation of the consensus test, the consensus approach would still flout the normative requirements stemming from the point and purpose of the ECHR. This stronger claim, associated with the moral reading of the Convention, is at the heart of debunking criticisms of the consensus approach. It has been most forcefully and clearly articulated by Letsas (Letsas, 2004; Letsas, 2007, chapter 2).

In order to better grasp why Letsas argues that the consensus approach is not merely superfluous but, in fact, incompatible with the moral reading of the Convention, recall the Court's reasoning in such cases, outlined earlier. By resorting to consensus, the Court apparently abstains from providing any substantive normative reasons about the point, purpose and moral value of ECHR rights. Instead, it seems to merely defer to what it thinks the common European standard is with respect to the human rights issue at hand, identified by a vague reference to the practices of the majority of contracting states (or even practices of non-contracting states and other international institutions). This is, for example, exactly the way in which the Court appears to have recently proceeded in the particularly controversial *Lautsi v. Italy* and *S.A.S. v. France* cases.[7] From the vantage point of the moral reading of the ECHR, the Court's choices raise two broad kinds of concern. First, by resorting to common understandings of contracting states through the comparative law study of the solutions adopted by these states on a given Convention right issue, the Court would aggregate solutions determined by the beliefs of political majorities about the content of Convention rights. It would thus appear to presuppose that these beliefs, and not independently identifiable moral values, determine the content of ECHR rights (Letsas, 2004). Letsas contends that this attitude is in tension with Strasbourg's 'interpretive ethic' (Letsas, 2010), which gives pride of place to the idea that the moral values underpinning ECHR rights are objective and, as such, irreducible to the beliefs that contracting states hold about them (see also Benvenisti, 1999). Second, adding insult to injury, the consensus approach would also appear to flout the very normative *raison d'être* of Convention rights: the fact that they are rights purporting to protect individuals and minorities from the hostile preferences of majorities. In particular, as already indicated, it would seem to follow from the anti-majoritarian nature of ECHR rights that the Court has the mission to provide an independent moral check on member states with respect to Convention rights issues. The Court arguably fails to do this when it merely mirrors or upholds member states' majoritarian current practices and beliefs.

Letsas's forceful critique seems to present proponents of the consensus approach that also subscribe to the moral reading of the Convention with a harsh dilemma. If they want to stick to some version of the consensus requirement, they should either abandon their commitment to objective moral truth by espousing a conventionalist view to the effect that the contracting states' concurring beliefs figure among the determinants of the content of Convention rights, or else they should adopt a theory of ECHR rights that makes majoritarian preferences the determining moral factor. Both of those alternatives are unattractive. On the one hand, moral conventionalism seats uneasily with the universalist ambition of human rights, as well as with the Court's own method of 'autonomous concepts' (Letsas, 2004). On the other hand, consequentialist moral conceptions such as utilitarianism, which roughly make maximisation of aggregate preference-satisfaction an objective criterion of moral rightness, are widely believed to be traditional enemies of human rights and could hardly be considered as natural candidates for an attractive conception of Convention rights (Letsas, 2007, chapter 5).

Decisions for complex normative systems: the problem of uncertainty

Do the previously mentioned considerations exhaust what sense there is to be made of the consensus approach under a moral reading of the ECHR? I beg to differ. In other work (Tsarapatsanis, 2015), I have developed an institutional account of the Court's interpretive practice, defending the margin of appreciation doctrine by appealing to normative considerations pertaining to shared responsibility and subsidiarity in the implementation of the Convention. It is a significant virtue of institutional accounts that they purport to explain and justify doctrines of judicial deference and self-restraint, such as the consensus approach or the margin of appreciation, without abandoning the ambition of reading the ECHR morally. Institutional accounts supplant the moral reading of the Convention: they hold that institutional reasons about the proper division of labour between the Court and national institutions, and not merely substantive ones about the moral point or value of human rights, are relevant to the determination of judicial outcomes. Unlike substantive reasons, which abstract from the identity of the Court *qua* court and refer only to the merits of the individual case, institutional reasons apply specifically to the Court as an enforcing institutional agent, by determining the Court's powers and responsibilities within a wider scheme of institutional cooperation. Such reasons may justify the Court's responsibility to defer to contracting states' shared understandings of Convention rights irrespective of the fact that these understandings are potentially at odds with the content of Convention rights seen from the perspective of an ideal moral theory of human rights. Moreover, these considerations are not *ad hoc*, applying only within the narrow context of the ECHR legal order, but pervasive in public law more generally (Kyritsis, 2015). Institutional approaches to judicial decision-making usually highlight issues of judicial competence and legitimacy as factors justifying both judicial restraint and deference to the choices made by the political branches of government. Judicial duties of deference to democratically legitimated institutions, underscored by institutional accounts, thus seems to cohere particularly well with the general structure of the consensus approach used by the ECtHR.

Nevertheless, in this chapter I contend that there is one additional question that should be asked with regard to the nature and role of the consensus approach and that insights from complexity theory are absolutely crucial to answering it. To begin with, recall that, by adopting a moral reading of the ECHR, I assume that pertinent moral reasons, both substantive and institutional, together with whatever empirical facts are made relevant by these reasons, jointly determine the truth values of propositions of Convention rights. The question, then, is whether the earlier-stated reasons and facts are epistemically accessible to judges given the judges' actual (as opposed to ideal) cognitive and, more generally, epistemic capacities. And, if so, what are the conditions and costs of such accessibility? This further question is particularly important, not least because efficient and reliable decision-making by the Court is not a theoretical, but an eminently practical, enterprise. That real, flesh-and-blood judges be able to reliably discover normative and empirical facts determining the truth values of particular propositions of Convention rights at an acceptable cost is what really matters in the collective enterprise of interpretation and enforcement of the Convention. Thus, even if there were, abstractly speaking, determinate objective right answers to all possible questions posed by the application of the Convention under a moral reading, their sheer existence would be utterly useless for the purposes of the administration of an effective regional system of human rights protection, should it turn out that these answers are epistemically inaccessible to real, as opposed to ideal, judges or accessible only at very high costs by comparison to the benefits delivered. I shall call this the *epistemic challenge* to the moral reading of the ECHR. I shall also claim that complexity theory is particularly important to framing and understanding the depth of the challenge before moving on to suggest that consensus may provide an acceptable solution to it.

The epistemic challenge helps bring into sharper focus the decision problem that judges of the Court face in hard cases. Succinctly put, the nature of the problem results from the combination of two sets of factors. First, under a moral reading of the Convention the considerations that provide reasons for individual judges to decide cases are frequently not just complicated but also *complex* in the specific sense outlined earlier. Second, individual judges are boundedly rational.

As a result, judges deciding in good faith are often unsure about the best course of action. In what follows, I shall begin by substantiating the first part of the claim. Grasping the role of complexity as a systematic generator of uncertainty in the functioning of the ECHR normative system is crucial. Then, in the next sub-section I focus on a number of features that constrain the epistemic capacities of judges.

Complexity and the decision problem faced by ECtHR judges

We can sharpen our initial grasp of the decision problem by tentatively distinguishing between three kinds of factors that the ECtHR must take into account when deciding cases: those relating to individual justice, those relating to the effect that the Court's case law has on the wider system of protection of rights under the ECHR and those relating to strategic considerations, widely conceived. The first have to do with granting appropriate relief to the particular individual complaining of a violation of a Convention right by a state party. The second revolve around the impact the case law of the Court has on the ECHR system of protection of human rights as a whole. The third concern different and complex contexts of interaction between, on one hand, individual members of a collegial Court among themselves and, on the other hand, interaction of the Court as a whole with states parties taken individually or *in tandem*. When consulting the proposed tripartite list, it is helpful to keep in mind two things. First, the distinction between different kinds of factors introduced here is not intended to reflect any deep properties of the factors themselves. It merely serves the tentative aim of helping us organise our thinking about the specification of the decision problem in hard cases. Thus, and to take an example, the reader should feel free to subsume the category of strategic considerations under either of the first two categories, if she perhaps thinks that these are not independent enough. Second, no controversial claim is made regarding the relative force of the reasons that the factors generate. This depends entirely on a fuller and more detailed specification of the point and purpose of the ECHR, which falls squarely outside the scope of this chapter. As a result, the list proposed here will have served its function if it can be plausibly accepted by people that otherwise reasonably disagree on the point and purpose of the ECHR and, accordingly, on the different weights to be assigned to the normative reasons stemming from the indicated factors.

We may begin with individual justice. It is uncontroversial that one of the ECHR's most dazzling achievements to date has been the initial recognition and gradual reinforcement of a right to individual petition, especially after the adoption of Additional Protocol 11 (Greer, 2006, pp. 1–59). The function of providing redress for alleged Convention rights violations by states parties for the benefit of specific individuals is one of the Court's most important tasks. Uncertainty with regard to the proper resolution of hard cases can occur at this level, without any need to take into account the wider effects of the case law of the Court on the ECHR system as a whole or of strategic considerations. In particular, judges can, for example, be uncertain or in reasonable disagreement about the best moral theory of human rights or about specific conceptions of such rights even from within an agreed-on general theory or about how to balance the protection of individual rights with institutional considerations such as the democratic legitimacy of the decisions taken by member states. In all these cases, nuanced judgment would be appropriate and reasonable judges could disagree about how best to exercise it. However, it is important to note that, if hard cases and uncertainty under a moral reading of the ECHR were only to do with the administration of individual justice in the preceding sense, then the decision problem that the Court faces would perhaps be difficult, but not necessarily complex in the sense specified earlier. In particular, there would not be any need to take into account adaptive interactions between the ECHR and domestic legal systems, since these factors would just be irrelevant with regard to the Court's mission.

I thus submit that what really makes the decision problem that the ECtHR faces not just difficult but complex, in the sense specified earlier on in the chapter, is the importance of taking into account, first, the impact that its decisions have on the wider system of protection of Convention rights, and, second, strategic considerations about the interaction of the Court with various institutional actors whose help and cooperation is vital in effectively enforcing the ECHR. Beginning with the first issue, it is important to note that, in many cases, the Court's judgments do not just resolve an issue relating to an individual claim of alleged ECHR violation following a petition, but, rather, set the minimum threshold of Convention rights protection across all contracting states. Thus, a judgment by the Court finding a violation of the Convention in a specific case often entails not only that the respondent state ought to modify its legislation but also that all other member states that have similar legislation ought to change it. Using examples from the case law of the Court may helpfully bring out the point. Thus, when the Court takes a position on issues such as same-sex marriage,[8] abortion,[9] closed-shop agreements[10] or prisoner voting rights,[11] it is *de facto* if not *de jure*, defining the minimal level of protection that will have to be accorded by all states parties with regard to the issue that is decided. However, as already noted, it is one of the major insights of complexity theory that the solutions provided by complex domestic systems to various issues are emergent properties of deeper interactions between adaptive agents reflecting dynamic temporary disequilibria. At the very least, then, and given phenomena of path-dependence, the Court has to be particularly careful when governing complex domestic systems, since the costs of imposing an erroneous unique solution to a given issue may make it subsequently infeasible to return the system to a previous state. Besides, this problem can be particularly acute when the correctness of the solution to some Convention issue depends heavily on empirical parameters as appears, for example, to be the case in *Sørensen*, mentioned in the earlier section 'Setting the Stage (b)', where the question was whether the legitimate aims pursued through closed-shop legislation could have been achieved in the absence of such legislation.

Complexity phenomena are also at play with regard to a variety of strategic interactions that are pertinent to deciding ECHR issues. In fact, the Court has to pay heed to the way other institutional actors are likely to apply its case law and, insofar as it does, it must have some view about the outcomes of such future interactions. The Court is part of a wider system of protection of ECHR rights, governed by the principle of subsidiarity and marked by an institutional partnership with domestic actors, which are under a duty to make their distinctive contributions within the system. This is the whole point, for example, of the rule of the exhaustion of domestic remedies before an application to the Court is deemed admissible (Tsarapatsanis, 2015, p. 686). Moreover, similar concerns also arise from the practical problems that the Court faces in the effective implementation of the Convention. Insofar as the Court has limited capacity, a large part of the role of handling implementation issues will inescapably be played by domestic authorities, which comprise but are not limited to courts. Issues of strategic interaction can also stem from the fact that the Court is an international court, with the result that political reactions to the implementation of its judgments by domestic authorities are harder to overcome than those faced by domestic courts. Thus, securing effective state compliance, either with regard to the behaviour of a single state when it comes to the implementation of a particular judgment[12] or, more generally, with regard to the patterned behaviour of states that appear to systematically challenge the legitimacy of the Court on any number of issues, can be an important source of normative considerations. Here again, complexity theory delivers crucial insights since it holds that states of (dis) equilibria of complex systems are always temporary and subject to disruption by adaptive behaviours of agents. This suggests that compliance and cooperation by contracting states should not be taken for granted, but, rather, should be seen as the emergent and potentially fragile property of past interactions between the Court and domestic legal actors. Perceived legitimacy of the Court by domestic agents can go some way towards addressing those issues, since it may stabilise and streamline expected behaviours. Still, there might be a real sense in which the complexity of the ECHR system and the multiplicity of agents' interactions make the impact of certain outcomes genuinely uncertain.

Besides, considerations stemming from patterns of strategic interaction are also at play at the level of decision-making by the Court itself. In fact, as Vermeule has forcefully pointed out (Vermeule, 2012, chapter 5), in multi-member courts, such as the ECtHR, individual judges favouring a particular optimal solution to a decision problem have to take into account the fact that they are sitting on a panel with other judges who may disagree with their views. Accordingly, and to take a hypothetical example, a judge who adheres to the moral reading of the Convention will have to make do with the fact that she is sitting on a panel with other judges who may not share her first-best interpretive approach. Other judges might be positivists or adopt some other approach. Interactions among judges are thus complex in the specific sense that the outcome of those interactions (the final judgment) is an emergent property of the Court, which does not necessarily reduce to the actions or intentions of individual judges. Moreover, states of doctrinal or interpretive stability on the part of the Court are also temporary and potentially fragile since they reflect the underlying complex emergent patterns of interactions among individual judges.

Now, complex situations such as these present judges with an important decision problem. The rational judge that finds herself in the minority will have to opt for a sub-optimal solution by her own lights. As Vermeule puts it (Vermeule, 2012, pp. 156–160), at this point she has a number of different strategic choices at her disposal. For example, she might adopt an 'evangelist' approach, disregarding the consequences of her behaviour in particular cases with the hope that, in the long run, she may convert other judges by the sheer force of her example. But she might also settle for a second-best view by strategically using her resources in some other way, for example by influencing the formulation of the reasons provided if she agrees with the outcome. Whilst this approach might initially seem opportunistic, it is far from evident that it is inconsistent with the very idea of the moral reading, since at a minimum it leads to acceptable outcomes by the judge's own lights. Be that as it may, the more general claim is to the effect that all cases of strategic interaction sketched earlier involve versions of the same core problem: how to cope with the fact that, once a moral reading is adopted as the best interpretive approach towards the ECHR, the judge adopting the approach has to cooperate with actors that do not necessarily share that view and which have the power to influence the real-world effects of the implementation of the Convention.

Epistemic constraints and reasoning strategies

As already observed, complexity with regard to systemic effects or strategic interactions is only one of the sources of judicial uncertainty. Uncertainty with regard to interpretation and application of the ECHR can be also exacerbated because of a number of familiar cognitive and epistemic constraints akin to what Christopher Cherniak (Cherniak, 1986, p. 8) has called the '*finitary predicament*' of human epistemic agents, to wit, the fact that their cognitive resources are limited. As a result of the finitary predicament, human agents' rationality has been called resource-dependent or bounded (Bishop and Trout, 2005). Bounded rationality approaches focus on how agents with limited information, time and cognitive capacities ought to make judgments and decisions (Tsarapatsanis, 2015, pp. 689–691). The approaches became particularly prominent since the 1970s, when an impressive array of experimental results in social psychology consistently showed that, under certain circumstances, human agents reason and decide in ways that systematically violate the formal canons of rationality (Bishop and Trout, 2005). At least part of the explanation for these shortcomings is attributed to the lack of cognitive resources available to human agents. Charting the actual limits of these resources is an important part of cognitive science and empirical psychology. Both conceptualise the mind as a finite information-processing device, strictly limited with regard to its memory, attention and computation capacities. These general considerations, which apply to judges insofar as the latter are human epistemic agents like any other, are complemented in a straightforward way by specific constraints on how the Court functions. Thus, pressures involving the capacity to efficiently process applications and deliver judgments in a timely fashion can turn out to be significant, insofar as the Court has a backlog of tens of thousands of pending cases[13] and is also committed to issuing decisions promptly as a matter of human rights (for more on these constraints see Tsarapatsanis, 2015, p. 690).

Now, as already indicated, under the moral reading of the ECHR, certain kinds of normative and empirical facts determine the truth values of propositions of Convention rights. In order to achieve the epistemic goal of apprehending these facts, judges need to deploy the appropriate epistemic means. I shall refer to these means as 'reasoning strategies'. Two kinds of normative constraints may be reasonably imposed on the selection of these strategies. First, they ought to be reliable, that is such as to allow agents to systematically track the relevant facts. This follows directly from the fact that, under the moral reading of the ECHR, the epistemic goal of judges is objective truth and not some other aim, such as justifiability or reasonableness. Second, they ought to be tractable, that is suitable for judges as epistemic agents endowed with finite cognitive resources.

Tractability brings immediately into play the epistemic constraints sketched above, especially bounded rationality. Thus, bounded rationality accounts ask which reasoning strategies agents with constrained cognitive resources ought to follow in order to reliably attain sets of specified epistemic goals for different kinds of environments. Accordingly, the reasoning strategies identified for boundedly rational agents are resource-relative: they are tailored to the actual cognitive abilities and resources of agents. Resource relativity as a constraint on the selection of reasoning strategies can be justified in various ways. To begin with, one can appeal to the 'ought-implies-can' norm: no judge should use a reasoning strategy that is clearly intractable. Moreover, and more controversially, reasoning strategies are also constrained by cost/benefit considerations. Suppose, for example that, if judges of the Court had infinite time, they could score better on the reliability dimension. However, ECtHR judges do not have infinite time, and in fact, they are under relentless time pressure, amplified by the ever-increasing volume of their caseload. It follows that, depending on the circumstances in which they are placed, judges could sometimes reasonably trade off marginal increases in reliability for speed, by following appropriately economical reasoning strategies, such as a more deferential and less fine-grained standard of review, through the margin of appreciation doctrine, if they have reason to trust the judgment of national authorities (Tsarapatsanis, 2015, p. 685). Generalising the point, we might say that it is not enough that reasoning strategies score high on the reliability dimension: it is important that they also come at an acceptable cost with regard to the finite epistemic resources of judges.

The upshot for the purposes of the present discussion is that reasoning strategies that take the formulation of objectively true propositions about the ECHR under a moral reading as their epistemic goal ought to take account of the judges' epistemic resources limitations. Even if the relevant moral and empirical facts could in principle be accessible to resource-independent agents, such as agents with normal perceptual capacities that perform no reasoning mistakes, have sets of completely consistent beliefs and infinite memory and time, we still ought to ask, first, whether they are they also in principle accessible to resource-dependent judges sitting at the ECtHR and, second, at what cost. Moreover, the epistemic constraints briefly alluded to in this section exacerbate the problem of responding appropriately to the uncertainty that the complexity of the ECHR system of rights protection generates in hard cases. This is so even without supposing that disagreement about the identification of the relevant moral facts poses any special epistemic problem *per se*. Whilst disagreement among epistemic peers can sometimes cause considerable uncertainty, my primary focus here is on the resource-restrained epistemic abilities of judges, which would be important even if disagreement were completely absent.

The consensus approach as a reasoning strategy

To summarise the argument thus far, I have stressed that in hard cases judges of the Court can be uncertain about the best solution, owing to the non-exhaustively specified combination of sheer difficulty, complexity due to systemic impact and strategic interaction and epistemic constraints. This is the case even if we assume, as I did throughout the chapter by adopting the moral reading of the ECHR, that there are objective right answers to hard cases. Given the pervasive epistemic issues I mentioned, it is possible and, in fact, desirable to distinguish between objective solutions to hard cases and reliable and tractable reasoning strategies for actual (as opposed to ideal) boundedly rational ECtHR judges. Whereas the former, under the moral reading of the ECHR, make objective morality a constitutive part of what makes propositions about the ECHR true, the latter aim at articulating ways for non-ideal agents to gain access to this truth by maximising the chances of correct outcomes or minimising the risks of incorrect ones. It could thus be the case that judges might be more likely to arrive at the objectively correct outcome under a moral reading in an *oblique* way rather than via direct moral reasoning. From the point of view of criteria of assessment of reasoning strategies, unconstrained moral reasoning shall have to be compared with a number of contenders and evaluated across a number of dimensions, chief among which figure reliability and tractability.

I thus submit that the consensus approach should be understood as just one of those contenders, that is as an oblique reasoning strategy which bypasses unconstrained moral reasoning and is apt to aid ECtHR judges in arriving at correct decisions whilst attempting to effectively govern a complex system. There are two points that should be noted here. First, classifying the consensus approach as a reasoning strategy and not as a criterion of truth of propositions about the ECHR straightforwardly avoids Letsas's criticism of the consensus approach outlined earlier. In particular, on this understanding consensus does not determine the truth about ECHR rights and therefore does not amount to majoritarianism about human rights. It is merely a heuristic device that may provide epistemic help to judges under conditions of uncertainty generated by complexity. Second, the argument of this chapter is that heuristics such as consensus can provide such an epistemic help under conditions of uncertainty and as a means to address it. No claim is thus made that consensus would be relevant to the moral reading even in the absence of uncertainty. However, it is indeed argued that, once the insights from complexity theory are fully taken into account, uncertainty proves to be a particularly pervasive phenomenon with regard to ECHR adjudication.

At this point, an important caveat should be underscored. It is one thing to indicate the possibility of understanding the consensus approach as a reasoning strategy, which was my main aim in this chapter, and it is quite another to argue in favour of the overall plausibility of such an approach *qua* reasoning strategy. Moreover, a fuller account should compare the relative merits of the consensus approach with those of other approaches, such as unconstrained moral reasoning, potentially specifying contexts in which the use one or the other could be more warranted. Since I do not have the space for a detailed treatment, I only provide a number of tentative considerations in favour of such plausibility, which open venues for future research, along with a number of critical comments.

I have already said that reasoning strategies for non-ideal epistemic agents should be assessed along the twin dimensions of reliability and tractability. Complexity theory provides an important source of considerations in favour of using consensus as a reasoning strategy along these dimensions. As already observed, consensus takes account of solutions to ECHR issues that have emerged spontaneously from the interactions of agents composing domestic legal systems and which it could be perilous to upset absent very strong reasons. Moreover, imposing a completely novel solution in the absence of consensus could carry significant costs in case the wrong decision is made, due to path-dependence. These considerations at the very least suggest that concerns deriving from aspects of complexity theory can inform a prudential attitude on the part of judges, especially when it comes to assessing solutions to ECHR issues that could have a wider systemic impact. Likewise, and with respect to tractability, the identification of common solutions adopted by contracting states may be much more economical than the calculation of complex normative and empirical factors.

Moreover, *qua* reasoning strategy, consensus may also be useful as a collective intelligence device. Collective intelligence arguments can take many different forms, but they revolve around the core notion that, under certain conditions, collectives may epistemically outperform individuals. Among the arguments advanced is the Condorcet Jury Theorem (henceforth 'CJT'), which roughly states that aggregating the beliefs of sincerely voting independent individuals on some subject, at least when the individuals are more likely to be right than wrong, increases the likelihood of choosing the right answer (for an informal presentation of CJT see Landemore, 2013, pp. 70–75). Interestingly, Eric Posner and Cass Sunstein have explicitly used CJT to argue that, under certain conditions, aggregating the solutions that relevantly similar states have provided to some issue can provide a good reason to believe that the majority solution is correct (Posner and Sunstein, 2006). While the proposal is far from uncontroversial and faces a number of technical challenges that cannot be addressed in detail here, it provides, if plausible, a clear justification for some (disciplined) form of the consensus inquiry *qua* reasoning strategy. In the same vein, it could be possible to explore the plausibility of models of cognitive diversity (Page, 2007). Roughly, the core idea of such models is that aggregating views based on different inter-pretations of how the world works maximises the chances that the median answer will be right as opposed to the one provided by a randomly chosen individual from the group (Page, 2007, p. 197). As in the application of CJT to solutions provided by domestic legal systems to various issues, it bears further exploring whether cognitive diversity models could also provide a justification for the con-sensus approach.

Here again, the notion of collective intelligence connects to insights from com-plexity theory. The solutions to various human rights issues provided by contract-ing states could be understood as emergent properties of complex national legal systems, resulting from the interactions of the agents composing those systems and not reducible to their individual actions or intentions. Common patterns of such emergent properties could thus be harnessed to enhance the cognitive capacities of ECtHR judges under conditions of uncertainty. Moreover, disciplin-ing and streamlining the use of consensus through more formal models such as CJT or cognitive diversity could help augment the legitimacy of the Court as well as the predictability of its reasoning, with beneficial effects with respect to both compliance and cooperation. Last, insofar as the consensus inquiry provides clear results due to common patterns of solutions, it can help stabilise the behaviour of diverse judges in a way that unconstrained moral reasoning, which frequently leads to disagreement, perhaps cannot. In this way, it could help address the issue of strategic interaction of judges in multi-member courts. The flipside of this, with regard to tractability, is that access to reliable information about the solu-tions adopted by different contracting states and aggregation of those solutions to provide a (rebuttable) guide for decision under uncertainty carries its own important costs. Whether such a reasoning strategy is ultimately more frugal, relative to gains in reliability, than is deciding by unconstrained moral reasoning will thus depend on specifics that cannot be touched on here.

Conclusion

In this chapter, I argued that, understood as a reasoning strategy, the consensus approach is compatible with a moral reading of the ECHR. My main argument consists in distinguishing between criteria of truth of propositions about Conven-tion rights and reasoning strategies. The function of the latter is merely epistemic. Using insights from complexity theory, I claimed that some reasoning strategies could be rational responses to the decision problem judges of the Court face in hard cases. In these cases, uncertainty is not accidental but the combined product of the bounded rationality of judges with often-complex factors, such as strategic interactions and systemic effects of the Court's judgments. I then contended that a number of heuristic devices used by the Court, such as the consensus approach, should be best understood as reasoning strategies and not as criteria of truth of propositions about the ECHR. Last, I briefly sketched the possibility that insights from complexity theory could also be used to justify using the consensus approach as a collective intelligence device. However, a lot more would need to be said in order to fully assess the relative merits of the consensus approach compared to other approaches, such as unconstrained moral reasoning. I therefore conclude by submitting that more systematic exploration of specific reasoning strategies for courts facing hard cases in complex normative environments could yield high theoretical and possibly practical payoffs.

Notes

1 For a particularly clear example in this and other respects, see *Sørensen and Rasmussen v. Denmark*, Application Nos. 52562/99 and 52620/99, Judgment of 11 January 2006.
2 *Ibid.*
3 *Ibid.*
4 *Ibid.*
5 See, for example, *Vo v. France*, Application No. 53924/00, Judgment of 8 July 2004.
6 See, for example, *Pretto and Others v. Italy*, Application No. 7984/77, Judgment of 8 December 1983.
7 See *Lautsi and others v. Italy*, Application No. 30814/06, Judgment of 18 March 2011; *S.A.S. v. France*, Application No. 43835/11, Judgment of 1 July 2014.
8 See *Schalk and Kopf v. Austria*, Application No. 30141/04, Judgment of 24 June 2010.
9 See *A, B and C v. Ireland*, Application No. 25579/05, Judgment of 16 December 2010.
10 See Sørensen judgment, n.1 above.
11 See Hirst v. the United Kingdom (No.2), Application No 74025/01, Judgment of 6 October 2005.
12 See, as a salient example, the non-compliance (yet) by the United Kingdom with regard to the Hirst case (above n.11).
13 There were 64,850 pending cases as of 31/12/2015. See the ECtHR's 2015 report at p. 187 available here: www.echr.coe.int/Documents/Annual_report_2015_ENG.pdf (last accessed 1/4/2017).

Bibliography

Ambrus, Mónika (2009), 'Comparative Law Method in the Jurisprudence of the European Court of Human Rights', 2(3) *Erasmus Law Review* 353.
Benvenisti, Eyal (1999), 'Margin of Appreciation, Consensus, and Universal Standards', 31 *Journal of International Law and Politics* 843.
Bishop, Michael and J. D. Trout (2005) *Epistemology and the Psychology of Human Judgment* (Oxford: Oxford University Press).
Cherniak, Christopher (1986), *Minimal Rationality* (Cambridge, MA: MIT Press).
Dworkin, Ronald (1996), *Freedom's Law: The Moral Reading of the American Constitution* (Cambridge, MA: Harvard University Press).
Dzehtsiarou, Kanstantsin (2015), *European Consensus and the Legitimacy of the European Court of Human Rights* (Cambridge: Cambridge University Press).
Greer, Steven (2006), *The European Convention on Human Rights: Achievements, Problems and Prospects* (Cambridge: Cambridge University Press).
Helfer, Laurence R. (1993), 'Consensus, Coherence and the European Convention on Human Rights', 26 *Cornell International Law Journal* 133.

Kyritsis, Dimitrios (2015), *Shared Authority: Courts and Legislatures in Legal Theory* (Oxford: Hart Publishing).

Landemore, Hélène (2013), *Democratic Reason: Politics, Collective Intelligence and the Rule of the Many* (Princeton, NJ: Princeton University Press).

Letsas, George (2004), 'The Truth in Autonomous Concepts: How to Interpret the ECHR', 15(2) *European Journal of International Law* 279.

Letsas, George (2007), *A Theory of Interpretation of the European Convention of Human Rights* (Oxford: Oxford University Press).

Letsas, George (2010), 'Strasbourg's Interpretive Ethic: Lessons for the International Lawyer', 21(3) *European Journal of International Law* 509.

Letsas, George (2013), 'The ECHR as a Living Instrument: Its Meaning and Its Legitimacy', in Andreas Føllesdal et al. (eds.) *Constituting Europe: The European Court of Human Rights in a National, European and Global Context* (Cambridge: Cambridge University Press), 106.

Page, Scott (2007), *The Difference: How the Power of Diversity Creates Better Groups, Firms, Schools and Societies* (Princeton, NJ: Princeton University Press).

Page, Scott (2010), *Diversity and Complexity* (Princeton, NJ: Princeton University Press).

Posner, Eric and Cass Sunstein (2006), 'The Law of Other States', 59 *Stanford Law Review* 131.

Ruhl, J. B. (2008), 'Law's Complexity: A Primer', 24(4) *Georgia State University Law Review* 885.

Tsarapatsanis, Dimitrios (2015), 'The Margin of Appreciation Doctrine: A Low-Level Institutional View', 35(4) *Legal Studies* 675.

Vermeule, Adrian (2012), *The System of the Constitution* (Oxford: Oxford University Press).

Wheatley, Steven (2016), 'The Emergence of New States in International Law: The Insights from Complexity Theory', 15 *Chinese Journal of International Law* 579.

7 Prospects for prosecuting non-state armed groups under international criminal law

Perspectives from complexity theory

Anna Marie Brennan

Introduction

This chapter examines a complicated question: Should we hold non-state armed groups as a collective entity accountable under international criminal law, and if so, how? The prevalence of non-state armed groups in the 21st century would suggest that international criminal law should have already grappled with this question. However, from its inception, international criminal law has had very little, if anything, to do with collective entities, how to think about them and how to hold them accountable for international crimes (Werle, 2005, p. 35). This is without doubt because of international criminal law's focus on individuals and is even more important now as questions emerge about how the non-state armed group as a collective entity can be brought within international criminal law's ambit (Clapham, 2008; Bellal et al., 2011, and Kleffner, 2009). Many non-state armed groups are structurally complex consisting of dynamic networks of interactions with a loosely organised command structure that is in a constant state of flux. As a result, it can be difficult to pinpoint which individuals within the group are responsible for the planning and commission of international crimes. This is in direct contrast to other types of collective entities such as state armed forces that are generally determinant closed systems with a clear-cut hierarchical command structure. The lack of a hierarchical command structure and a centralised leadership exercising direct control over subordinates enables the non-state armed group to change its *modus operandi* quite easily and thus complicates the task of bringing members of these groups to justice (Brennan, 2018).

This raises questions about whether it is still appropriate for international criminal law to remain focused on the pursuit of individual members or whether we should also be pursuing the accountability of the collective entity as well. We need then to clarify our thinking on the issue and this chapter proposes to examine the methodologies developed by sociologists to make sense of this question – specifically that we should look to a facet of systems theory called complexity theory to inform our analysis of prospects for holding the non-state armed group as a collective entity accountable under international criminal law. In particular, this chapter examines whether it makes a difference for international criminal law if its applicants are complex adaptive systems, a concept derived from complexity theory. Complexity theory aims to explain the behavioural pattern of objects, actors and agents that are loosely organised (Wheatley, 2016). It proposes that interactions between a sizeable number of individuals can lead to phenomena that are not presaged from the individual's behaviour itself (Chinen, 2014, p. 704). As a form of self-organisation, the non-state armed group is more than the aggregate of its members, distinct from its environment but at the same time not impervious to it. What would be the implications for the individuals in such a system if the system itself can be held accountable for international crimes?

This chapter makes one key claim: that the non-state armed group is not always reducible to the individuals who ostensibly are members. Instead, it can be regarded as a nascent phenomenon that has emerged from complex interactions among individuals and the conceptual tools and structures that individuals use within their physical and social environments (Yezdani et al., 2015, p. 305). This implies that there is a gap between the collective entity that is the non-state armed group and the intrinsic interactions of the individual members from which the group emerges. So the non-state armed group stands apart from its individual members. Consequently, therefore, complexity theory frustrates the central premise of international criminal law that the natural person as such is the most appropriate bearer of responsibility. In other words, the theory posits that the non-state armed group, as an emergent phenomenon, is not a proxy for the individual members who act on its behalf, nor is it merely an extension of the individual. It is instead a distinct entity, which can perpetrate international crimes. So for the most part crimes do not occur because of the criminal propensity of one particular person but instead stem from the distinct phenomenon, which has resulted from the multiple interactions of individuals.[1]

My conclusions can be stated in a few lines: the majority of the chapter is dedicated to defending these ideas and to highlighting their implications for international criminal law. Building on existing accounts of international criminal justice, I delineate an ontology of the non-state armed group that undermines the individual-centric focus of international criminal law. I argue that if international criminal law is to be effective in achieving its central goal to suppress impunity for international crimes it must have a legal means of investigating and prosecuting collective entities such as the non-state armed group as well. Extending accountability to non-state armed groups as collective entities can be justified to laypersons with reference to the goals international criminal justice seeks to pursue, but such measures and their effects would also be compounded with difficulty. Thus, I argue that other efforts should be made to enhance non-state armed groups' compliance with international law and their accountability for breaches of it. Accountability for international crimes would therefore perhaps mean something different, depending on the type of measures considered. By taking this dilemma seriously we can make sense both of our varied intuitions about the scope of justice, and of persisting problems within international criminal law.

The non-state armed group and the dilemma of collective entity responsibility in international criminal law

The dilemma of juggling a non-state armed group's responsibilities for its members' conduct with its obligations towards other subjects of international law is not new. It would seem that the drafters of the International Law Commission's (ILC) Articles on the Responsibility of States for Internationally Wrongful Acts[2] were somewhat cognisant of this issue; in 1956 the first special rapporteur for the ILC, Garcia-Amador contended that

> [i]t is no longer true, as it was for centuries in the past, that international law exists only for, or finds its sole raison d'etre in, the protection of the interests and rights of the States; rather its function is now also to protect the rights and interests of its other subjects who may properly claim its protection.
>
> (1956, para. 57)

Garcia-Amador's comments formed part of a discussion that states have a responsibility to guarantee that foreign nationals have the same rights as their own citizens. However, they also suggest that the commission was of the view that the articles were to be construed in their broader context in which all subjects enjoy rights and protections under international law. This wider perspective makes room for considering not only the rights and interests of non-state armed groups but also their duties and the extent to which they can be held accountable for international crimes.

Indeed, the ILC articles distinguish between primary obligations that derive from specific treaties, customary international law and other sources of international law, on one hand, and secondary obligations that stem from the law of state responsibility, on the other. Reconciling a non-state armed group's responsibilities under international law could happen at either a primary or a secondary level. Their obligations could be accounted for and enshrined within international law among the primary rules of state responsibility, as international courts and tribunals demarcate the specific obligations of non-state armed groups and construct remedies for their wrongful acts. There is some evidence that this is beginning to happen: international courts and tribunals have given consideration to the obligations of non-state armed groups under international law. Judges have harnessed the Geneva Conventions[3] as they considered the minimum conditions for ascertaining the duties of non-state armed groups in armed conflict situations: these include amongst others defining the degree of organisation a non-state armed group must possess in order for the Geneva Conventions to apply to their conduct, their control over territory, their capacity to gain access to weapons and training, their ability to develop a coherent military strategy and their capability to generally act as one unit and negotiate agreements such as ceasefires and peace accords.[4] As a result, judges have been cognisant that non-state armed groups have not only "recognised and respected the international treaties of the United Nations and conventions on war"[5] but also "issued communiqués assuming responsibility for . . . attacks."[6]

In particular, Bilkova argues that the *Case Concerning Military and Paramilitary Activities in and Against Nicaragua* supports her claim that the rules on responsibility for internationally wrongful acts can be extended to non-state armed groups; she points out that "violations of primary rules binding on . . . [non-state armed actors] by a conduct attributable to them would constitute internationally wrongful acts entailing responsibility with its classical content" (Bilkova, 2015, p. 76). However, Bilkova's argument is flawed since she does not outline exactly how a principle of collective entity responsibility could be introduced in international law or how international courts and tribunals, in particular, the International Criminal Court, should approach claims against non-state armed groups. The *Nicaragua* case may imply that non-state armed groups must comply with international humanitarian law but the decision also supports a separate argument that the decision-maker declined to adjudicate on the precise parameters of the duties of non-state armed groups in international law perhaps for lack of competence or because he or she is limited by terms of reference or by the primary rules themselves.

This is a trend that has been followed by other international courts and tribunals as well.[7] While they have readily accepted that non-state armed groups must comply with international humanitarian law (IHL) they have so far parked any further elaboration about their precise obligations. It is perhaps understandable that decision makers would be apprehensive about applying rules that were originally designed for states. The legal personality of non-state armed groups is likely to be limited since they generally only exist for a short time because they are suppressed by the state and are disbanded, they succeed in seizing power and form a new government or they create a new state. Such groups thus cannot enjoy rights akin to the permanent nature of international actors but also perhaps do not have the capacity to fulfil legal obligations either. So what would be the impact of a finding that a non-state armed group as a collective entity has committed an international crime not only on the group but also on its members? Bilkova posits that non-state armed groups should most probably be held responsible for the conduct of its members and constituent organs acting on its behalf, but she does not outline how to implement this or acknowledge that the organisational structure of some non-state armed groups can be so ambiguous that it is difficult to ascertain whether these individuals are members of the group in the first place (Bilkova, 2015, p. 76). Thus, determining whether the individual member's conduct can be attributed to the non-state armed group will at times be difficult to ascertain.

A decision-maker could follow an approach suggested by the ICRC called the "continuous function" test; he or she could take into account whether "a person assumes a continuous function for the group involving his or her direct participation in hostilities" as a decisive criterion for ascertaining individual membership of the group (Melzer, 2009, p. 33). This test, however, relegates to the background the contemporary realities of non-state armed groups and the fact that the ambiguous organisational structure of some non-state armed actors means that it can be difficult to pinpoint whether an individual member performs a continuous function for the group. Under this approach, the principle of individual criminal responsibility could run the risk of being ignored as a matter of substance or, at best, becoming part of a spasmodic mishmash of law in which the genesis of how international crimes are perpetrated plays a significant role in some modes of liability and lesser role in others. No doubt the challenges of consistency and legitimacy caused by multiple competing legal principles are much broader than the particular concerns raised here, but in this context a prosecutor could find him- or herself stuck between a regime that prioritises individual criminal responsibility and another that impedes it, thus running the risk of not only slowing down but also even stalling investigations into alleged crimes and prosecutions.

Attributing responsibility to collective entities and ethics theory: prospects for international criminal law

The issue of distributing responsibility for international crimes to collective entities has been somewhat under-explored in international criminal law, but there is emerging scholarly work on this issue in ethics, which may provide some guidance here (Darcy, 2006; French, 1984; Feinberg, 1970; Lukes, 2005; Connolly, 1974; Arendt, 1987; Martin, 1976, Copp, 1984; Kutz, 2000; and Levinson, 1974. Much of the literature focuses on whether a collective entity can be the subject of moral responsibility or whether the collective entity's members are the proper bearer instead. Other questions arise about what type of collective entity can bear responsibility or if it is even pertinent for criminal responsibility to be distributed to individual members at all? Moreover, if an individual member is held responsible for an international crime, but there is evidence to suggest that it was instigated by the group why are the consequences of that crime only borne by the individual members and not by the non-state armed group which facilitated its perpetration?

Although this question has repercussions for group responsibility, the literature tends to agree that "judgments about the moral responsibility of [a collective entity's] members are not logically derivable from judgements about the moral responsibility of a collectivity" (Held, 1970, p. 475, and Miller, 2006). So moral judgements about a collective entity can only be made after an individual member must have committed a wrongful act. This implies that the best justification for having individual members shoulder the responsibility of the collective entity is that they are inculpated in the collective entity's wrongful act. A number of reasons have been put forward to support this proposition. First of all, some commentators are of the view that if the individual members have committed themselves to the objectives of the group, then they must bear responsibility for its acts (Chinen, 2014, p. 716). Others argue that membership itself is a basis for holding an individual directly responsible for what the collective entity does (Sepinwell, 2011, p. 241). This is because membership insinuates a commitment to the collective entity and its members and a commitment to face the consequences for the collective entity's wrongful acts in recognition that the collective entity "is his as well as theirs" (Ibid).

However, an alternative approach would be to focus on the collective entity instead of the individual member. The collective entity would be responsible for wrongful acts committed by each member as a matter of course. So when one member of the collective entity commits a wrong, the collective entity as a whole is held responsible for that wrong since "each member of the community is an expression of its moral center" (Reiff, 2008, p. 209). Pursuant to this approach, if the collective entity is our fundamental concern, any misgivings about individual members must be confined to the sidelines. But, whether or not a non-state armed group should bear the consequences of a wrongful act committed by individual members will ultimately depend on the purpose of the punishment and whether the punishment fulfils that purpose (Hardin, 1968, p. 1243).

Criminal punishment is often justified for reasons of deterrence, retribution and rehabilitation. In theory, establishing what type of punishment should be imposed on groups and whether it will be effective in fulfilling its purpose will be particularly challenging in view of the characteristics of the non-state armed group and the fact that it is only individual members who have thus far borne the brunt of punishments for the perpetration of international crimes. Therefore, it is necessary to strike a balance between the aim of punishing the non-state armed group as a collective entity and the consequences this may have for individual members' future behaviour within the group. The type of punishment enforced on the non-state armed group is of considerable importance. Realistically, the group could only be subjected to a pecuniary punishment; it is not conceivable to imprison the entire membership of the group by virtue of the fact that they are members (Sriram, 2010, p. 66). Liability would therefore not be based on the culpability of the group but instead on the sense that an individual who has suffered harm as a result of its wrongdoing should be provided compensation or another form of reparation or restitution (Pasternak, 2011, p. 213).

Research on moral responsibility can provide some guidance to international criminal law on this dilemma. This can be readily observed by considering the rationales for holding non-state armed groups responsible for the wrongdoing of its members, the possible arbitrariness of the individual being the sole focus of international criminal law, the group's enjoyment of the benefit derived from the individual member's wrongful conduct and the group as a system in condoning and inciting the individual to carry out crimes. Could all these constitute rationales for holding the non-state armed group as a collective entity accountable under international criminal law? The next sub-section considers each of these rationales in turn.

The individual as the principal unit of concern for international criminal law

A contention can be made that it is fully conceivable that the individual be regarded as the main focus of international criminal law, and it therefore follows logically that collective entities are not prosecutable before the ICC. Therefore, it could be argued that the individual-centric focus of international criminal law should be retained in its current form. When theorising the dilemma of distributing responsibility for international crimes, it could be said that questions about the collective entity fade into oblivion because as discussed earlier, without the principle of individual criminal responsibility international criminal law would have no arms or legs to function. This argument has undeniable merit; but even if the natural person continues to be the only subject of international criminal law it does not follow as either a moral or legal matter that the non-state armed group as a collective entity should be excluded from international criminal justice. Of course, to retain the individual as the only unit of concern in the distribution of international criminal justice has been deeply considered and is based on well-established principles from domestic law, but it also confines responsibility for international crimes to individuals. This leads to the rather odd situation that the role of the collective entity in the planning and commission of international crimes is ignored in the international criminal justice regime with the implication that individuals take precedence in the pursuit of accountability.

This approach engenders illegitimacy. If international criminal law is supposed to concern itself with the dynamics of how international crimes are committed in reality, it is unclear why it does not concern itself with the accountability of collective entities or support reparations for their crimes or at a minimum concede to them. Malle posits that: "humans have no trouble reasoning about the actions and minds of groups and have the desire to blame and punish them when they act immorally" (Malle, 2010, p. 136). So it would seem that the same theories are utilised when attributing blame to individuals as they do to collective entities. At the same time, there would appear to be considerable repugnance to guilt by association, holding collective entities responsible for wrongdoing seems to be widely acceptable but less so for individual members of collective entities. But the capacity and effectiveness of international criminal law to punish groups are very limited (Ibid). Prior to the Nuremberg trials there was a clear decision to eschew collective responsibility, but in doing so, international criminal law perhaps runs the risk of being ineffective if it is not cognisant of these normative values and fundamental conceptions of morality. Without additional clarification, it is hard to elucidate why collective entities should not bear at least some of the burden of responsibility for crimes that emanate from the group itself.

The non-state armed group as the beneficiary of the individual member's act

One further justification for distributing responsibility for international crimes to non-state armed groups as a collective entity is that an entity, which benefits from the acts of individual members should shoulder the burden of accountability. There are persuasive arguments that non-state armed groups benefit from the conduct of individual members because the crime is often perpetrated in pursuance of the aims and objectives of the group. If an individual's conduct brings about these benefits, it is perhaps reasonable to assume that the group as a collective entity should bear some responsibility as well. This contention is further bolstered if the group's recruitment of the individual as a member of the group involves some sort of acquiescence to be responsible for that member's actions.

This argument is persuasive for a number of reasons. The benefits of imposing accountability on the non-state armed group as a collective entity may far outweigh the costs of putting an accountability mechanism in place to prosecute every individual member who played a role in the commission of the international crime, and in most cases the non-state armed group directly benefits from the conduct of the individual member since they were formed to accomplish a particular strategic goal which is achieved through the individual member's conduct. Moreover, by virtue of the individual's membership the non-state armed group has an expectation of return. But in view of the differences in the genesis of non-state armed groups the benefits accrued from an individual member's conduct may not slot very easily into a non-state armed group individual member benefit–burden paradigm. In addition, given differences in the aims and objectives of non-state armed groups, it is difficult to compare the benefits enjoyed by one non-state armed group from the conduct of an individual member with those of another group. Last but not least, the content of a non-state armed group's obligations under international law is highly contested, and in any event, the non-state armed group does not ordinarily commit itself to being held jointly responsible for its individual member's conduct.

Even if this rationale for rerouting responsibility from individual members to the group did not have such shortcomings, in order to be enforceable in law it would need to be developed into a body of legal principles used to ascertain how and when a non-state armed group should be held accountable as a collective entity for international crimes. This raises the dilemma of commensurability: it is not clear-cut whether the benefits of an individual member's conduct in fulfilling the aims and objectives of the group should extend liability to the non-state armed group as a collective entity. Moreover, a body of legal principles based on the proposition that recruitment by the non-state armed group necessitates a commitment to some form of joint liability with the individual member would perhaps require international law to move beyond its focus on the criminal responsibility of individuals to the duties the non-state armed group owes to them. Of course, some international instruments already make some reference to the responsibilities of non-state armed groups, and perhaps some of the dilemmas raised in this chapter will incentivise their articulation.[8] Even if this was the case, it is not clear whether an elaboration of a non-state armed group's responsibilities at the international level should include a principle that a non-state armed group as a collective entity be held responsible alongside its individual members for that individual's actions. This would be tantamount to committing a non-state armed group to a form of direct criminal liability, even though in some instances it would be difficult to prove a link between the crime and the non-state armed group because as discussed in the following, some groups have such an ambiguous organisational structure that it can be difficult to pinpoint whether the individual member received orders from the centralised command the act.

The non-state armed group as a system in the planning and commission of international crimes

Some scholars would argue that a non-state armed group, which provides material support for the wrongdoing of a member should be held accountable for that member's conduct (McConnell, 2016; Bellal, 2015, p. 305). However, international criminal law does not account for this: a non-state armed group which infringes an international norm cannot be held responsible for that infringement, even if an individual member was acting in an official capacity with its authority. More important is the question of whether the distribution of responsibility to the collective entity is necessary when it is the individual members who ultimately commit the international crime in the first place. In a group, members are deemed to act on its behalf. So when a non-state armed group commits an international crime, this is not a question of collective responsibility because individual criminal responsibility can be invoked instead. Nevertheless, the organisational structure and *modus operandi* of non-state armed groups vary greatly. On one hand, there are a small number of groups that function as quasi-states, while on the other hand, there are groups which do not have state-like features but, nevertheless, are highly organised and have the capacity to assert control over vast swathes of territory and the civilian population within it (Frowein, 1992).

Nevertheless, that the non-state armed group can comprise the environment in which violations of IHL can be committed is recognised by the legal framework governing the conduct of hostilities. IHL has not merely bound non-state armed groups since the adoption of Common Article 3 to the Geneva Conventions 1949, but breaches of this law also entail individual criminal responsibility of the group's members. A snapshot of ongoing hostilities in Syria, Afghanistan, Iraq, the Democratic Republic of Congo, Nigeria and Mali showcases how war crimes are not just the purview of states. This is abundantly established by the systematic operation orchestrated by ISIS in Syria and Iraq of deliberately spreading terror amongst the civilian population, the perpetration of sexual slavery, rape, mutilations, collective punishment and the recruitment and use of child soldiers and so on; the widespread killing of civilians and taking of hostages by Boko Haram in Nigeria; the policy of the Union of Congolese Patriots and Patriotic Forces for the Liberation of Congo to forcefully recruit and use children in armed conflict; the indiscriminate use of landmines, booby traps and other improvised explosive devices intended to cause widespread devastation and civilian casualties by the National Union for the Total Independence of Angola; or the summary executions of civilians by the Revolutionary United Front in Sierra Leone.

Besides war crimes, the law on crimes against humanity has also been expanded to encapsulate non-state armed groups. Although it had been contended for a number of decades that such crimes could only be committed in pursuance of a governmental policy this is no longer the case as acknowledged in the Rome Statute of the International Criminal Court, the case law of the *ad hoc* tribunals and customary international law. In particular, the Trial Chamber of the International Criminal Tribunal for the Former Yugoslavia in *Prosecutor v. Duško Tadić* held that crimes against humanity can be perpetrated by "forces which, although not those of the legitimate government, have *de facto* control over, or are able to move freely within, defined territory."[9] However, Werle has posited that territorial requirement is not necessary and that this category of international crime can be perpetrated by any group of individuals if it has the capacity to commit a widespread and systematic attack against a civilian population (Werle, 2005, p. 228). In any event, it is generally recognised that the system which provided the material support for the perpetration of the crimes may either be a state or a non-state armed group. The convictions of Alex Brima, Brima Kamara and Santigie Kanu, who were members of the Armed Forces Revolutionary Council, for crimes against humanity before the Special Court for Sierra Leone provide ample evidence for this actuality.[10]

There is also nothing within the law on genocide which implies that this crime can only be perpetrated by a state and its organs and officials. The Genocide Convention 1948 is silent on the matter of non-state armed groups since it is only concerned with state responsibility and individual criminal responsibility. Although the most prominent instances of genocide in the 20th century, the holocaust perpetrated by the Nazis and the Rwandan genocide of 1994 concerned state systems the judgement of the International Court of Justice in the *Case Concerning the Application of the Convention on the Prevention and Punishment of the Crime of Genocide (Bosnia and Herzegovina v Serbia and Montenegro)* demonstrate that acts committed by non-state armed groups which can be attributed to a state can be classified as genocide if the conditions laid down in Article II of the Genocide Convention 1948 are fulfilled.[11] This further strengthens the argument that non-state armed groups can perpetrate the crime of genocide.

Indeed, in recent months it has been alleged that ISIS has deliberately killed members of the Yazidi group, a Kurdish ethno-religious group, with the intent to destroy the group in whole or in part. It may very well be the case that non-state armed groups will more frequently perpetrate crimes of genocide since they may have the material and personnel capacity necessary to do so. Thus, non-state armed groups can be regarded as systems that can plan, commit and repeatedly perpetrate system crimes whether they are war crimes, crimes against humanity or genocide. For now, however, international criminal law does not take into account whether a non-state armed group has the capacity to perpetrate international crimes, despite the overwhelming empirical evidence to the contrary. The idea that only members of a non-state armed group should bear responsibility for those crimes is abstract, given the sociopolitical realities of how such crimes are planned and committed. The ambiguous organisational structure of some non-state armed groups provides a platform through which individual members can argue that their prosecution is arbitrary and unjust. If this is true, the argument that the non-state armed group has approved an individual member's actions and, therefore, has consented to bear the cost of that responsibility becomes tenuous.

Complexity theory: the repercussions for the responsibility of non-state armed groups for international crimes

There are considerable theoretical and empirical questions about complexity theory, especially whether it can clarify how non-state armed groups work or if the concept can only apply by analogy. If correct, the theory provides an insight into the non-state armed group and its consequences for the matters examined here. The theory suggests that the non-state armed group is a complex adaptive system that has emerged from the interactions of individuals and the variety of conceptual and social tools such as law, ethics, language, culture and religion that bolster cohesion among significant numbers of individuals that may share a common objective. The non-state armed group as a complex adaptive system has casual effects in its environment, on its members and the interactive processes from which it has emerged (Fresard, 2004, p. 46). Although a non-state armed group can on occasion be influenced by a particular individual member, it tends to persist despite them (Haer, 2015, p. 35). As such, the non-state armed group is greater than its total members since it has the capacity to achieve far more than one individual acting alone, so there is no definitive connection between the two.

Conceptualising the responsibility of the non-state armed group as a collective entity

The ontology of complex adaptive systems has considerable ramifications for the responsibility of the non-state armed group as a collective entity in general and for the principle of individual criminal responsibility more importantly. Since it enforces the actuality of complex adaptive systems, complexity theory arguably supports the claim that collective entities are ontologically distinct and separate from their individual members. So the degree to which responsibility necessitates that there is an actor on whom responsibility can be imposed, complexity theory implies that the non-state armed group can be regarded as such an actor because it has causative effects that cannot be devolved into integral components. Complexity theory is therefore inconsistent with the central premise of international criminal law: that the non-state armed actor is not an appropriate entity to which criminal responsibility should be attributed on the grounds that it is simply an extension of the individual.

That said, the ontology of the complex adaptive system indicates that the aperture between the non-state armed group and the individual is still significant for two reasons. First, complexity theory substantiates that rather anodyne conduct has superfluous consequences at group level. In particular, Bella asserts that

> [t]o merely blame individuals . . . is to avoid . . . the essential claim of emergence: that the character of wholes should not be reduced to the character

of parts . . . [Therefore] outcomes can emerge through the efforts of normal competent and well adjusted people much like ourselves.

(Bella, 2006, p. 103)

Take for example, the members of an armed opposition group whose role is to follow the instructions of the centralised command and manufacture improvised explosive devices. The interactions of individuals following straightforward rules or instructions could result in the catastrophic loss of civilian life. If Bella's contention is correct, then complexity theory helps to support the idea that the non-state armed actor must be a focal point of responsibility under international criminal law as a collective entity. This is because individual acts that may contribute to the emergent conduct of the non-state armed group may be inconspicuous.

Nevertheless, if the actions of an individual are indeed inconspicuous but international crimes can still occur at group level then the distributive dilemma becomes even more acute. The

> non-linear relationship between the complex interactions of individuals and the phenomena that emerge from those interactions means that in most cases it will be impossible to trace direct connections between an individual and the complex adaptive system from which it is a part and the impacts that system may have in the world.
>
> (Chinen, 2014, p. 723)

Considering how complexity theory would posit that the perpetration of an international crime generally stems from the interactions of multiple individuals it is difficult to delineate why an individual member of a non-state armed group should shoulder responsibility for that crime alone when numerous other individuals may have contributed towards its commission as well? This implies that even if an individual member conforms with the complex system's behaviour, there may still be no consequential link between that conformity and the crimes that system has committed. The nonlinearity between the individual members and the non-state armed group infers that the distribution of responsibility requires a link no matter how tenuous between the group and its members. So complexity theory can perhaps help explain why the distribution of responsibility for international crimes can be difficult to elucidate.

Complexity theory also has consequences for the concept of responsibility. Under international criminal law, it is deemed pertinent that individuals should bear responsibility for a crime where he or she understood the potential consequences and nevertheless, committed the act anyway. However, if an act can lead to unforeseen consequences in the future, how can a person be considered responsible for them. Juarrero posits that "[s]elf-organization . . . counsels for a wider denotation for the term cause, one reconceptualised in terms of 'context-sensitive constraints' to include those causal power that incorporate circular causality, context-sensitive embeddedness, and temporality" (Juarrero,

2008, p. 280). Klaus Mainer also supports this argument, stating that in a linear model causation "the extent of an effect is believed to be similar to the extent of the cause. Thus, a legal punishment of a punishable action can be proportional to the degree of damage effected" (Mainzer, 1997, p. 435).

However, according to complexity theory phenomena can emerge from random events, so the principle of proportionate responsibility can be questioned. Complexity theory presupposes that most issues have nonlinear characteristics. Indeed, Juarrero asserts that "[a]s the ecological, economic, and political problems of mankind have become global, complex, nonlinear, and random, the traditional concept of individual responsibility is questionable" (Juarrero, 2008, p. 280). The same can be said of non-state armed groups: the acts they commit in the short term can result in consequences that are felt by the wider international community in the long term. Thus, regard should be had to the context in which responsibility is being applied. While in the short term the individual member of the non-state armed group will retain some element of control over the effects of their actions for which they can be held accountable they are also restricted by their environment and the internal dynamics which lead to the emergent behaviour of the group (Frederick, 2003, p. 5). These limitations have led some commentators to be sceptical about whether a member of a non-state armed group has much autonomy. For instance, Kelly contends that (individuals)

> are shaped by environment, genetics, and experience in a way that affects what we perceive as reasons and narrows the horizon of possibilities for action. Environmental, genetic and psychological factors all shape what count as reasons for a person. Recognising this should challenge our confidence that a given wrongdoer was morally capable of doing better.
>
> (Kelly, 2011, p. 194)

This is precisely why some scholars who wish to retain some semblance of the concept of responsibility argue that it should be constructed. In particular, Schütz states how responsibility "does not exist prior to its assignment" (Schütz, 1994, p. 160).

For the purposes of this chapter, the question whether a non-state armed group has enough autonomy or some other characteristics to rationalise the assignment of responsibility to it does not need to be immediately determined. But it is my view that the ontology of complex systems is such that international criminal law which disperses the responsibility of a complex system like the non-state armed actor downwards to individuals who are members of that group will inevitably be the result of construction. This is haphazard since there is not always a linear connection between the non-state armed group and the individual. The ontology of the non-state armed group implies that when the entity commits an international crime, there is no way to sidestep the criticism that its individual members are unfairly bearing the brunt of accountability for it.

The non-state armed group: an ontology or a legal concept?

It would appear that complexity theory and its ontology of the complex adaptive system has not been able to answer all our questions. If this is the case, why should we pay any attention to it especially since international law seems to yearn for us to set questions about the ontology of the non-state armed group to one side? Indeed, the ILC's Articles on the Responsibility of States for Internationally Wrongful Acts would suggest that "[c]ontemporary international law is clearly the work of states, every last word conforms to their wishes. States jealously safeguard their constitutional attribute of independent sovereignty and thereby affirm their monopoly over both national and international matters" (Lejbowicz, 1999, p. 292). So there would seem to be very little leeway for non-state armed groups in this quandary. At the same time, prosecuting the non-state armed group as a collective entity under international criminal law would imply that they are international legal subjects, but as discussed in the second section of this chapter, there is no clear-cut definition of non-state armed groups, and this is clearly illustrated in the myriad of terms that have been used to describe them.

Zegveld traces the legal concept of the non-state armed group on the one hand and the sociopolitical accounts on the other by delineating whether they should satisfy a minimum set of criteria as to their size and power to qualify as legal subjects (Zegveld, 2002, p. 134). However, contemporary international law would posit that the non-state armed group does not have "a legal status attaching to a certain set of affairs by virtue of certain rules or practices" (Crawford, 2006, p. 5). Portmann contends that international practice is cognisant of, first, a formal conception of international personality where legal rules delineate the criteria for legal personality and, second, an 'actor conception' in which legal personality is composed of actors who take part in and contribute to the creation of international law (Portmann, 2010, p. 176). He suggests that it is futile to "recognise certain entities as international persons: [because] such status in itself does not entail any international legal rights, duties or capacities whatsoever" (Ibid). It is for international norms to bestow legal rights and obligations on entities and not the mere fact of international legal personality. On the basis of these amalgamated constructs, even if non-state armed groups could be described as having a legal status they "do not automatically possess so-called fundamental rights and duties nor are they automatically authorised to formally contribute to the creation of international law" (Portmann, 2010, p. 177).

Of course, the idea that international law should regard the non-state armed group as a legal concept would be quite appealing, if only because it sidesteps the difficulties with construing the non-state armed group as a sociopolitical object. It has further significance for the issue of responsibility itself. As discussed in the previous section, the majority of the literature has strived to answer whether a collective entity can ever be subject to moral judgement since it is unclear whether the tools we utilise to make moral judgements about individuals can also be applied to entities such as non-state armed groups. Developing a legal conception of the non-state armed group would do very little to alleviate the problem since non-state armed groups are regarded as incapable of fulfilling their duties under international law, so it could be said that international law is still not comfortable with regarding a non-state armed group as having the capacity to bear legal responsibility for its wrongful conduct, preferring instead to distinguish them from states under the law and treat its members as mere domestic criminals. As Portmann provides, "the only consequence directly stemming from international personality is the capacity to invoke international responsibility and to be held internationally responsible" (Portmann, 2010, p. 275).

Classifying a non-state armed group as a legal concept would also have ramifications for the principle of individual criminal responsibility under international criminal law. A non-state armed group needs individuals to act for it. In turn, if the non-state armed group could be held accountable under international criminal law as a collective entity the member's conduct could be imputed to the collective entity. As a result, the member would not always face the consequences for their role in the perpetration of the international crime. This was also a dilemma recognised by Roberto Ago, the second rapporteur for the ILC, with regard to states. However, he averred that the entity to which the individual's actions are imputed must be a legal one and not a socio-political entity. The same can be contended with regard to non-state armed groups:

> There are no activities of the . . . [non-state armed group] which can be called 'its own' from the point of view of natural causality as distinct from that of legal attribution . . . In describing the . . . [non-state armed group] as a real entity . . . like any other legal person-one must nevertheless avoid the error of giving an anthropomorphic picture of the collective phenomenon, in which the individual-organ would have his personality absorbed and annulled in the whole world and would be an inseparable part of it, rather like an organ of the human body.
>
> (Ago, 1970, para. 39 note 4)

So there would be conceptual difficulties with holding a non-state armed group accountable as a collective entity under international criminal law.

In particular, there would be difficulties in utilising the principle of individual criminal responsibility under the Rome Statute if the non-state armed group is regarded as an entity that subsumes the individual member. Indeed, how would the principle of individual criminal responsibility exist alongside a principle of collective entity responsibility since from an evidential point of view pursuing the non-state armed group as a collective entity would perhaps be a more straightforward option for the Office of the Prosecutor than pursuing individual members? As a result, members of the group who played a crucial role in the planning, ordering and commission of international crimes would potentially evade justice, thereby fostering a culture of impunity. It would be essential for international criminal law to preserve some degree of autonomy and responsibility for individuals who act on behalf of non-state armed groups, in particular, high-ranking members who may have ordered or failed to prevent the commission of international crimes. Realistically, the only way such autonomy could be pre-served is to dismiss the concept of the non-state armed group as a collective entity that can subsume the individual member's conduct.

Thus, analysing the non-state armed group in legal terms has its benefits, but also its limitations. The idea that a non-state armed group becomes a legal sub-ject of international law by meeting certain pre-existing legal criteria which it had no role in creating or developing would be a major bone of contention for them. Indeed, many non-state armed groups refuse to be bound by laws that were created by what they perceive to be the "enemy state." One could retort by contending that at least some fundamental international legal principles preceded the existence of states so why not for non-state armed groups as well? Law existed before the Treaty of Westphalia in 1648, but at the same time it is "shorthand for the processes in which law and social-political entities interacted well before 1648 and thereafter, not just a single event" (Chinen, 2014, p. 727). Of course, the debate about whether law or sociopolitical entities existed first is incessant, but modern-day international practice has certainly taken the stance that one precedes the other.

So perhaps it is crucial to re-consider James Crawford's use of the phrase "state of affairs" to figure out its legal implications. As Chinen points out "[t]erri-tory, population, government, and the ability to engage in international relations are legal terms . . . but they also purport to say something meaningful about the socio-political and geographic state of affairs to which legal analysis applies" (Ibid). One could posit that as an autopoietic system international criminal law views the world with reference to its own terms. But even so, the state of affairs should not be relegated to the back burner. As with the relationship between law and ethics, if legal categorisations depart from reality a number of challenges ensue. There is a danger that international criminal law could become ineffective either because it cannot sufficiently address this reality or it hazards becoming pointless for this same reason. Both levy a price on international criminal law's legitimacy. As explored earlier, the reality is that individual members bear the brunt of responsibility for international crimes. But if we recognise that the state of affairs is an important component in any analysis of international criminal law, ontology can perhaps be brought within the fold. With regards the sociopolitical entity that is the non-state armed group, complexity theory would suggest that ontology is one of an entity with a membership, individuals who can be held accountable for international crimes but as a result of being part of a complex adaptive system are not linearly connected to that non-state armed group. It is long overdue for international criminal law to embrace this reality.

Conclusion

As stated in the introduction, international criminal law has so far had very lit-tle to do with the accountability of the non-state armed group as a collective entity and this chapter has examined a fundamental consequence of that actuality. It has explored how certain ideas from complexity theory impact international criminal law, in particular, the challenges that arise from how the distribution of criminal responsibility within a non-state armed group to individual members conflicts with how international crimes are planned and orchestrated within the group in reality. Complexity theory contributes to this discussion by providing an ontology of the non-state armed group that contradicts international criminal law's focus on the individual; it is an entity that is not differentiable from its individual members. So if international criminal law is to be effective and achieve its objective to thwart impunity and deter future crimes it must deal with the non-state armed group itself as well. However, the theory also helps to elucidate that any attempt to interlink individual and group conduct is rife with difficulty: as a complex adaptive system, the non-state armed group could not exist without the interactions of its individual members, but the non-state armed group is also more than the aggregate of its members. Thus, the relationship between the individual member and the non-state armed group cannot be clearly delineated. Unfortunately, international criminal law continues to overlook this ontology thereby risking incoherence. This conclusion does not exclude group responsibility as such but does make it difficult to account for it and makes the job of distributing accountability upwards from the individual members to the non-state armed group more difficult.

Efforts to hold the non-state armed group accountable for the behaviour of its individual members have been and will continue to be challenging at best because of the nonlinear connection between the group and the individual member. This does not imply that efforts should not be made: forging a law of general responsibility and encouraging international engagement with non-state armed groups to enhance their compliance with international law would help to overcome some of the dilemmas discussed in this article. However, we should not hope for any of these approaches to be completely effective. Laws are ignored and engagement often yields little by way of behavioural change. This article could thus seem quite dispiriting: like a lot of research it endeavours to illustrate why a particular dilemma in international criminal law recurs perennially and why any possible strategies to overcome it are likely to fail. Nevertheless, the consequences of only distributing responsibility for international crimes to individual members of the non-state armed group are far-reaching especially since, for the most part, it is the group as a whole who may be condoning and inciting the individual member to carry out crimes. Therefore, it is at the very least useful to consider whether the benefits ensuing from holding individual members of the non-state armed group accountable outweigh those negative consequences and whether there is any means to overcome those impacts, in particular, because there may be no rational means to vindicate them.

Notes

1 Prosecutor v. Duško Tadić, Case No. IT-94-1-A, Judgment, 15 July 1999, para. 191.
2 For the text of the Articles and commentaries see Official Records of the General Assembly, Fifty-Sixth Session, Supplement No. 10 (A/56/10). Ch. V.

3 Geneva Convention for the Amelioration of the Condition of the Wounded and
the Sick in Armed Forces in the Field, 75 UNTS 31; Geneva Convention for the
Amelioration of the Condition of Wounded, Sick and Shipwrecked Members of
Armed Forces at Sea, 75 UNTS 85; Geneva Convention Relative to the Treatment of Prisoners of War, 75 UNTS 135; Geneva Convention Relative to the
Protection of Civilian Persons in Time of War, 75 UNTS 287. All four Conventions entered into force on 21 October. See also Protocol Additional to the
Geneva Convention of 12 August 1949, and Relating to the Protection of Victims of International Armed Conflict, 1125 UNTS 3; 16 ILM 1391 (entered
into force 7 December 1978); and Protocol Additional to the Geneva Convention of 12 August 1949, and Relating to the Protection of Victims of Non-
International Armed Conflicts, 1125 UNTS 609; 16 ILM 1442 (entered into
force on 7 December 1978)
4 Prosecutor v. Ramush Haradinaj, Case No. IT-04–84–84-T, Judgment (Trial
Chamber), 3 April 2008, para. 60; Prosecutor v. Jean Paul Akayesu, ICTR-96–
4-T, Judgment (2 September 1998), para. 619.
5 Ibid., para. 69.
6 Ibid., para. 91.
7 Prosecutor v. Ramush Haradinaj, Case No. IT-04–84–84-T, Judgment (Trial
Chamber), 3 April 2008, para. 60; Prosecutor v. Jean Paul Akayesu, ICTR-96–
4-T, Judgment (2 September 1998), para. 619.
8 See, in particular, Common Article 3 to the Geneva Conventions as supplemented
by Additional Protocol II.
9 Prosecutor v. Duško Tadić, IT-94–1, Judgment, Trial Chamber (7 May 1997),
para. 654.
10 Prosecutor v. Alex Tamba Brima, Brima Bazzy Kamara and Santigie Borbor Kanu,
SCSL-04–16-T, Trial Judgment (20 June 2007).
11 Bosnia and Herzegovina v. Serbia and Montenegro, ICJ Case Concerning the
Application of the Convention on the Prevention and Punishment of the Crime
of Genocide, Judgment (26 February 2007) General List No. 91, paras. 278–97.

Bibliography

Ago, Robert (1970), 'Special Rapporteur Second Report on State Responsibility', ii
Yearbook of the International Law Commission 177.
Arendt, Hannah (1987), 'Collective Responsibility', in J. Bernhauer (ed.) *Amor
Mundi* (Dordrecht: Nijhoff).
Bella, David (2006), 'Emergence and Evil', 8(2) *Emergence: Complexity and Organi-
zation* 102.
Bellal, Annyssa (2015), 'Establishing the Direct Responsibility of Non-State Armed
Groups for Violations of International Norms: Issues of Attribution', in N. Gal-
Or, C. Ryngaert and M. Noortmann (eds.) *Responsibilities of the Non-State Actor
in Armed Conflict and the Market Place: Theoretical Considerations and Empirical
Findings* (Leiden: Brill), 305.
Bellal, Annyssa and Stuart Casey-Maslen (2011), 'Enhancing Compliance with Inter-
national Law by Armed Non-State Actors', 3(1) *Goettingen Journal of Interna-
tional Law* 175.
Bilkova, Veronika (2015), 'Armed Opposition Groups and Shared Responsibility', 62
Netherlands International Law Review 69.
Brennan, Anna (2018), *Transnational Terrorist Groups and International Criminal
Law* (Oxford: Routledge).

Chinen, Mark (2014), 'Complexity Theory and the Horizontal and Vertical Dimensions of State Responsibility', 25(3) *European Journal of International Law* 703.

Clapham, Andrew (2008), 'Extending International Criminal Responsibility beyond Individuals to Corporations and Armed Opposition Groups', 6 *Journal of International Criminal Justice* 899.

Connolly, William (1974), *The Terms of Political Discourse* (Lexington: D.C. Heath and Company).

Copp, David (1984), 'What Collectives Are: Agency, Individualism and Legal Theory', 23 *Dialogue* 253.

Crawford, James (2006), *The Creation of States in International Law* (Cambridge: Cambridge University Press, 2nd ed.).

Darcy, Shane (2006), *Collective Responsibility and Accountability under International Law* (The Hague: Martinus Nijhoff).

Feinberg, Joel (1970), 'Collective Responsibility', in J. Feinberg (ed.) *Doing and Deserving: Essays in the Theory of Responsibility* (Princeton, NJ: Princeton University Press), 222.

Frederick, William (2003), 'Emergent Management Morality: Explaining Corporate Corruption', 5 *Emergence* 1.

French, Peter (1984), *Collective and Corporate Responsibility* (New York: Columbia University Press).

Fresard, Jean-Jacques (2004), *The Roots of Behaviour in War: A Survey of the Literature* (Geneva: International Committee of the Red Cross).

Frowein, Jochen (1992), 'De Facto Regime', 1 *Max Planck Encyclopedia of Public International Law* 966.

Garcia-Amador, F. (1956), 'Special Rapporteur Report on International Responsibility', ii *Yearbook of the International Law Commission* 174.

Haer, Roos (2015), *Armed Group Structure and Violence in Civil Wars: The Organizational Dynamics of Group Killing* (London: Routledge).

Hardin, Garrett (1968), 'The Tragedy of the Commons', 162 *Science* 1243.

Held, Virginia (1970), 'Can a Random Collection of Random Individuals Be Morally Responsible?' 67 *Journal of Philosophy* 471.

Juarrero, Alicia (2008), 'On Philosophy's 'To Rethink' List: Causality, Explanation and Ethic', 20 *Ecological Psychology* 278.

Kelly, Erin (2011), 'Reparative Justice', in T. Isaacs and R. Vernon (eds.) *Accountability for Collective Wrongdoing* (Cambridge: Cambridge University Press), 194.

Kleffner, Jann (2009), 'The Collective Accountability of Organized Armed Groups in System Crimes', in A. Nollkaemper and H. van der Wilt (eds.) *System Criminality in International Law* (Cambridge: Cambridge University Press), 238.

Kutz, Christopher (2000), *Complicity: Ethics and Law for a Collective Age* (Cambridge: Cambridge University Press).

Lejbowicz, Agnes (1999), *Philosophie du Droit International: L'Impossible Capture de l'Humanite* (Paris: PUF) (ICRC translation).

Levinson, Sanford (1974), 'Responsibility for Crimes of War', in M. Cohen et al. (eds.) *War and Moral Responsibility* (Princeton, NJ: Princeton University Press), 104.

Lukes, Steven (2005), *Power: A Radical View* (New York: Palgrave Macmillan, 2nd ed.).

Mainzer, Klaus (1997), *Thinking in Complexity: The Computational Dynamics of Matter, Mind and Mankind* (Berlin, Springer, 5th ed.).

Malle, Bertram (2010), 'The Social and Moral Cognition of Group Agents', 19(1) *Journal of Law and Policy* 95.

Martin, Benjamin (1976), 'Can Moral Responsibility Be Collective and Non-Distributive?' 4 *Social Theory and Practice* 93.

McConnell, Lee (2016), *Extracting Accountability from Non-State Actors in International Law* (London: Routledge).

Melzer, Nils (2009), *Interpretive Guidance on the Notion of Direct Participation in Hostilities under International Humanitarian Law* (Geneva: International Committee of the Red Cross).

Miller, Seumas (2006), 'Collective Moral Responsibility: An Individualist Account', XXX *Midwest Studies in Philosophy* 176.

Pasternak, Avia (2011), 'The Distributive Effect of Collective Punishment', in T. Isaacs and R. Vernon (eds.) *Accountability for Collective Wrongdoing* (Cambridge: Cambridge University Press), 213.

Portmann, Roland (2010), *Legal Personality in International Law* (Cambridge: Cambridge University Press),

Reiff, Mark (2008), 'Terrorism, Retribution and Collective Responsibility', 34 *Social Theory and Practice* 209.

Schütz, Anton (1994), 'Desiring Society: Autopoiesis Beyond the Paradigm of Mastership', 5 *Law and Critique* 149.

Sepinwell, Amy (2011), 'Citizen Responsibility and the Reactive Attitudes: Blaming Americans for War Crimes in Iraq', in T. Isaacs and R. Vernon (eds.) *Accountability for Collective Wrongdoing* (Cambridge: Cambridge University Press), 241.

Sriram, Chandra (2010), *Globalizing Justice for Mass Atrocities: A Revolution in Accountability* (Abingdon: Routledge).

Werle, Gerhard (2005), *Principles of International Criminal Law* (The Hague: TMC Asser Press).

Wheatley, Steven (2016), 'The Emergence of New States in International Law: The Insights from Complexity Theory', 15(1) *Chinese Journal of International Law* 579.

Yezdani, Omer, Loius Sanzogni and Arthur Poropat (2015), 'Theory of Emergence: Introducing a Model-Centred Approach to Applied Social Science Research', 33(3) *Prometheus: Critical Studies in Innovation* 305.

Zegveld, Liesbeth (2002), *The Accountability of Armed Opposition Groups in International Law* (Cambridge: Cambridge University Press).

Section IV

Complexity and business and finance regulation

8 Governing complexity

Mark A. Chinen[1]

Introduction

A major concern of the international financial crisis of 2008 and 2009 was the damage a collapse of the derivatives market would cause to the financial system and to the wider economy. Because of the vast size of that market, the web of contractual relationships among the counterparties to those instruments, and the difficulties in understanding the terms of the derivatives themselves and in appraising them, the financial community was concerned that the failure of one party to honour the terms of a single agreement would trigger a system-wide cascade of failures, with enormous negative impacts. This danger was serious enough, but the size and complexity of that market were just two of several factors that contributed to the contraction. Imprudent borrowing and lending practices in the United States and elsewhere, ill-designed financial products, mismatched incentives created by executive compensation policies and collusion between credit-rating agencies and debt issuers have all been blamed for the crisis.[2]

Accounting for the failure of governance to anticipate and prevent that crisis is of course part of preventing similar ones. However, as the international community tried to understand and draw lessons from these events, one had the sense not only of moving targets but of changing targets too and, further still, of a shift in our way of seeing targets. How much each factor recited earlier contributed to the crisis is subject to debate, just as what measures should be adopted to prevent future crises. For example, derivatives themselves are solutions to the problem of managing risk (Ayadi and Behr, 2009, p. 190). Risk is a fact of life, and nothing prevents other forms of risk management from evolving and spreading to the point where they themselves pose systemic risks. Moreover, for at least some observers, the crisis itself has called into question the very framework through which international finance is viewed and justified. If we assume *arguendo* that framework is flawed, how do we proceed if there is no viable alternative?

This chapter discusses complexity theory and its implications for governance in these circumstances. Complexity theory is of relatively recent vintage, with yet-to-be resolved methodological issues, and thinking about its meaning for governance is newer still. The theory suggests that if international finance, the global economy, and law are complex systems, then they are either uncertain by nature or have uncertain features and thus are never subject fully to forecast and therefore are never subject fully to control.

Perhaps this says nothing new. Complexity theorists do not urge we abandon planning altogether. Planning is possible over the short term, and one can anticipate how actors are likely to behave under complex conditions. It may be the best governance can do in the long term is to promote stability or resilience to systemic threats. In any event, complexity theory justifies scepticism about any set of policies that purport to be universally applicable. These "lessons" from complexity theory are worth exploring further, and I do so here. But the theory provides at least one more insight. Complexity theory does not dictate an ethic or set of values, including the values of procedural and substantive inclusion implicit in the criteria of good governance: participation, accountability, effectiveness, responsiveness, and fairness. However, once we adopt those values, complexity theory might shed light on how they are "structured" and implemented. It tells us that those values will take on a different cast and be subject to their own dynamism as they manifest themselves in different contexts.

Governing complex systems

The features of complexity theory have been described in more detail in earlier chapters. The one most important for this chapter is the idea of complex systems: that individual actions at the micro-level can lead to emergent phenomena and self-organisation not predicted at the micro-level. Such systems are highly dependent on initial conditions that give rise to such emergent phenomena and are always evolving, although they may appear to be in states of equilibrium for significant amounts of time. Thus, such systems cannot be described completely, and their evolution cannot be predicted over the long term.

Complexity theory has been applied in a wide range of fields, including mathematics, physics, biology, economics, sociology, and cognition.[3] Of particular interest has been work in economics. Most have described the economy as a linear system in various states of equilibrium, subject to exogenous shocks. The appeal of nonlinear, dynamic accounts of the economy is that bubbles, financial crises and so on can be explained from within the system itself, a system that is constantly in process (Serletis and Gogas, 1999, p. 83; Arthur, 2014, p. 5). It could be argued, however, that thus far, the impact of complexity theory on economic policy has been modest. The issue is whether complexity can be detected in the available economic data. Several studies have concluded, primarily through statistical analyses of time-series data, that nonlinearity does exist to some extent in parts of the economy. This includes the stock market (Park et al., 2007), stock returns (Kanas, 2003 and 2005), the housing market (Dieci and Westerhoff, 2011), commodity futures (Chen and He, 2010) and the relationship between inflation and commodity prices (Kyrtsou and Labys, 2006). Others, however, have argued that nonlinearity in the economy has not been proved because of weaknesses either in the data or in the methodology. (Serletis and Shintani, 2006; Brooks, 1998).

Complexity theory is thus a work in progress, with more to be done in determining whether and how the theory can be applied to the social world. This chapter discusses the possible implications for governance *if* the financial system, the broader economy, and other social systems such as the law are in fact complex systems that interact and impact the other.

One such implication goes to forecasting and its impact on policy. Here the lessons are arguably quite underwhelming: first, short-term planning is possible, but long-term planning is not, and second, one size does not fit all. This is not earth-shattering. Geyer (2003, p. 253) writes,

> A complexity perspective . . . would argue that there is no clear linear policy answer to all situations. . . . [B]eyond creating a stable fundamental order within which individuals can learn, interact and adapt, there is little a state can do with linear certainty.

According to Cilliers (1998, pp. 139–40), policymaking therefore involves, among other things, gathering as much information on an issue as possible, even though it is understood that it is impossible to obtain all information; considering as many potential outcomes as possible while understanding that it is impossible to consider all of them; and ensuring ways to re-evaluate and revise a judgment if there are flaws. This approach to policymaking thus begins to resemble those supported by deliberative democracy (e.g., Habermas, 1996, p. 300) or the social realisations approach proposed by Sen (2009). Those approaches also appreciate the need for large, albeit finite amounts of information in the decision-making process and for reviewing previous policy decisions. So it can be said that as a practical matter, complexity theory offers nothing new to policymaking. This might be particularly true with the economy. Mainzer (2007, p. 327) argues that even if it were proved conclusively that the economy is a chaotic nonlinear system, it would still be possible to engage in "local predictions."

Yet even now, complexity theory changes our orientation toward governance, or confirms suspicions we have had about theories of governance. First, it provides grounds for scepticism of any rule or norm that purports to be universally applicable. This is because a rule's impact will depend on local conditions. Furthermore, any rule of governance is itself the result of a complex adaptive process as people perceive their environment and respond to it by choosing from an array of possible rules. Under complexity theory, the very assessment of the environment is necessarily incomplete, meaning that any response will be too (see, e.g., Romano, 2014), and since the rule eventually selected has "competed" with other solutions to the same problem, it is often only marginally better than others that were bypassed. There is no guarantee that such a rule is optimal.

In addition to cautioning us against sweeping solutions, complexity theory provides a more nuanced way to make short-term predictions about how a complex system might develop and how actors would be expected to act within such a system and thereby have an impact on the system itself. Some scholars have tried to show how the economy might be governed *qua* nonlinear complex system, but for purposes of this chapter, the more interesting work has focused on the behaviours of agents who interact within an economic system to determine what kinds of phenomena might emerge from such interaction. In one such study, Westerhoff (2008) uses computer simulations to model boundedly rational heterogeneous agents who follow three simple investment strategies (follow rising prices, trade on the fundamental price of the asset, or stand pat) in an artificial financial market. Westerhoff finds that interactions among them can lead to bubbles and crashes endogenous to the system, which is driven by one investment strategy coming to dominate the others and then receding as another comes to dominate, although no strategy completely disappears. He then models the impact of forms of financial regulation imposed exogenously, such as transaction taxes, central bank interventions, and trading halts, and finds, for example, that a small tax on transactions stabilises markets (*ibid.*, p. 208). If these models can be validated empirically, they could help us in regulatory policymaking.

At present, however, most analysis of this kind is more conceptual or metaphorical. Chiu (2010a and 2010b) engages in such an analysis in her description of the regulatory space of the financial industry. She sees a "decentred" landscape, "populated by resourceful, competent and powerful industry participants, and on the other hand consisting also of agencies of authority that have a public character" (Chiu, 2010a, p. 171). Chiu assesses the strengths and weaknesses of various forms of regulation that range from self-regulation by private actors to pure "regulator-led governance" (*ibid.*, pp. 173–81). Towards one end of the spectrum, for example, is "transactional governance," in which parties set private rules through contract. Private ordering has the benefit of better reflecting party preferences. Chiu warns, however, that contractual terms can often become standardised; not wrong in itself but leading to systemic risk if the standards are sub-optimal. Moreover, contracts rarely provide strategies for dealing with systemic failures and do not account for externalities (*ibid.*, p. 174). On the other end of the spectrum is "meta regulation," the attempt to link private ordering with public values that are provided by regulators. Here, the benefit is enabling individuals to engage in private ordering but ensuring they are informed by and take into account public values. However, Chiu argues that such values can be eclipsed easily by the technical demands of a particular system, what she terms "capture by technocracy." This, in turn, leads to lack of accountability because it is difficult for non-expert outsiders to monitor financial activities (*ibid.*, pp. 179–80).

Chiu's regulatory space is far more complex than Westerhoff's artificial market. In this space, participants are engaged in a dynamic relationship, "co-producing and co-enforcing norms of governance" (*ibid.*, p. 170) so that the various forms of regulation are no longer exogenous to the financial system but emerge from the participants themselves, each with particular interests and concerns, some overlapping and others potentially conflicting. Moreover, as Chiu argues, it is possible for any regulatory space to be dominated by one or more participants and thus not only determine which regulatory "heuristic" will be used but also in theory come to dominate all heuristics; hence, her argument that other stakeholders in the financial services industry, such as accountants, credit rating agencies, and attorneys, be strengthened so that they can participate in the formation of these norms of governance (*ibid.*, p. 301).

Other scholars have tried to anticipate how actors might operate within complex systems on the international level and how that environment might change as a result. Alter and Meunier (2009) argue the increasing density of interlocking international legal regimes has led to a complex system of rules that exhibits much of the uncertainty described earlier. One can expect five consequences from such complexity. First, since the outcome of complex networks of rules is hard to predict, there will be greater emphasis on the implementation of those rules by either administrators or adjudicators. Second, one can expect actors to engage in "chessboard politics," in which international actors pursue various goals in several institutions to forum-shop within a particular legal region, to shift of regimes themselves, and to engage in strategic inconsistency: creating contradictory rules in a parallel regime to undermine rules in another (*ibid.*, pp. 16–17). Third, actors are likely to use the decision-making tools of bounded rationality, among them a reliance on experts and the use of heuristics as such experts make the first rough cuts at assessing and responding to various policy issues (*ibid.*, pp. 17–18). Fourth, international regime complexity and its reliance on experts result in the generation of small-group environments, as the same experts find themselves asked to address different international issues and the number of international venues increases, thus bringing into play the positive and negative aspects of small-group dynamics (*ibid.*, pp. 18–19). Finally, the feedback effects inherent in complex systems can mean competition among various institutions, unintentional reverberations in other regimes, a lack of accountability because feedback and secondary effects make it hard to identify who might be responsible for a particular issue, a greater emphasis on loyalty because of acts of a state in one area will affect how it is perceived in others, and, last, greater ease of exit from international regimes through non-compliance, regime shifting, or withdrawal from international organisations (*ibid.*, pp. 19–21).

As actors operate within this complex system of rules, one can see how the structure itself will be impacted, just as financial actors have the potential to impact Chiu's regulatory landscape. For example, if, as Alter and Meunier predict, complexity promotes competition among institutions as they "vie" for resources and prestige, it is quite likely we would see such institutions under pressure to recast themselves to attract international actors. This is one reason why targets are recognisable, as such, but are moving nonetheless. The work of describing complex governance structures and institutions, the ways in which people and other agents might be expected to interact within that system, and the ways in which the two co-evolve leads to what Baxter (2011, pp. 264–68) describes as a shift in focus of governance from direction to adaptation. Governance will continue to steer individual and corporate behaviour and to set out the basic structures for interaction but more with a view towards enhancing the resilience of complex social systems (Kavalski, 2008) and certainly not towards establishing some kind of stasis.

Governing values

The last section focused on more "pragmatic" implications of complexity theory for governance, but the discussion of Chiu's work suggests that the theory might also address the values that underlie such governance. The criteria of good governance have been identified as participation, accountability, effectiveness, responsiveness, and fairness, with the implication that governance is not legitimate unless all five requirements are met. Implicit in several of these are norms of procedural and substantive inclusion such that we would question any form of governance that excludes people from either its processes or the benefits it secures (Lubell, 2013, p. 548).

Some complexity theorists argue that the theory leads precisely to that kind of inclusion, albeit framed in terms of diversity. The terms are not synonymous, but I use the term *inclusion* here to mean making room for difference. Cilliers explored deeply the relationship between complexity theory and an ethic of diversity. For him, ethical decisions are embedded within our existence in complex systems: because such a system cannot be completely defined or described, it must be interpreted and evaluated, that is, given meaning, in necessarily incomplete ways. Such interpretive acts have two aspects. First, for Cilliers (2010), meaning in a complex system requires what he terms "constrained difference" and "repeatable identity." For a thing to have meaning it must be comparable to something else, and at the same time, there must be some, albeit sometimes permeable, boundary so that the various components of a system can interact. The more relationships, the more constrained a component individual will be, but the richer it will be by virtue of those relationships (*ibid.*, p. 10). Second, Cilliers argues that articulating the meaning of a complex system is an act of ethical judgment. As Preiser and he write, "[s]ince we cannot have complete knowledge of complex things we cannot 'calculate' their behavior in any deterministic way. We have to interpret and evaluate. Our decisions always involve an element of choice which cannot be justified objectively" (Preiser and Cilliers, 2010, p. 274).

Preiser and Cilliers suggest a four-part, provisional moral imperative in respect of complex systems that does not itself provide the content of an ethical system but, like Kant's more famous categorical imperative, helps evaluate ethical decisions in light of the reality of such systems, a reality in which our knowledge is limited, any decision we make is inherently provisional, we have no choice but to act in any event, and, as just discussed, meaning emerges through interactions with others (ibid., p. 275):

1   justify your actions only in ways which do not preclude the possibility of revising that justification,
2   make only those choices which keep the possibility of choice open,
3   your actions should show a fundamental respect for difference, even as those actions reduce it,
4   act only in ways which will allow the constraining and enabling interactions between the components in the system to flourish (ibid., pp. 275–76).

Diversity and flexibility are obviously important parts of this imperative, as seen in maxim 3's admonition to show respect for difference even as one curtails it, and in the premise that meaning itself derives from constrained diversity. Other complexity theorists follow suit and urge that without diversity and the ability for various components to interact, human societies will not be robust (Geyer and Rihani, 2010, p. 184; Carline and Pearson, 2007, p. 79).

The remainder of this chapter relies much on the work of Cilliers and other complexity theorists who advocate diversity as necessary for either a robust complex system or a meaningful one. Before proceeding, however, it seems appropriate to discuss what might seem like a theoretical quibble. Although complexity theory might help us better understand both how societies thrive and how diversity is important to that success, in my view, it cannot tell us we *ought* to value such success and such inclusive diversity. As Mainzer (2007, p. 434) writes, complexity theory is "a mathematical, empirical, testable, and heuristically economical methodology. . . . It is not an ethical doctrine in the traditional sense . . ." (see also Preiser and Cilliers, 2010, p. 274). Here is another form of the Humean problem, whether significance can be attached to observed phenomena, even if they are complex. One can argue that the Humean conception unjustifiably privileges observed phenomena over epistemology and ethics, and indeed, Kunneman (2010, pp. 161–62) believes that one of complexity theory's contributions is that it places the three on a level plane. In this sense, constrained difference might be a fundamental requirement for the human ability to perceive reality and to attach significance to it, and it may be that ethics may be far more complex and more deeply embedded within human consciousness and behaviour than is contemplated by more traditional, but theoretically problematic, conceptions of ethics based on reason or theology. Furthermore, it might be self-evident that human thriving is to be valued, and thus, aspects of human interaction, cognition, and interpretation that contribute to such thriving should too (hence the reason these remarks might seem quibbling). But it is less clear that complexity theory itself tells us why such values are inevitable. However, if we choose the values of procedural and substantive inclusion, complexity theory might be able to provide some insight into the form such values take and how they might be implemented.

Law and inclusion

The governance of major aspects of economic globalisation poses a daunting challenge in this regard. Boulle (2009, p. 363) writes that the "law" of globalisation is an emergent phenomenon "comprising international treaties, national laws and court decisions; sub-national laws and regulations; standards of business groups and industry bodies, and the practices of private corporations." As Boulle notes, "[t]his is a mixture of hard law enforceable by courts and soft law that influences state and corporate behaviour but lacks judicial enforceability" (*ibid.*). Such norms can be understood as emerging out of interactions among private actors, the state, and international organisations. From the perspective of legitimacy, what is remarkable about this mixture of hard and soft law is how little input most people appear to have in their creation. Under widely accepted norms of democracy, it is unclear why people or states should honour rules they had no hand in developing, particularly when it is perceived that the rules have failed to protect them.

However, complexity theory is wary of any rule that purports to be universal in application, whether it is a financial regulation or even as important an idea as democracy. This should come as no surprise, given the various instantiations of the Western strand of democracy, from its roots in Athens to the capitals of modern, bureaucratic states (see Dunn, 2005). We would expect the norms of procedural and substantive inclusion to play themselves out in different ways depending, in part, on the systems in which they operate so that such inclusion will take on a particular cast in the governance of a state and another in the regulation of the financial system. Cilliers and others recognise that the theory might therefore serve as grounds for a type of situational ethic. He urges we take "present ethical (and legal) principles seriously – to resist change – but to be keenly aware of when they should not be applied, or have to be discarded" (Cilliers, 1998, p. 139). In his view, to follow rules responsibly "may imply that the rules must be broken" (*ibid.*). This does not mean the rule would be invalidated: "if it is a quasi-rule emerging from a complex set of relationships, part of the *structure* of this kind of rule will be the possibility *not* to follow it" (*ibid.*).

There are at least three ways one might construe this last statement. First, the word *structure* brings to mind the work of Balkin (1998), who borrows concepts from Claude Levi-Strauss to argue that human cognition and human social organisation are made possible through mechanisms that have a tool-like character. Such tools are the "abilities, associations, heuristics, metaphors, narratives and capacities that we employ in understanding and evaluating the social world" (*ibid.*, p. 6). They also comprise human customs and institutions. According to Balkin, these tools are subject to "bricolage," meaning that cultural tools are the cumulative result of other tools. As a result, cultural bricolage has four characteristics: first, it is cumulative in that tools can only be created from the materials at hand; second, it has multiple and sometimes unintended uses; third, it is recursive in that tools can lead to the development of other tools that are then applied to former tools; and, finally, it has unintended consequences because of its multiple uses (*ibid.*, p. 32).

A second way in which the structure of norms might be such that they must take different forms, sometimes to the point where they might be "disobeyed," is that there may be times when the application of an otherwise valid principle or rule will not withstand the scrutiny of the provisional ethical imperative discussed earlier, and thus, the rule or principle should not be followed in that instance. Take, for example, the question whether a democracy may choose to become a dictatorship. The issue appears to be paradoxical because, on one hand, democratic principles would seem to indicate that mature, free people have the right to choose the kind of government they prefer. On the other hand, such a choice would undermine choice itself. Under the provisional imperative, it could be argued that a democracy should not be allowed to choose dictatorship because, at a minimum, the switch to an authoritarian regime would violate maxim 2: "make only those choices which keep the possibility of choice open."

A third, closely related way in which the structure of quasi-rules is such that they can be disobeyed stems from the fact that it is the relationships between people are what give rise to the "quasi-rules" we identify as ethical norms or norms of governance. Under this reading of the imperative, people will always be preferred over the rules and other structures that emerge from them. To see why this might be so, it is important to remember that complexity theorists view social phenomena as emerging from the complex interactions of individuals. "The essential feature is that of the co-evolution of successive layers of interacting elements both horizontally and between levels" (Allen et al., 2010, p. 42). Sawyer's emergence paradigm (Sawyer, 2005, pp. 210–23) provides a helpful picture of what this might look like. Sawyer argues that the social reality with which we are interested can be understood as emergent phenomena that Sawyer groups into five levels, as shown in Figure 8.1.

As can be seen from Sawyer's table, emergent phenomena originate from the individual to interactions with others through language (Levels A and B). Out of such interaction emerge the "communicative tools" such as topic, context, etc. that make such communication possible (Level C). Eventually, more stable

| **Social Structure (Level E)** |
| Written texts (procedures, laws, regulations); material systems and infrastructure (architecture, urban design, communication, and transportation networks) |
| **Stable emergents (Level D)** |
| Group subcultures, group slang, and catchphrases, conversational routines, shared social practices, collective memory |
| **Ephemeral emergents (Level C)** |
| Topic, context, interactional frame, participation structure; relative role and status assignments |
| **Interaction (Level B)** |
| Discourse patterns, symbolic interaction, collaboration, negotiation |
| **Individual (Level A)** |
| Intention, agency, memory, personality, cognitive processes |

Figure 8.1 Sawyer's Emergence Paradigm (2005, p. 211)

phenomena emerge: according to Sawyer, this includes among other things group subcultures, shared social practices and collective memory (Level D). Finally, Sawyer sees social structures such as written texts and the physical aspects of a society, such as architecture, urban design, and so on as resulting from the interactions taking place at lower levels (Level E.)

One need not adopt Sawyer's paradigm wholesale (one can get lost trying to parse the various levels) to appreciate, first, that such phenomena have "both constraining and enabling effects on individuals" (*ibid.*, pp. 216–17) and, second, that the emergent phenomena at each level are themselves in a process of adaptation and transformation as they interact with other phenomena at the same and different levels (which confirms the work of Chiu and of Alter and Meunier). The state is, of course, the primary example. Its enabling function is obvious. Boucher (1998, p. 387) argues in this regard that the state continues to be "the sustainer of all our cultural, social, and political institutions and practices" and "the predominant agent through which citizens are collective actors on the world scene." Its constraining effect is equally clear. And yet, as I discuss in the following in connection with international finance, it, too, is undergoing transformation because of interactions between states and multilateral organisations; states and other states; and states and sub-state actors.

This schema helps us appreciate better Cilliers and others' claim that an ethic that takes complexity theory seriously, including one that informs a system of governance, must pay attention to the relationships between people, to "act only in ways which will allow the constraining and enabling interactions between the components in the system to flourish." One could thus interpret the maxim so that in human societies, people are more important than the ideas and structures that spring from them. Thus, the perennial question for governance, particularly in its democratic forms is whether the norms of hard and soft law that result from the myriad interactions at the micro-level enables individual flourishing, with the caveat that under complexity theory, we can make that assessment only in the short term. The norms of procedural and substantive inclusion make the claim that such well-being is not possible without either, for they enable and constrain interactions at any level. As discussed earlier, we would be suspicious of any form of governance in which people are excluded from its processes or from the benefits it secures. Yet in the end, even these norms would cede to human welfare.

Participation in financial governance

In the remainder of this chapter, I discuss how the norms of procedural and substantive inclusion play out in financial regulation on the international level. The conundrum of financial regulation is well known. On one hand, finance is highly technical, and thus, it seems natural to delegate its regulation to experts. On the other hand, the negative impact of financial crises on the wellbeing of ordinary people is obvious. Because most people have little knowledge of the financial system, let alone means to influence it, yet bear the brunt of system failures, the result can be great disaffection with financial governance. As argued earlier, under the norms of procedural and substantive inclusion, it is unclear why people or states should honour financial rules they had no hand in developing, particularly when it is perceived that the rules have failed to protect them.

We must remind ourselves, however, that this analysis approaches the limits of complexity theory. As Miller and Page (2007, pp. 221–22) point out, under the current analytic tools of agent-based interaction, it is relatively straightforward to model the actions of two agents or an infinite number of them. However, those analytic tools cannot yet account for agents who interact with a relatively limited number of actors. Miller and Page argue that the way forward is to pay attention to the networks of communications agents use to interact (*ibid.*, p. 222). For a complex systems approach to financial regulation, this means paying attention to international and domestic institutions and other means through which financial actors and governments interact and develop governing norms within the regulatory landscape described by Chiu.

A complete study of who participates in the creation of the "law" of global finance is beyond this chapter's scope, but it could be argued that the "system" which has emerged from the interactions of the various regulatory agencies and other actors has the hallmarks of complex systems, including the epistemic challenges such systems pose, and each component of the system has given rise to a large literature. It is possible to give only a cursory treatment of two, albeit important aspects of financial governance: the participation of states in the formation of international financial norms and the influence of civil society.

The participation of states in financial rule-making – the G20

Before the financial crisis, to the extent states were interested in the international coordination of financial policy, rule-making was dominated by the rich states of North America, Europe, and Japan, through unilateral domestic policy, their dominant positions in multilateral organisations such as the International Monetary Fund (IMF), and the exclusive nature of groups such as the G7 and Financial Stability Forum.[4] That "equilibrium" was undergoing change even before the crisis. The growing economic strength of economies such as Brazil, Russia, India and, in particular, China, was already putting pressure on the status quo, as illustrated by the calls for quota and voice reforms in the IMF to better reflect that growth. The IMF, in turn, was responding to challenges to its legitimacy and relevance, as countries like China began to retain vast foreign currency reserves of their own, in part, to avoid the structural adjustment policies that accompany IMF assistance. These developments were also influenced by criticism of international institutions, punctuated by protests against globalisation, in general, that shut down the 1999 World Trade Organization's ministerial meeting in Seattle and that accompany meetings of the IMF, the World Bank, and the G20 itself.

If nothing else, the crisis of 2008 and 2009 made it difficult if not impossible to exclude the emerging economies from the table (Torres, 2009). The G20 was identified as the organisation within which major decisions about the international financial system are to be made. The Financial Stability Forum was disbanded and replaced by a larger Financial Stability Board. Furthermore, the G20 supported the quota and voice reforms contemplated by the IMF, the most recent of which came into effect in 2015. Finally, the G20 evolved from a gathering of finance ministers to a meeting of heads of state.

What does such wider participation mean ten years after the crisis? The G20 is said to represent two-thirds of the world's population, four-fifths of global gross domestic product, and three-quarters of world trade (G20, 2016).[5] It is fair to say that the G20 has been recognised as the primary forum for international cooperation in international financial governance, if only because it is not tenable for the G7 or Organisation for Economic Co-operation and Development (OECD) countries to fashion rules without the participation of China (Jorgensen and Strube, 2014, p. 9). Beeson and Bell (2009, pp. 72–73) are sympathetic to arguments that the G20 may have begun as a sop to emerging economies, and that the G7 countries, particularly the United States, will continue to try to exert its influence in this forum, but they point out that the heightened sense of vulnerability in the financial system during the crisis and structural change in the wider economy, in particular, changing economic power among nations, will create incentives for cooperation. Indeed, Cammack (2012) argues that the emergence of the G20 marks a shift in the centre of gravity of the global economy from the north to the south and east.

Greater participation, however, may not be without its costs. As the G20 was emerging as the group that would address the crisis, Mosely and Singer (2009, p. 425) worried that "the sheer size of the group, as well as its diversity of interests, domestic environments, and development levels, will render agreements difficult" (*ibid.*). Failure to reach consensus would, in turn, call into question the G20's effectiveness, with negative impacts on legitimacy. This fear has been borne out to some extent, at least according to critics such as Gür (2015). Deadlock at the G20 level could encourage the development of regional standards and bilateral negotiations. (Mosely and Singer, 2009, p. 425). The issue then becomes determining the circumstances under which the proliferation of institutions "fosters regulatory convergence, and the circumstances under which this proliferation generates centrifugal pressures that lead to regulatory fragmentation" (*ibid.*, pp. 424–45). This, in turn, raises the question when convergence is desirable and when fragmentation is.

Another question is whether the G20 is vulnerable to the same attacks on legitimacy experienced by the G7 (Postel-Vinay, 2014, pp. 24–47; Ciceo, 2010, p. 124). In this respect, Payne (2008) argued that the exclusive nature of the G7 meant that it failed the three tests of political legitimacy proposed by Sohn, 2005, p. 1) (1) "inclusiveness," referring to wide participation of a relevant group in decision-making; (2) "rule-governance," that policymaking and implementation be guided by rules; and (3) "fair return," that there be a fair sharing of costs and benefits by all who are expected to comply" (Sohn, 2005, pp. 489–90, cited by Payne, 2008, p. 527). Sohn himself argued five years after the global response to the Asian debt crisis that the G7 and G20 were too exclusive, had unclear rules of decision-making, and often came up with solutions that focused more on the domestic economies of Asia rather than reforming international aspects of the system that contributed to the Asian crisis (Sohn, 2005, pp. 493–95).

Some ten years later, it is unclear whether these perceived shortcomings have been overcome. To be sure, a strong argument can be made that the G20 has now replaced the G7 as the primary venue for the deliberation of international financial policy. But the rules for membership in the G20 remain unclear, as are its processes for decision-making. Far more states are not members of the G20 than members so that many states continue to be rule-takers instead of rule-makers. This issue has become more pressing, as the G20 has expanded its portfolio to include issues such as climate change, gender equality, migration, and food security (G20, 2017). Yet as Boulle (2009, p. 107) points out, those very states are essential to international governance because eventually, it is the legislatures of those states which implement international norms through domestic law. Can they be expected to do so if they were not at the table in the first place?

The development of newer organisations such as the Global Governance Group, or 3G, underlines these concerns and demonstrates that the dynamics set in motion by inclusion and exclusion cannot be ignored. The 3G comprises 30 small and medium countries not members of the G20 (Global Governance Group, 2016).[6] Cooper and Momani (2014) argue that the 3G has emerged as a response to the exclusive nature of the G20. However, in their view, instead of serving as a venue for resistance to the G20, the 3G serves as a bridge between the G20 and the rest of the world's states, resulting in greater legitimacy for the G20. Cooper and Momani note that in the years following the crisis, several of the 3G countries had been identified by the G20 and OECD countries as tax and corporate havens, unfairly, in those countries' view. Those countries succeeded in changing the terms of the debate by insisting on the primacy of the UN as the institution for international financial cooperation, whilst inserting itself into the G20 process by consulting with G20 countries and proposing technical solutions to the issue (*ibid.*, pp. 221–22). Since then, the G20 continues to consult with the 3G about issues of common concern.

Even this cursory discussion of the issues facing the G20 underscores the complex and changing nature of the norms of inclusion as they play out on the international level. The G20 is newer version of an older tool for cooperation, the G7, which, in turn, was formed with the hope that decisions could be facilitated by the advantages of relatively small size. The G20 can be seen as a response to the problem of exclusivity, and a vindication of the value of inclusion, but ironically retains enough of the features of the former mechanism that concerns about its legitimacy persist, primarily because of the exclusion of the rest of the international community. Moreover, that very inclusion could undermine the nimbleness enjoyed by smaller groups. The emergence of organisations such as the 3G highlight the irony that every "solution" to the problem of international governance raises the same set of issues. And it is not just that any action will cause an equal and opposite reaction, or that any system of decision-making will give rise to tensions between inclusion and efficiency. The G20 is just one manifestation of even deeper, cross-cutting impulses at the international level that also implicate norms of inclusion: on one hand, that under international law, states are equal and independent and, on the other, that some states are more powerful than others. These two impulses almost guarantee that the interactions of states will be nonlinear and complex.

Citizen feedback on the impacts of international financial rules

States do not find it easy to participate in the formation of international financial norms, and more participation is not without cost. However, a complexity theorist would likely argue that state inclusion alone is not enough to create a resilient financial system. The technical nature of finance and the potential for industry capture ensures that the number of actors who influence financial policymaking will be low. However, if policy decisions are to be evaluated properly, there must be ways ordinary people can provide feedback on how financial norms such as a capital adequacy standard or an austerity measure, have an impact on their lives, even if such feedback is blunt.

There is a long way to go. Baker (2009) argues that financial policymakers are members of a closed, epistemic community that has difficulty hearing and responding to voices from outside who have wider interests (*ibid.*, p. 196). States themselves are represented by persons who share the same ethos (*ibid.*, p. 211). Since this cuts across countries it is not enough that groups such as the G20 or even the 3G include more states since this only results in adding persons who share the same world view.

To be sure, groups including the G20 have begun to consult with segments of civil society, and in this vein, Germain (2010) is somewhat hopeful that broader public participation in international financial regulation is possible. He argues that a *prima facie* case can be made that the decentralisation of power represented by the shift to the G20 has enabled a nascent public sphere in the area of financial governance. This sphere is constituted by an interstate institutional framework, financial markets, an active media, and the international activities by civil society (*ibid.*, p. 504). Germain finds evidence of such a public sphere in acknowledgements of heads of international financial organisations that the world financial system is a public good; the Highly Indebted Poor Countries Initiative, a product of strong pressure from civil society; and in the way in which public participation was invited in the development of the Basel II capital adequacy standards (*ibid.*, pp. 504–505). This public sphere remains fragile (*ibid.*, p 508; see also Scholte, 2013; Prache, 2015), but Germain believes the public will continue to find ways to make its voice heard (Germain, 2010, pp. 508–509). In a similar vein, Kastner (2013) argues that the adoption of consumer-friendly policies, resulting in the Consumer Financial Protection Bureau in the United States and initiatives from the European Supervisory Authorities in the European Union, can only be explained by a greater willingness of international and domestic financial institutions to receive input from civil society.

Recent political developments in the United States and elsewhere show how fitful progress can be, but if Germain and Kastner are right that a nascent public sphere is emerging in international financial governance, this is another example of how complex systems provide both challenges to and new opportunities for furthering the norms of inclusion, as well as how those norms might manifest themselves. On one hand, to the extent the financial regulatory landscape is decentralised, that very landscape could make it hard for persons and institutions outside the transnational financial community to participate in decision-making. On the other hand, the same landscape allows Germain's public sphere to emerge and thus perhaps have an impact on financial decisions taken at the international level.

This, however, says nothing about the adequacy of such modalities for public participation. One could argue that to the extent there is any public sphere in international finance, it raises its own issues of inclusion and exclusion, such as the ad hoc nature of some debates, the ability of developing countries to participate in civil society, and so on (Kelley, 2011, pp. 543–46). Indeed, to refer to Balkin's concept of bricolage, we would expect that the multifunctional "uses" of the public sphere itself would have unintended consequences that will shift the way that sphere is understood and structured. Complexity theory would argue that any structure for participation and our ways of conceiving of such participation are always in process and thus must always be reassessed.

Conclusion

Complexity theory has not made the job of governance any easier. If the theory is correct, the great challenges to governance, including the international finance system and the broader economy, are the emergent results of myriad interactions among people and the institutions they have created that both constrain and enable social life. Such phenomena are incapable of complete description and their paths are unpredictable, thus making governance possible only in the short run and requiring frequent re-evaluation. To be sure, complexity theory does not consign society to inaction – people have been successful members of dynamic, complex systems long before any theory emerged to articulate them. But the theory gives theoretical grounds for scepticism of any policies or structures of governance that purport to be applicable for all time and in all places. It might also serve as the basis for a context-specific way of understanding and implementing the norms of procedural and substantive inclusion that have come to inform our ideas of legitimate governance. We would expect those norms to play themselves out differently at the different levels of organisation and in different contexts, even though we would expect to see a family resemblance between the various instantiations of those norms. The theory tells us the environments to which governance responds and the mechanisms, conceptual or otherwise, that governance uses in such responses, are always interacting and co-evolving. Thus, as discussed earlier, the theory suggests that we should be prepared to engage in a constant re-evaluation and renegotiation of the terms of inclusion in governance at every level.

Notes

1 An earlier version of this chapter appeared in Boulle (2011). My thanks to Maria Luisa Hernandez Juarez for her helpful research assistance.
2 There are many accounts. Among them are Financial Crisis Inquiry Commission (2011), Rajan (2010), and International Bar Association (2010).
3 One of the best discussions of the applications of complexity theory in various scholarly fields is found in Mainzer (2007).
4 Again, the literature here is enormous. For a discussion of the power of Western countries in impacting IMF policy, see (Copelovitch, 2010). Payne (2008) discusses the G7 and the challenges to its legitimacy. Somewhat ironically, the G20 now plays an important role in endorsing policies developed by multilateral financial institutions like the IMF. (Eccleston et al., 2015).
5 The G20 membership consists of Argentina, Australia, Brazil, Canada, China, France, Germany, India, Indonesia, Italy, Japan, Mexico, Russia, Saudi Arabia, South Africa, South Korea, Turkey, the United Kingdom, the United States, and the European Union. (G20, 2017).
6 As of the 2016 Ninth Ministerial Meeting, the members of the Global Governance Group are Bahrain, Barbados, Botswana, Brunei, Chile, Costa Rica, Finland, Guatemala, Jamaica, Kuwait, Liechtenstein, Luxembourg, Malaysia, Monaco, Montenegro, New Zealand, Panama, Peru, the Philippines, Qatar, Rwanda, San Marino, Senegal, Singapore, Slovenia, Switzerland, the United Arab Emirates, Uruguay, and Viet Nam (ibid.).

Bibliography

Allen, Peter M., Mark Strathern and Liz Varga (2010), 'Complexity: The Evolution of Identity and Diversity', in P Cilliers and R Preiser (eds.) *Complexity, Difference and Identity: An Ethical Perspective* (Dordrecht: Springer), 41.
Alter, Karen J. and Sophie Meunier (2009), 'The Politics of International Regime Complexity', 7(1) *Perspectives on Politics* 13.
Arthur, W. Brian (2014), *Complexity and the Economy* (Oxford: Oxford University Press).
Ayadi, Rym and Patrick Behr (2009), 'On the Necessity to Regulate Credit Derivatives Markets', 10(3) *Journal of Banking Regulation* 179.
Baker, Andrew (2009), 'Deliberative Equality and the Transgovernmental Politics of the Global Financial Architecture', 15(2) *Global Governance* 195.
Balkin, Jack M. (1998), *Cultural Software: A Theory of Ideology* (New Haven: Yale University Press).
Baxter, Lawrence G. (2011), 'Adaptive Regulation in the Amoral Bazaar', 128(2) *South African Law Journal* 253.
Beeson, Mark and Stephen Bell (2009), 'The G-20 and International Economic Governance: Hegemony, Ccollectivism, or Both?' 15(1) *Global Governance* 67.
Boucher, David (1998), *Political Theories of International Relations: From Thucydides to the Present* (Oxford: Oxford University Press).
Boulle, Laurence (2009), *The Law of Globalization: An Introduction* (The Netherlands: Kluwer Law International).
Boulle, Laurence (ed.) (2011), *Globalisation and Goverance* (Capetown: Syber Ink).
Brooks, Chris (1998), 'Chaos in Foreign Exchange Markets: A Skeptical View', 11(3) *Computational Economics* 265.
Cammack, Paul (2012), 'The G20, the Crisis, and the Rise of Global Development Liberalism', 33(1) *Third World Quarterly* 1.

Carline, Anna and Zoe Pearson (2007), 'Complexity and Queer Theory Approaches to International Law and Feminist Politics: Perspectives on Trafficking', 19(1) *Canadian Journal of Women and the Law* 73.

Chen, Shu-Peng and Ling-Yun He (2010), 'Multifractal Spectrum Analysis of Non-linear Dynamical Mechanisms in China's Agricultural Futures Markets', 389(7) *Physica A: Statistical Mechanics and Its Applications* 1434.

Chiu, Iris H-Y. (2010a), 'Enhancing Responsibility in Financial Regulation – Critically Examining the Future of Public – Private Governance: Part I', 4(2) *Law and Financial Markets Review* 170.

Chiu, Iris H-Y. (2010b), 'Enhancing Responsibility in Financial Regulation – Critically Examining the Future of Public – Private Governance: Part II', 4(3) *Law and Financial Markets Review* 286.

Ciceo, Georgiana. (2010), 'Reshaping the Structures of Global Governance. What Lessons Are to Be Learned from the Latest Financial Crisis', 2 *Analele Universității din Oradea. Relații Internaționale și Studii Europene* 116.

Cilliers, Paul (1998), *Complexity Theory and Postmodernism: Understanding Complex Systems* (New York: Routledge).

Cilliers, Paul (2010), 'Difference, Identity and Complexity', in Paul Cilliers and Rika Preiser (eds.) *Complexity, Difference & Identity: An Ethical Perspective* (Dordrecht: Springer), 3.

Cooper, Andrew F. and Bessma Momani (2014), 'Rebalancing the G-20 from Efficiency to Legitimacy: The 3G Coalition and the Practice of Global Governance', 20(2) *Global Governance* 213.

Copelovitch, Mark S. (2010), 'Master or Servant? Common Agency and the Political Economy of IMF Lending', 54(1) *International Studies Quarterly* 49.

Dieci, Roberto and Frank Westerhoff (2011), 'A Simple Model of a Speculative Housing Market', 22(2) *Journal of Evolutionary Economics* 303.

Dunn, John (2005), *Democracy: A History* (New York: Atlantic Monthly Press, 1st American ed.).

Eccleston, Richard, Aynsley Kellow and Peter Carroll (2015), 'G20 Endorsement in Post Crisis Global Governance: More Than a Toothless Talking Shop?' 17(2) *British Journal of Politics and International Relations* 298.

Financial Crisis Inquiry Commission. (2011), *The Financial Crisis Inquiry Report* (Washington, DC: U.S. Government Printing Office).

G20.(2016),'Prioritiesofthe2017Summit'<www.g20.org/Content/DE/_Anlagen/G7_G20/2016-g20-praesidentschaftspapier-en.pdf?__blob=publicationFile&v=2>, accessed 11 July 2017.

G20. (2017), 'G20 Germany 2017 Hamburg: Shaping an Interconnected World' [online]. Berlin: Press and Information Office of the Federal Government <www.g20.org/Content/DE/_Anlagen/G7_G20/2016-g20-broschuere-bpa-en.pdf?__blob=publicationFile&v=4>, accessed 11 July 2017.

Germain, Randall (2010), 'Financial Governance and Transnational Deliberative Democracy', 36(2) *Review of International Studies* 493.

Geyer, Robert (2003), 'Beyond the Third Way: The Science of Complexity and the Politics of Choice', 5(2) *British Journal of Political Science* 237.

Geyer, Robert and Samir Rihani (2010), *Complexity and Public Policy: A New Approach to 21st Century Politics, Policy and Society* (Abingdon, OX: Routledge).

Global Governance Group. (2016), *Press Statement by the Global Governance Group (3G) on Its Ninth Ministerial Meeting in New York* [press release] 22

September 2016 <www.mfa.gov.sg/content/mfa/media_centre/press_room/pr/2016/201609/press_20160923.html>, accessed 11 July 2017.

Gür, Nurullah (2015), *The G20 and the Governance of Global Finance* (Istanbul: SETA Foundation for Political, Economic and Social Research).

Habermas, Jürgen (1996), *Between Facts and Norms*, trans William Rehg (Cambridge, MA: MIT Press).

International Bar Association. (2010), 'Interim Report: Preliminary Views on the Financial Crisis', 4(3) *Law and Financial Markets Review* 251.

Jorgensen, Hugh and Daniela Strube (2014), 'China, the G20 and Global Economic Governance' [pdf] (Sydney: Lowry Institute for International Policy) <www.lowyinstitute.org/publications/china-g20-and-global-economic-governance> accessed 11 July 2017.

Kanas, Angelos (2003), 'Nonlinear Forecasts of Stock Returns', 22(4) *Journal of Forecasting* 299.

Kanas, Angelos (2005), 'Nonlinearity in the Stock Price-Dividend Relation', 24(4) *Journal of International Money and Finance* 583.

Kastner, Lisa (2013), *Transnational Civil Society and the Consumer-Friendly Turn in Financial Regulation* [pdf] (Paris: Max Plank Sciences Po Center on Coping with Instability in Market Societies) <www.maxpo.eu/pub/maxpo_dp/maxpodp13-2.pdf>, accessed 11 July 2017.

Kavalski, Emilian (2008), 'The Complexity of Global Security Governance: An Analytical Overview', 22(4) *Global Society* 423.

Kelley, Claire R. (2011), 'Financial Crises and Civil Society', 11(2) *Chicago Journal of International Law* 505.

Kunneman, Harry (2010), 'Ethical Complexity', in Paul Cilliers and Rika Preiser (eds.) *Complexity, Difference and Identity: An Ethical Perspective* (Dordrecht: Springer), 131.

Kyrtsou, Catherine and Walter C. Labys (2006), 'Evidence for Chaotic Dependence Between US Inflation and Commodity Prices', 28(1) *Journal of Macroeconomics* 256.

Lubell, Mark (2013), 'Governing Institutional Complexity: The Ecology of Games Framework', 41(3) *Policy Studies Journal* 537.

Mainzer, Klaus (2007), *Thinking in Complexity: The Computational Dynamics of Matter, Mind, and Mankind* (Berlin: Springer 5th ed.).

Miller, John H. and Scott E. Page (2007), *Complex Adaptive Systems: An Introduction to Computational Models of Social Life* (Princeton, NJ: Princeton University Press).

Mosely, Layna and David A. Singer (2009), 'The Global Financial Crisis: Lessons and Opportunities for International Political Economy', 35(4) *International Interactions* 420.

Park, Joongwoo Brian, Jeong Won Lee, Jae-Suk Yang, Hang-Hyun Jo and Hie-Tae Moon (2007), 'Complexity Analysis of the Stock Market', 379(1) *Physica A: Statistical Mechanics and its Applications* 179.

Payne, Anthony (2008), 'The G8 in a Changing Global Economic Order', 84(3) *International Affairs* 519.

Postel-Vinay, Karoline (2014), *The G20: A New Geopolitical Order* (New York: Palgrave Macmillan).

Prache, Guillaume (2015), 'The Role of Civil Society in EU Financial Regulation', in Pablo Iglesias-Rodríguez (ed.) *Building Responsive and Responsible Financial Regulators in the Aftermath of the Global Financial Crisis* (Cambridge: Intersentia), 185.

Preiser, Rika and Paul Cilliers (2010), 'Unpacking the Ethics of Complexity: Concluding Reflections', in Paul Cilliers and Rika Preiser (eds.) *Complexity, Difference and Identity: an Ethical Perspective* (Dordrecht: Springer), 265.

Rajan, Raghuram G. (2010), *Fault Lines: How Hidden Fractures Still Threaten the World Economy* (Princeton, NJ: Princeton University Press).

Romano, R. (2014), 'Regulating in the Dark and a Postscript Assessment of the Iron Law of Financial Regulation', 43(1) *Hofstra Law Review* 25.

Sawyer, R. Keith (2005), *Social Emergence: Societies as Complex Systems* (Cambridge: Cambridge University Press).

Scholte, Jan A. (2013), 'Civil Society and Financial Markets: What Is Not Happening and Why', 9(1) *Journal of Civil Society* 129.

Sen, Amartya (2009), *The Idea of Justice* (Cambridge, MA: Harvard University Press).

Serletis, Apostolos and Periklis Gogas (1999), 'The North American Natural Gas Liquids Markets Are Chaotic', 20(1) *Energy Journal* 83.

Serletis, Apostolos and Mototsugu Shintani (2006), 'Chaotic Monetary Dynamics with Confidence', 28(1) *Journal of Macroeconomics* 228.

Sohn, Injoo (2005), 'Asian Financial Cooperation: The Problem of Legitimacy in Global Financial Governance', 11(4) *Global Governance* 487.

Torres, Heitor F. S. (2009), 'Relacionando o G-20 à governança global e à ordem mundial [Relating the G20 to Global Governance and to the World Order]', 47(106) *Meridiano* 52.

Westerhoff, Frank H. (2008), 'The Use of Agent-Based Financial Market Models to Test the Effectiveness of Regulatory Policies', 228(2–3) *Jahrbücher fur Nationalökonomie und Statistik* 195.

9 Complex regulatory space and banking

Michael Leach

The central purpose of this chapter is to consider how complexity might usefully be applied to the study of regulation and regulatory systems. As with any field, one must be cautious when cross-applying a new and unfamiliar perspective and its associated methodologies to something like regulation. One of the central claims made here is that in order for this to be a meaningful exercise, complexity needs to piggyback onto existing concepts from regulation scholarship in order to distinguish and make sense of important patterns of relationships within a regulatory environment. It proposes that the concept of the 'regulatory space' can be a useful handmaiden for this purpose and offers a new conceptual framework of the 'complex regulatory space' as the outcome of their union. It explores what a complex regulatory space would look like and uses banking as an example to demonstrate how it can highlight the relationships, dynamics and tensions that give shape to patterns of banking actor behaviours and their relationship to systemic phenomena. It also locates the place of law in regulation, one that is necessarily 'decentred' as a contextual feature of actor behaviour rather than a core structural feature of regulation itself. Overall, the tone of this chapter is meant to be both exploratory and circumspect. Applying complexity to something like regulation is far from straightforward, but what it might sacrifice in simplicity it can exchange for greater nuance and new insights into regulation as a complex social phenomenon.

Complexity and regulation

Complexity

Any exploration of complexity's applicability to new fields needs to start with some conceptual delineation of what 'complexity' is. Complexity research studies large-scale, multi-bodied composite structures that have interacting, networked components whose organisation is neither random nor centrally controlled and whose manifold and variable relational behaviours collectively produce observable, large-scale patterning effects throughout the whole (Mitchell, 2009, p. 12; Schelling, 2006, p. 21, Auyang, 1998, p. 4). Studying complex systems involves explaining the large-scale dynamics of such entities in terms of the patterning and structuring effects of the interacting behaviours of their component parts. Explanations of complex phenomena try to identify linkages between systemic phenomena and the dynamic interactions of individual constituents operating within a given environment. Doing so appreciates different scales and levels of analysis, requiring sensitivity to different concepts depending on what is being described (Auyang, 1998, p. 5). Being able to see the forest for the trees requires understanding how they relate to one another at different levels of organisation and interaction. Thus, complexity scholars distinguish between system phenomena at the micro-level of individual interactions; at the middle or meso-level, looking at interactions of collective entities; and at the macro-level where the properties of a system as a whole are evident and describable in terms of system concepts (Auyang, 1998, p. 15).

These levels are connected by processes of 'emergence' and 'feedback'. Complex system properties are said to 'emerge' when the aggregated effects of micro-level interactions produce observable patterns at the meso- and macro-levels, which may feed back to the micro-level to further alter micro-contexts of future individual component behaviours (Arthur, 1999, p. 108). 'Risk vulnerability' in banking, for instance, can be understood as either a meso-level emergent property of the actions of groups of actors within a banking system, or as a macro-level feature of the system itself, emerging from the actions and interactions of all actors and collectives in a given banking system. This emergent property can be seen as feeding back into the system as micro-level banking actors get an impression of systemic risk vulnerability and change their behaviour and interactions with one another as a result. Complex systems are said to change when the aggregate effect of sufficient changes in the interactive behaviours of system components alter the meso- and macro-level characteristics of the whole.

This multilayered study of complex systems does not imply hierarchical structures or normative orderings of a given system. Rather, it appreciates that complex phenomena can be studied at different degrees of abstraction. It tries to capture the whole by tracing the mechanisms that connect them, explaining macro-level system phenomena in terms of the micro-level mechanisms that produce them, not in terms of linear causation but in terms of causal relations between aggregate and variable component behaviour and system characteristics. Complex social systems like regulation are usually prone to great behavioural variability because of the agency of actors to choose from a range of possible behavioural responses to their contextual environments. This they do by employing variable forms of reason and logic, belief and faith and gut feelings to craft uncertain hypotheses about the world around them which they test, retain, alter and/or discard over time through their experience (Durlauf and Young, 2001, p. 1), accepting that any full accounting of the states of their manifold components at any given moment is empirically intractable (Merali and Allen, 2011, p. 32; Auyang, 1998, p. 236). Because of this variability, complexity studies accept that any full accounting of the states of a system's manifold components at any given moment is empirically difficult (if not impossible), but also unnecessary to explain the whole (Merali and Allen, 2011, p. 32; Auyang, 1998, p. 236).

Bringing complexity to regulation

Given the preceding, a study of regulation as a complex phenomenon, therefore, would seek out multi-layered descriptions that link the characteristics and processes of regulatory systems to interactive individual behaviour of constituent actors within them.[1] There is nothing inherent in complexity or regulation to indicate how exactly this might be done. Much regulation scholarship provides relatively narrow, un-complex descriptions of centralised bureaucratic structures operating along hierarchical normative and accountability frameworks that seek to achieve specific behavioural controls in society (Morgan and Yeung, 2007, p. 3; Ogus, 1994). A complexity perspective, though, demands a broader lens. Some regulation scholars have used it to explain regulation policy as an emergent outcome of co-evolving policymaking and economic systems (Cherry, 2007, p. 379). Others have seen regulation as external, state-led interventions that manage complex, nonlinear natural and social natural resource 'systems' according to some linear cause–effect rationality (Mahon et al., 2008; Rammel et al., 2007). In both cases, regulation's reality is described not in terms of its normative or authority structures but in terms of patterns of interaction among a regulatory system's composite actors and institutions.

This chapter takes a similar path, demonstrating how banking can be can be understood as a complex phenomenon that 'emerges' from the systemic effects of the aggregate behaviours of interconnected and interdependent banks, regulators and other actors operating in a given marketplace or economy. Whether law in such a construct is an emergent characteristic of those same constituent actors or is an external framework imposed by the state, it is analytically 'decentred' (Black, 2001), becoming one of several contextual features that has variable effects on the idiosyncratic behaviours and interrelationships of many actors that constitute them. Indeed, while the business of banking may be subject to linear regulatory controls, banking systems themselves do not behave in linear fashions. While laws might give some structure to the system, they are not the system itself, and many other factors besides law bear on actor behaviours and interactions.

However, simply declaring that banking is a regulated complex system does not explain how banking works. While complexity draws our attention to how the whole relates to the individual in a regulatory system, by itself it offers little guidance about which relationships matter or why. It is unwise to assume that complexity will easily translate to the study of regulation. Regulation is a particular type of social phenomenon, and although complexity may shed new light on it, doing so without the benefit of concepts suited to the study of regulation might risk meaningless results. As stated earlier, though, much regulatory theory does not share the same ontological base of complexity and any fit would therefore be awkward at best.

There is, however, a somewhat niche concept in regulation scholarship, namely the 'regulatory space', that does closely align with complexity and that has potential to be grafted to it in useful ways. The 'regulatory space' is an intentionally open-ended and flexible concept. Unlike more dominant models of regulation that project a dichotomy between public authority and private interests, it instead appreciates regulation's interconnected, networked composition (Hancher and Moran, 1989). Regulatory space studies are typically critical of centralised, hierarchical notions of regulatory power, seeing instead power as something that is fluid and dispersed in ways that problematise the otherwise neat juridical division between public and private authority (Hancher and Moran, 1989, pp. 273–275). Regulatory space allows for descriptions of actor behaviour that go beyond mere analyses of deviance/compliance by describing regulatory behaviour as strategic responses to interactive environmental contexts that are conditioned or mediated by power distributions (Canning and O'Dwyer, 2013, pp. 174–175). These power relations shape a regulatory space in terms of how it is distributed and affects actor relations and behaviours that are driven by a strategic pursuit of control over resources (Scott, 2001, p. 331, 333). All activity in a space revolves around certain dominant regulatory 'issues', creating discursive practices that generate systemic regulatory outcomes (Young, 1994, p. 86). Regulatory space studies also differentiate themselves from most regulation scholarship by describing discursive practices that generate systemic outcomes in the form of regulatory 'problems' and their solutions, both arising out of regulatory space actor interrelationships (Young, 1994, p. 84). While law is appreciated as having some ordering effect in these processes, its role is ambiguous, and regulatory space scholars have approached it in a variety of ways: as something that is culturally determined (Hancher and Moran, 1989, p. 280); or bounded by political and economic context (Haines, 2003);[2] or as 'a normatively closed, but cognitively open, reflexive legal discourse', only one of many discourses that influence regulatory processes and relationships' (Lange, 2003, p. 419).

As a concept, regulatory space does have limitations, though. It tends to describe spatial arrangements and interactive processes only in terms of how they are produced by the arrangement of a regulatory space at a given period. It therefore does not easily lend itself to change descriptions and has difficulty handling questions of agency within regulatory spaces. Seeing actor behaviour as reproducing or being conditioned by the power and relational structures of the regulatory space leaves relatively little room to explain how actors themselves contribute to those same arrangements, why their behaviour can be variable, or how this produces change in regulatory spaces since doing so requires explaining how spatial structures change independently of the actors that compose them.

Complexity and regulatory space therefore have much to offer each other to meet their respective limitations. While regulatory space can provide some recognisable conceptual footing to complexity for the study of regulation, complexity, in turn, can offer regulatory space a multi-layered approach that can link variable individual actor behaviours to larger regulatory system-wide phenomena to explain regulatory change processes as resulting from the aggregation of variable and highly contextualised micro-level actor behaviour. In the next section, this is demonstrated in the example of a 'complex regulatory space' for banking that makes sense of banking as a dynamic, changing and evolving regulatory phenomenon, described using both complexity and regulatory space in terms of actor behaviour and relationships within contexts of power, discourse and resources.

Complex regulatory spaces for banking

This complexity and regulatory space combination can be guided by a shortlist of research questions: (a) how regulatory actors interact and to what effect; (b) how actor behaviour responds and contributes to discourses around regulatory 'issues'; (c) how aggregate alterations to actors, relationships, resources, and issue discourses produce patterns of change visible at the middle- or macro-levels of the regulatory space; (d) how such changes, in turn, produce further feedback change effects at different levels of the space; and (e) how those interactions are structured by actors' respective capacities to access and control resources. The remainder of this chapter takes these questions and sketches out what a complex regulatory space for banking might look like. It identifies three key discursive 'issues' that are of central importance for banking – 'risk', 'liquidity' and 'confidence' – and will highlight how banking's main actors, institutions and their interrelationships engage with and are structured by them, mindful of how these actors typically rationalise their contexts as they consciously or unconsciously choose how to behave and interact with one another (Schelling, 2006, p. 14). It then explores the regulatory space concept of 'resources' from a complexity perspective as things that regulatory actors engage or pursue in meaningful ways. The resulting depiction of banking as a complex regulatory space will mostly reflect a basic model of 'traditional' banking based on deposit collecting and lending. Modern banking practices, however, involve much more than that, especially when one brings finance into consideration. There is insufficient room in this chapter for a lengthier and more sophisticated modelling, so its purpose must be understood as being demonstrative rather than comprehensive.

Issues

Joni Young describes regulatory 'issues' as the 'problems' that are constructed discursively by actors within a regulatory space that shapes their behaviour in response (Young, 1994, p. 86). In a complex regulatory space these 'issues' are the topics around which most regulatory activity revolves. As such, they are analogous to 'attractors' familiar to complexity theorists, in the sense that banking actor behaviour can be understood as constantly negotiating these concerns (Goertzel, 1997, p. 5). This chapter argues that there are three primary attractor 'issues' that underlie all relevant activity within a complex banking regulatory space – 'Risk', 'Liquidity' and 'Confidence' – which, while normatively neutral, are given normative content by actors when they interpret their respective contextual environments through them. It is important to note that this treatment of the three will be rather different from how economists typically discuss them. While it is very common for economists to measure systemic 'risk' (Acharya et al.,

2017), 'liquidity' (Cetina and Gleason, 2015) or 'confidence' (Dailami and Masson, 2009) in banking, such measurements tend to equate quantification of the behaviours to their systemic effects. As will be seen, a complex regulatory space approach understands their systemic role in very different ways.

Risk

Risk is a concern for any regulated industry or sector (Black, 2010), but it is especially important for banking since the fundamental premise of banking is to engage risk in pursuit of future gain (Scholtens and Van Wensveen, 1999, p. 1247). The business of banking consists primarily of managing a risky balance between short-term liabilities of liquid deposit funds and long-term, illiquid loan portfolios.[3] Bank profitability traditionally arises from the ability to maintain solvency while leveraging liquid depositor funds out as interest-generating, but risky, loans. Depositors provide banks with funds in the form of current accounts in return for a (usually low) rate of interest in exchange for bank promises to return deposit funds as currency upon demand and to facilitate payments and financial transactions. This very basic three-way relationship among banks, depositors and borrowers is indicative of how financial assets in banking are fundamentally relational: they are essentially commitments by people to pay cash to others at certain moments in the future. Uncertainty and risk are inherent to any such commitment since any future cash payout will be contingent on the ability and willingness of those with a liability to pay at that time.

Risk is a regulatory 'issue' in the sense that banking actors interpret the scope and relative desirability of choices of action available to them in terms of risk. Risk structures and delimits the scope of possible action based on how actors interpret what is acceptable or undesirable risk and their variable willingness to engage it in pursuit of uncertain future gains. Banks are perpetually exposed to a variety of risks, such as: liquidity risk (the risk of having insufficient funds to finance their short-term liabilities), credit risk (the risk that their loan assets will default), operational risk (the risk of their employees behaving improperly), market risk (the risk that the market will go into a downturn and decrease future returns) and foreign exchange risk (the risk that the value of foreign currencies will fluctuate to the detriment of assets denominated in those currencies), to name a few. Risk management involves balancing the trade-off between risk and profit. Banks routinely calculate their exposure to risk, but how much risk they will accept is subjective and contingent on that bank's business strategy and sense of security (Heffernan, 2005, p. 105). For example, a cautious bank that chooses to lower its credit risk by lending only to borrowers who are extremely unlikely to default effectively shrinks the pool of available borrowers and denies itself the opportunity to generate higher returns by lending at high interest rates to riskier borrowers (Minsky, 2008, p. 265). Bank choices on how to strike this balance will be informed by their desire for profits, their understanding of their particular contexts, socially prescribed behavioural conventions, mandatory legal requirements, and past experience.

Risk informs the relationships of banks with one another as well as with state regulators, depositors, and borrowers. By depositing funds in a bank, depositors implicitly accept the risk of bank failure, however small it might be. Borrowers risk the consequences of not being able to repay their loans when they are due, usually by losing what they put up as collateral. These relationships will vary in different regulatory contexts, and different market conditions will provide different incentives to actors to engage in various ways with others, which, in turn, will produce different aggregate patterns of relations in an economy that can feedback to different effects in a complex regulatory space (Boot and Thakor, 2000, pp. 681–682).

The central concern of bank regulators is to manage risk by taking measures to ensure that both private and state-owned banks do business in ways that are good for the economy and in line with government policy (World Bank, 2012, p. 45). How regulators choose to pursue those objectives, though, can depend on variable political and economic power relations within a regulatory space. In general, though, governments are very concerned about banking risk for many reasons, not least because when banks lose their gambles and miscalculate their risky profit balancing act the political, social and economic effects can be immensely deleterious. It is because of this that governments charge their regulators with enforcing policy, legislation and regulatory measures to control bank behaviour (Minsky, 2008, p. 349). At the same time, though, regulators that are too heavy-handed and overly police a regulatory space can also push firms out by making banking un-lucrative (Murshed and Subagjo, 2002, pp. 251–252). Thus, a key policy dilemma for legislators and regulators is how to best manage systemic risk while striking the balance between surveilling an economy and fostering its competitive moneymaking potential (Harris et al., 2014; Gorton and Winton, 2014).

The behavioural ordering effect of risk as an attractor issue can also be appreciated at the meso- or macro-levels of the complex regulatory space. In this sense, though, rather than being a behavioural issue, risk becomes a system characteristic that emerges from the aggregate effect of the collective decisions made by banking actors as they act on their individual risk calculations. When aggregated, these give the system as a whole an emergent general risk footing. Thus, a banking system filled with banking actors who generally engage in high-risk lending will, as a whole, be exposed to more risk than one whose actors are more restrained (or 'prudent' in banking terminology). This emergent systemic risk can also feed back to the micro-level, such as the recognised phenomena of 'disaster myopia', where a banking system projects enough of an impression of health and low risk that it encourages banking actors to ignore or neglect risk contingencies (Honohan, 2000, p. 88). Government regulators are concerned about systemic risk in this aggregated sense but appreciate that its source lies in actor behaviour, and thus, regulatory restrictions can be imposed to limit the scope of permitted actor behaviour to prevent systemic exposure to risk. The 'Capital Asset Ratio', for instance, is a prudential regulatory measure that requires banks to hold a prescribed minimum level of liquid capital in reserve at all times, designed to prevent banks from lending excessively.

Liquidity

In traditional economic theory, liquidity is a concept that captures the ease by which a financial asset can be exchanged for something else without losing much or any of its value. Markets are considered 'liquid' when the financial assets that flow through them can be exchanged quickly and easily. Banks have long been seen as playing a critically important liquidity-providing intermediary role in economies by facilitating the exchange of financial assets (Minsky, 2008, p. 349). From a complex regulatory space perspective, though, liquidity is an attractor issue in the same way that risk is. It is a perpetual concern about which banking actors must make constant and continuous calculated business choices. And like risk, liquidity is an emergent meso- or macro-level characteristic of a complex banking regulatory space because of the aggregate effects of banking actor liquidity-based choices. Actors with too little liquidity (ie. little access to easily exchangeable assets) will find it difficult to find cost-effective financing to manage their payments and transactions. If many actors in a space are simultaneously unable to finance their payments, this can cause the economy as a systemic whole can slow downt. On the other hand, too much liquidity can be systematically problematic as well. Banks with access to too much cheap liquid capital can collectively cause an economy to over-inflate or unduly increase its overall risk exposure by allocating abundant loan capital to too many low-quality, risky investments.

The primary liquidity concern for state regulators is to ensure that bank practices are structured in ways that ensure there is enough liquidity in a market for funds to flow freely and easily throughout it. Bank concerns with liquidity at the micro-level are somewhat different. 'Traditional' banking consists of banks using depositor funds to finance loans elsewhere in the economy while maintaining sufficient internal liquidity to ensure that they can repay all of their short-term liabilities, like depositor withdrawals, at all times. Depositor funds are 'liquid' because banks can easily pool them together and issue them out as loans but become 'illiquid' once they are loaned out because banks cannot do anything further with them until they are repaid. Banks must ensure not only that loans will generate returns when they are repaid with interest but also that they have enough currency held in reserve to meet all of their short-term liabilities (Heffernan, 2005, p. 3). Because cash held in reserve generates no future earnings, banks have an incentive to leverage out as much as possible, but banks that over-leverage without enough cash in reserve become insolvent if they are suddenly unable to satisfy their liabilities. If depositor withdrawal demands unexpectedly spike, this can force banks with insufficient liquid capital on hand to quickly liquidate other assets, like by selling investment securities at a loss or by borrowing from other banks. Banks fail once they run out of options to stay solvent (Brunnermeier et al., 2009, p. 14). Liquidity is therefore a key concern for banks as their survival and profitability depend on their ability to manage this uneven cash flow between long-term illiquid loan assets and short-term liabilities. Liquidity is an attractor issue in the sense that balancing liquidity concerns is a perpetual aspect of banking actor decision-making, although how such balances are struck can vary from actor to actor.

Furthermore, as with risk, while liquidity has a positive, measurable dimension, it is also an intangible systemic property that emerges out of the interrelations of banks, regulators and others actors. It is something that banking actors continually observe, interpret and make calculated decisions about as they try to strike a desirable balance between opportunities and dangers inherent to banking business. This systemic liquidity effect emerges mainly from bank interconnectivity through the correspondent deposit accounts that they hold with one another. When banks facilitate payments between their clients they do not exchange actual cash. Instead, the banks credit and debit the respective deposit accounts that they hold with each other. Doing so effectively converts cash-based deposits into a form of liquid and easily tradable 'money' that can be exchanged instantaneously through accounting practices (Gai et al., 2007). This payment system, along with short-term inter-bank lending, is the primary way in which banks are interrelated and interdependent within a banking regulatory space. It is the ease by which this interconnectivity can facilitate payments throughout an economy from which the liquidity characteristic of the whole regulatory space emerges.

If the numbers and degree of bank interconnectivity are large enough, losses suffered by one or more banks can cascade through inter-bank networks causing system-wide 'contagion' (Allen and Gale, 2000) as more and more banking actors deleverage by making loss-making sales, driving down asset prices and depressing hopes of future returns (Brunnermeier et al., 2009, p. 5, 11). Indeed, some observers have noted that it is a bank's interconnectedness, rather than its size, that in times of crisis can make it systemically dangerous and 'too connected to fail' (Haldane and May, 2011). Regulators try to prevent this and maintain systemic liquidity levels by using a number of regulatory tools at their disposal to intervene and affect bank choices and their interrelations, such as with interest rate changes, government bailouts, or controversial forced creditor bail-ins (Capello and Ervin, 2010).

Thus, from the perspective of state regulators, systemic liquidity is generally desirable because it facilitates economic transactions and investment. However, it can be sustained only as long as individual banks trust that each are sufficiently capitalised and that their promises to honour their liabilities to one another are credible. If one bank loses that credibility it will have a difficult time convincing other banks to lend to it or make payments with it. While this is certainly bad news for a bank in bad times, it is also systemically problematic because banks are highly inter-dependent. Overall, liquidity levels in an entire economy can decrease if one or more banks run into trouble because it can generate emergent meso-level signals that feed back to the micro-level as other banks interpret its troubles as a predictor that bad times are coming. This, in turn, encourages them to take precautions by hoarding capital and increasing their own internal liquidity, liquidating assets and drawing down their correspondent deposit accounts at other banks (Gai et al., 2007, p. 156). If enough banks in a banking regulatory space take such measures to protect themselves, the emergent collective effect is often described as liquidity evaporating or disappearing from a marketplace.

Confidence

The British financier Sir Evelyn de Rothschild once famously claimed that 'the single most important commodity traded in the City of London is confidence' (de Rothschild, 2013). Banks can only function when depositors, borrowers, government regulators and other banks have sufficient confidence in them and their solvency. Depositors will only deposit their disposable funds in banks they trust, while banks will only convert risk-free, liquid deposit capital into long-term, illiquid, risky loans if they have confidence in the creditworthiness of borrowers. Banks cooperate with one another through inter-bank lending and payments systems because they have confidence in each other's commercial viability. The ultimate fear around which much bank risk management revolves is a massive loss of confidence, the nightmare prospect of mass withdrawals caused by widespread and sudden depositor fears about the security of their bank deposits. While 'bank runs' can start at individual banks, bank interconnectedness can cause chain reactions that can cripple an entire banking system (Diamond and Dybvig, 1983, p. 404).

Like risk and liquidity, confidence has an inherently relational character and manifests as both a micro-level behavioural concern as well as a meso-/macro-level systemic characteristic (Llewellyn, 2014, p. 225). Systemic confidence is the intangible aggregate of all (micro-level) banking actors' confidence impressions of the (meso/macro) whole. One cannot get an impression of meso- or macro-level systemic confidence in the banking system by looking at a single, poorly performing bank. However, that bank's troubles can have ripple effects on other banks with which it cooperates, and can signal and contribute to meso- or macro-level impressions about the overall security of banking system. Relationships break down when banking actors lose confidence in one another and in the system as a whole, which can endanger the entire structure of banking if it happens on a mass scale.

This confidence is fragile since it is subject to the bounded rationality of actors. An economic downturn can turn into a widespread crisis if it sparks a general loss in confidence, causing liquidity to dry up as banking actors take measures to protect themselves, effectively slowing down the economy (Brunnermeier et al., 2009, p. 3). Given the systemic danger of a sudden loss of confidence, a state may intervene with mechanisms like deposit insurance schemes to ease depositor fears but in doing so risks removing the disciplining effect that the fear of a bank run can bring that keeps banks cautious and prudent in their lending and investment choices (Demigüç-Kunt and Detragiache, 2002). Thus, while governments and their regulators are keen to ensure that their banking and financial systems enjoy sufficiently high confidence such that their integrity and coherence can be maintained, too much confidence can also be problematic if it inspires actors to take excessive risks, collectively spawning investment 'bubbles' that can distort a market and make it vulnerable (Scheinkman and Xiong, 2003).

The challenge facing regulators and policymakers, therefore, is to determine how much and what kind of information is needed to generate sufficient market-wide confidence that will keep investors investing, depositors depositing, and banks lending to borrowers and cooperating with one another, without causing adverse effects on competition. The 2008 global financial crisis demonstrated how confidence's contingent intangibility makes it difficult to manipulate, such that that even pumping billions of dollars of liquidity into a banking systems cannot guarantee the restoration of system-wide confidence (Swedberg, 2013, pp. 514–515).

Resources and power

While attractor issues give us a sense of the central dilemmas around which actor behaviour revolves in a regulatory space and can give a sense of the scope of behavioural choices available to actors, attractor issues are not enough to explain why actors end up behaving the ways they do within it. A central concern for any complex regulatory space study is the question of why it is that actors in that space behave and interact the ways they do and, thus, why the space takes on the shape that it does. Regulatory space scholarship typically answers this by characterising regulatory actor behaviour as competition for 'resources' in pursuit of gain. Colin Scott describes the regulatory space as a fragmented array of four 'resources' over which regulatory actors compete and that they strategically deploy, namely: wealth, information, legal authority, and organisational capacity (Scott, 2001, p. 334). Actors have variable access to these resources and variable capacities to use them, which accounts for formal and informal power differentials within a regulatory space, and this uneven distribution of power in turn explains the variability of actor behaviours and interrelations with one another (Scott, 2001, pp. 336–338).

This chapter cautiously adopts Scott's taxonomy of 'resources' for the complex regulatory space framework. Cautiously, because while conceptually it might seem obvious that regulatory actors would compete for resources in ways that reflect how powerful they are, in practice, the analytical application of these explananda is not always so clear-cut and actor behaviour is highly variable. Complex descriptions of the dynamics of a banking regulatory space can reveal how relationships among actors, resources and power are contextual, nonlinear, and contingent and can offer some explanation for why the competitive pursuit of resources can produce behavioural patterns that are variable and unpredictable. By highlighting how relationships between actors and the aggregate effects of their collective behaviours can shape a regulatory space by producing emergent systemic effects at different levels of collective abstraction, a complexity analysis can provide an account of how spaces are shaped and change in terms of those relational dynamics. But it can also show how the very meaning of a 'resource' in a regulatory space itself is contextual and contingent. Like jet fuel, 'resources' can provide actors with power in some circumstances if they are used in particular ways, but they can also be problematic and sometimes dangerous if used inappropriately in others. Thus, the meaningfulness of whatever actions regulatory actors take in a competitive complex regulatory space is necessarily subjective, contextual and contingent. When aggregated, though, they can provide a link between the meaningfulness of regulatory actor behaviour and the emergent systemic qualities of risk, confidence and liquidity detailed earlier.

Wealth

While on the surface the regulatory space conceptualisation of competition for 'wealth' can offer some obvious explanations for actor behaviour, a sensitivity to the complex dynamics and interrelations of actors within a regulatory space reveals considerable contingency and ambiguity in how 'wealth' relates to power within one. At first glance, wealth might seem an obvious resource that motivates banking competition in a modern capitalist economy. The standard paradigm of such competition, after all, is that of largely private banks[4] pursuing wealth as profit by competing to attract depositors and pursue profit-generating lending opportunities within a risky marketplace that is overseen by state regulators who try to influence that competition through law and regulation (Independent Commission on Banking, 2011, 153). While the degree and nature of bank competition may vary from one banking environment to another, it will always affect individual bank decisions about how to strategically structure their operations in their pursuit of profit and therefore will also shape the nature and effectiveness of regulatory interventions to control their behaviour (Awrey, 2013, p. 414).

Yet, while understanding actor motivations in terms of the pursuit of 'wealth' as motivation might seem obvious in banking, the concept of 'wealth' as power is ambiguous. One might normally assume that regulatory actors with greater 'wealth' wield greater power (Scott, 2001, p. 337), but banks are only 'wealthy' in the sense that they have ready access to deposit capital that they can leverage out as risky profit-generating loans. That capital, however, rarely belongs to the banks themselves. Banks have access to 'wealth' as deposit capital only if they can convince depositors to deposit their disposable funds with them and to not withdraw them *en masse*. While depositor choices are also motivated by the pursuit of personal gain, they are determined by a variety of contextual factors that banks can only partly control. Depositors are informed not only by personal choices to save or consume but also by meso- and macro-level impressions that feed back to them as information about the health or security of the economy or the confidence and trustworthiness of the banking sector. Aggregate patterns of depositor 'wealth' in the form of disposable funds deposited in banks that result from that process then shape a regulatory space and banking practice. In emerging economies especially, the importance of the latter cannot be taken for granted (Hasan et al., 2013).

A bank's 'wealth' is also conditioned by risk. The more that banks lend out their available capital, the more they expose themselves to risk, and therefore a bank's access to capital only makes it 'wealthy' and powerful relative to how it is used and within the context of the economy it is in. Indeed, the bigger and more leveraged a bank is, the more vulnerable it can be to economic downturns when more and more of its loans underperform or default. In this sense, bank 'wealth' is also intimately connected to the 'wealth' of those to whom they issue loans. Regulators are generally interested in ensuring that banks use their 'wealth' in ways that maintain sufficient confidence and liquidity in the economy to ensure stable growth because consequences for the economy as a whole are dire if they use it imprudently and recklessly. Their ability to intervene, though, is partially dependent on their own financial and material resources, i.e. regulator 'wealth' especially in terms of state budget allocations that are determined by external bureaucratic process and politics.

Information

When one looks at 'information' as a regulatory resource from a complexity perspective, it too becomes more ambiguous than it might initially seem. Economists, finance analysts and journalists, as well as central banks and credit-rating agencies exercise power through the information they produce and disseminate. It makes intuitive sense that better access to reliable information about markets or technology can better prepare banking actors to gain advantages over their competitors (Marquez, 2002, pp. 901–903). Thus, one can say that asymmetries of information access and control can create power differentials in a competitive space (Dell'Aricca, 2001). Banks exercise power over their borrowers by monitoring their performance, although how much they do this is a business choice that reflects their tolerance of risk and their optimism about the future of the economy (Gorton and Winton, 2002, pp. 38–31, 51–53; Winton, 1999). Similarly, regulators use information that banks are required by law to supply them with to monitor their business activities and their compliance with law and their licences.

At the meso- and macro-levels, though, information as power means something quite different in a complex regulatory space. The quantity, quality, type and flow of aggregate information through a complex regulatory space is fundamentally related to all actor behaviour. Indeed, behaviour *is* information in the way that it can signal things to others within the space and can structure relational patterns throughout it. For instance, strong corrective action taken by a regulator against certain banks can signal to other banks what standards of strictness they should expect from the regulatory in the future, while inaction will have the opposite effect (Leach, 2015, p. 108). Furthermore, a bank's struggles to stay solvent can signal to others that the economy as a whole might be in trouble. Regulators are particularly aware of this and are especially concerned about ensuring that information flows through an economy in ways that maintain systemic confidence. Again, however, information is double-edged. Manipulating information flows does not always generate desirable results. For example, when times are good, regulatory requirements that banks publicly disclose certain types of business information can install greater public confidence in a banking system by removing market uncertainty, but when times are bad, detailed information about banking operations can also have the opposite effect (Bar-Gill and Warren, 2008).

Taken together, complexity helps us appreciate that information flows in a competitive regulatory space can be fragmented and asymmetrical in ways that can simultaneously create opportunities for business advantage at the micro-level but with the potential to sow confusion and uncertainty at the meso-level of a marketplace, which, in the worst scenarios, spawns dangerous herd activity in difficult, panicky times (Schwarcz, 2009, pp. 220–225). Aggregated micro-level individual actions, like panicky withdrawals, can also produce meso- or macro-level information flows that feed back in the form of impressions of the state of the market, which further shapes action responses by others elsewhere in the regulatory space. In this way, tracking information flows in a complex regulatory space can be challenging, not only because that flow is omnidirectional but also because the nature of how and where information flows depends on how regulatory actors are organised and interact, how they collectively process and redistribute it throughout the space, how they control access to it, and how subjectively receptive they are when it comes to them. As such, one cannot treat information in banking as a constant, or as being evenly distributed, or assume that actors have equivalent interests in or needs for it or will interpret it the same way.

Legal capacity

Unlike more legalistic accounts of regulation that assume legal compliance and therefore privilege law as the central structuring device in a regulatory system (Ogus, 1994), a complex regulatory space analysis would be interested in the contingency of law's behavioural effects in terms of the variable ways it shapes actor behaviour and relationships and how it can confer power to some over others. Law will structure a regulatory space only to the extent that regulatory actors consider it when determining how to behave. This is not a constant and will vary from person to person and from context to context depending on a number of factors, like legal culture, standards of practice, past experience, as well as the strength and willingness of regulators to enforce it (Scott, 2001, p. 344).

Regulation scholarship typically associates legal power with the authority of the state (Morgan and Yeung, 2007, p. 4), and legislative choices about how much and what kind of legal authority should be provided to regulators emerge from political dynamics that revolve around the dilemma of how best to balance the benefits and detriments that regulatory intervention can have. That balance can be struck in many ways. In the most basic sense, regulators exercise power over banks by controlling the banking licence as legal permission to operate. 'Banks', in their most basic sense, are those organisations licenced to receive deposit funds that they invest in long-term loan contracts (Diamond and Dybvig, 1983, p. 402). The constant threat of revoking a licence is what gives a regulator the power to subject banks to stringent reporting requirements to demonstrate that they continue to meet basic standards (Marcus, 1984, pp. 564–565). However, the power and effectiveness of this disciplinary measure will always be contingent on how credible actors feel a regulator's threat is (Gorton and Winton, 2002, p. 90). In this sense, the power of law in a complex banking regulatory space is never fixed or absolute but, rather, is dependent on the ability of regulators to exercise their legal authority, which, in turn, depends on the effectiveness of the legal system, in general, and on other actors understanding and respect of that authority. As such, regulators in some countries can be entirely powerless to exercise their authority regardless of how much it may be formally prescribed by law (Demigüç-Kunt and Detragiache, 1998, p. 87).

Outside of the bank–regulator relationship, a general (i.e. aggregate, meso- and macro-level) absence of confidence in a country's legal system can also limit the effectiveness of formal legal measures that actors use to structure their relationships with one another. In places where judiciaries are untrustworthy or are known to be corrupt, banks may have little power to enforce provisions of their loan contracts, regardless of whatever legal rights they may formally enjoy to do so (Weil, 2011). Furthermore, low confidence in regulatory authority and powers of oversight can encourage bank actors to collude to provide regulators and the public with false information, such as in the case of the recent Libor scandal in the UK (Hou and Skeie, 2014). The analytical challenge for a complex regulatory space analysis is to understand how contextual pressures like these relate to patterns of actor relationships and emergent systemic characteristics of the space as a whole.

In this way, understanding 'legal capacity' as a resource in a complex regulatory space analysis 'decentres' law as the primary structural consideration for understanding how a regulatory system is shaped (Black, 2001). Doing so recognises both its double-edged quality, as well as its contextual nature, such that it is insufficient to assume that the mere presence of legal frameworks and institutions in a regulatory space will explain how or why it functions the way it does absent any reference to their relation to how actors behave and pursue their interests. Nor is it possible to understand the effect of a law on a regulatory space without understanding aggregate patterns of how actors relate to legal authority and how they use law to structure their relations among each other.

Organisational capacity

Finally, it is again intuitively obvious how superior organisational capacity might empower an actor in a competitive regulatory space. Better organised and capable actors, whether banks, regulators or borrowers, who enjoy high-quality and efficient internal procedures will be better able to pursue their objectives and gain advantage within the regulatory space. Yet, again, attention to complexity places organisational capacity in a relational perspective, such that rather than being an absolute quality, an actor's organisational capacity instead will be beneficial or detrimental depending on their positionality vis-a-vis others in the regulatory space. It is important to understand not only which actors have more or less organisational capacity at any given time but also what effects the ways that they try to achieve greater organisational capacity has on the regulatory space as a whole. Some forms of organisation will be mandated by formal rules, like regulatory prescriptions on corporate governance structures, while others will be idiosyncratic to individual actors, reflecting their business culture and strategy, but they can also be related to contextual factors like history, competitive dynamics in the market and disposable wealth, among others. Investing resources in building internal organisational capacity or doing so by engaging in external cooperative ventures with others are business decisions that must respond to a given bank's strategy, needs and contextual circumstances. Aggregate patterns of organisational structures and capacities among actors within a complex regulatory space are therefore the result of a variety of factors specific to a given space. The systemic effect of those patterns can affect the general shape of a regulatory space, not only determining which actors might dominate but also by contributing to larger systemic phenomena, such as public confidence, systemic vulnerability to risk, and liquidity flows through a market.

Conclusion

The aim of this chapter was to consider what added value and challenges complexity can bring to the study of regulation. A key argument here has been that complexity on its own is agnostic about what matters for regulation and therefore should look for guidance in regulation theory. The proposal here is that complexity should graft itself onto the ontologically similar concept of the 'regulatory space'. Doing so can provide a robust platform upon which a complex analysis of regulation can be built, based on their shared concerns with how actor relationships and interdependencies structure large-scale social phenomena. Using banking as a demonstrative model or case study, this chapter showed that complexity can provide novel contributions to the regulatory space approach, most notably by fleshing out the structural implications of actor and institutional interrelatedness that is at its conceptual core. It does so, however, at the cost of increasing ambiguity around some of regulatory space's key analytical concepts. For one, it demonstrates how regulatory 'issues' act like attractors influencing actor behaviour at the micro-level, but distinguishable from emergent properties that manifest at the meso- and macro-levels, which require different conceptual tools to understand them. This explains how 'risk' can be both a behavioural concern for all banking actors in their day-to-day business affairs as well as an emergent systemic property of banking, but that the two require different approaches to properly understand their systemic role and signficance. While the former is understandable by grasping the variable contexts in which actors deal idiosyncratically with risk, the latter requires aggregation calculations and systemic change dynamics that complexity specializes in. Furthermore, while complexity benefits from the regulatory space understanding of competition for resources as explananda for regulatory actor choices and motivations, fitting the concept of the 'resource' into a complexity framework reveals how contextual resources are and how contingent is the power that can be associated with them. If one accepts the idea that competition over resources is a useful framework for understanding why regulatory spaces are shaped the way they are, from a complexity standpoint the role that resources play within the space is highly nuanced, and their value is double-edged depending on who uses them, how, where and when. Thus, while the marriage between the two is not seamless, bringing complexity theory together with regulatory space can be nevertheless fruitful for regulation research because the two complement each other's limitations in useful ways.

Notes

1 This analytical approach should not be confused with Julia Black's tri-layered model of regulatory regimes (Black, 2001). Where Black differentiates micro, meso and macro types of hierarchical governance, here the levels distinguish different degrees of aggregation within a regulatory space.
2 Haines describes her study as 'regulatory character' rather than 'regulatory space'; however, her understanding of regulation as arising out of the interactions of actors and norms in their local contexts is similar.
3 Modern banking, of course, involves far more than mere deposits and loans and deal with a plurality of different financial instruments, like guarantees, letters of credit or securitised debt financing, among others. See Johnson et al. (2003, p. 6) and Heffernan (2005, p. 102).
4 In the case of state-owned banks, loan capital is sourced from state allocations that are determined by political or bureaucratic processes, that may be equally competitive in different ways, but the pursuit of profit from loan interest is the same.

Bibliography

Acharya, Viral, Lasse Pedersen, Thomas Philippon and Matthew Richardson (2017), 'Measuring Systemic Risk', 30(1) *The Review of Financial Studies* 2.

Allen, Franklin and Douglas Gale (2000), 'Financial Contagion', 108 *Journal of Political Economy* 1.

Arthur, W. Brian (1999), 'Complexity and the Economy', 284 *Science* 107.

Auyang, Sunny Y. (1998), *Foundations of Complex-System Theories: In Economics, Evolutionary Biology, and Statistical Physics* (Cambridge: Cambridge University Press).

Awrey, Dan (2013), 'Towards a Supply-Side Theory of Financial Innovation', 41 *Journal of Comparative Economics* 401.

Bar-Gill, Oren and Elizabeth Warren (2008), 'Making Credit Safer', 157 *University of Pennsylvania Law Review* 1.

Black, Julia (2001), 'Decentering Regulation: Understanding the Role of Regulation and Self-Regulation in a "Post-Regulatory" World', 54 *Current Legal Problems* 103.

Black, Julia (2010), 'The Role of Risk in Regulatory Processes', in Robert Baldwin, Martin Cave and Martin Lodge (eds.) *The Oxford Handbook of Regulation* (Oxford: Oxford University Press), 304.

Boot, Arnoud and Anjan Thakor (2000), 'Can Relationship Banking Survive Competition?' 55(679) *Journal of Finance* 681–682.

Brunnermeier, Mark, Andrew Crockett, Charles Goodhart, Avinash D. Persaud and Hyun Shin (2009), 'The Fundamental Principles of Financial Regulation', Geneva Reports on the World Economy, International Center for Monetary and Banking Studies (ICMB) and the Centre for Economic Policy Research (CEPR), Geneva, 24 January 2009.

Canning, Mary and Brendan O'Dwyer (2013), 'The Dynamics of a Regulatory Space Realignment: Strategic Responses in a Local Context', 38 *Accounting, Organizations and Society* 169.

Capello, Paul and Wilson Ervin (2010), 'From Bail-Out to Bail-in', *The Economist* (London, 28 January 2010).

Cetina, Jill and Katherine Gleason (2015), 'The Difficult Business of Measuring Banks' Liquidity: Understanding the Liquidity Coverage Ratio', Office of Financial Research Working Paper 15–20, 7 October 2015.

Cherry, Barbara A. (2007), 'The Telecommunications Economy and Regulation as Coevolving Complex Adaptive Systems: Implications for Federalism', 59(2) *Federal Communications Law Journal* 369.

Dailami, Mansoor and Paul Masson (2009), 'Measures of Investor and Consumer Confidence and Policy Actions in the Current Crisis', World Bank Policy Research Working Paper 5007, July 2009.

de Rothschild, Evelyn (2013), 'Banking Must Pursue the Holy Grail of Confidence', *The Financial Times* (New York, 24 June 2013).

Dell'Aricca, Giovanni (2001), 'Asymmetric Information and the Structure of the Banking Industry', 45(10) *European Economic Review* 1957.

Demigüç-Kunt, Asli and Enrica Detragiache (1998), 'The Determinants of Banking Crises in Developing and Developed Countries', 45(1) *IMF Staff Papers* 81.

Demigüç-Kunt, Asli and Enrica Detragiache (2002), 'Does Deposit Insurance Increase Banking System Stability? An Empirical Investigation', 29 *Journal of Monetary Economics* 1373.

Diamond, Douglas and Philip Dybvig (1983), 'Bank Runs, Deposit Insurance, and Liquidity', 91 *Journal of Political Economy* 401.

Durlauf, Steven N. and H. Peyton Young (2001), 'The New Social Economics', in Steven N. Durlauf and H. Peyton Young (eds.) *Social Dynamics* (Washington, DC: Brookings Institution Press), 1.

Gai, Prasanna, Neil Jenkinson and Sujit Kapadia (2007), 'Systemic Risk in Modern Financial Systems: Analytics and Policy Design', 8 *Journal of Risk Finance* 156.

Goertzel, Ben (1997), *From Complexity to Creativity: Explorations in Evolutionary, Autopoetic, and Cognitive Dynamics* (New York: Plenum Press).

Gorton, Gary and Andrew Winton (2002), 'Financial Intermediation', National Bureau of Economic Research, Working Paper No 8928.

Gorton, Gary and Andrew Winton (2014), 'Liquidity Provision, Bank Capital, and the Macroeconomy' (Working Paper, University of Minnesota).

Haines, Fiona (2003), 'Regulatory Reform in Light of Regulatory Character: Assessing Industrial Safety Change in the Aftermath of the Kader Toy Factory Fire in Bangkok, Thailand', 12(4) *Social & Legal Studies* 461.

Haldane, Andrew G. and Robert M. May (2011), 'Systemic Risk in Banking Ecosystems', 496 *Nature* 351.

Hancher, Leigh and Michael Moran (1989), 'Organizing Regulatory Space', in Leigh Hancher and Michael Moran (eds.) *Capitalism, Culture and Economic Regulation* (Oxford: Clarendon Press), 272.

Harris, Milton, Christian C. Opp and Marcus M. Opp (2014), 'Macroprudential Bank Capital Regulation in a Competitive Financial System', University of Chicago Working Paper, 9 October 2014.

Hasan, Iftekhar, Krzysztof Jackowicz, Oskar Kowalewski and Lukas Kozlowski (2013), 'Market Discipline During Crisis: Evidence from Bank Depositors in Transition Countries', 37(12) *Journal of Banking & Finance* 5436.

Heffernan, Shelagh (2005), *Modern Banking* (New York: John Wiley & Sons).

Honohan, Patrick (2000), 'Banking System Failures in Developing and Transition Countries – Diagnosis and Prediction', 29 *Economic Notes* 83.

Hou, David and David Skeie (2014), 'LIBOR: Origins, Economics, Crisis, Scandal and Reform', Federal Reserve Bank of New York, Staff Report No 667, March 2014.

Independent Commission on Banking (2011), 'Final Report Recommendations' (September 2011).

Johnson, Neil F., Paul Jefferies and Pak Ming Hui (2003), *Financial Market Complexity* (Oxford: Oxford Scholarship Online).

Lange, Bettina (2003), 'Regulatory Spaces and Interactions: An Introduction', 12(4) *Social & Legal Studies* 411.

Leach, Michael (2015), 'From Recklessness to Prudence: A Study of Regulatory Space Change in Indonesian Banking from 1997 to 2008', MSt Thesis, University of Oxford, Oxford.

Llewellyn, David T. (2014), 'Reforming the Culture of Banking: Restoring Trust and Confidence in Banking', 1 *Journal of Financial Management, Markets and Institutions* 221.

Mahon, Robin, Patrick McConney and Rathindra N. Roy (2008), 'Governing Fisheries as Complex Adaptive Systems', 32(1) *Marine Policy* 104.

Marcus, Alan (1984), 'Deregulation and Bank Financial Policy', 8 *Journal of Banking & Finance* 557.

Marquez, Robert (2002), 'Competition, Adverse Selection, and Information Dispersion in the Banking Industry', 15 *Review of Financial Studies* 901.

Merali, Yasmin and Peter Allen (2011), 'Complexity and Systems Thinking', in Peter Allen, Steven Maguire and Bill McKelvey (eds.) *SAGE Handbook of Complexity and Management* (London: Sage Publications).

Minsky, Hyman (2008), *Stabilizing an Unstable Economy* (New York: McGraw-Hill, 2nd ed.).

Mitchell, Melanie (2009), *Complexity: A Guided Tour* (Oxford: Oxford University Press).

Morgan, Bronwen and Karen Yeung (2007), *An Introduction to Law and Regulation: Text and Materials* (Cambridge: Cambridge University Press).

Murshed, S. Mansoob and Djono Subagjo (2002), 'Prudential Regulation of Banks in Less Developed Economies', 20 *Development Policy Review* 247.

Ogus, Anthony (1994), *Regulation: Legal Form and Economic Theory* (Oxford: Clarendon Press).

Rammel, Christian, Sigrid Stagl and Harald Wilfing (2007), 'Managing Complex Adaptive Systems – A Co-Evolutionary Perspective on Natural Resource Management', 63(1) *Ecological Economics* 9.

Scheinkman, José A. and Wei Xiong (2003), 'Overconfidence and Speculative Bubbles', 111 *Journal of Political Economy* 1183.

Schelling, Thomas C. (2006), *Micromotives and Macrobehaviour* (New York: W. W. Norton & Company).

Scholtens, Bert and Dick Van Wensveen (1999, 'A Critique on the Theory of Financial Intermediation', 24 *Journal of Banking & Finance* 1243.

Schwarcz, Steven L. (2009), 'Regulating Complexity in Financial Markets', 87 *Washington University Law Review* 211.

Scott, Colin (2001), 'Analysing Regulatory Space: Fragmented Resources and Institutional Design', *Public Law* 329.

Swedberg, Richard (2013), 'The Financial Crisis in the US 2008–2009: Losing and Restoring Confidence', 11 *Socio-Economic Review* 501.

Weil, Laurent (2011), 'How Corruption Affects Bank Lending in Russia', 35(2) *Economic Systems* 230.

Winton, Andrew (1999), 'Don't Put All Your Eggs in One Basket? Diversification and Specialization in Lending', Financial Institutions Center Working Paper #00–16.

World Bank (2012), *Rethinking the Role of the State in Finance* (Washington, DC: World Bank).

Young, Joni J. (1994), 'Outlining Regulatory Space: Agenda Issues and the FASB', 19 *Accounting, Organizations and Society* 83.

10 Regulating for ecological resilience

A new agenda for financial regulation

Jamie Murray

Financial regulation has undergone a fundamental transformation in the wake of the 2007/8 systemic crisis in global financial systems. The global financial system is now understood as systemic, interconnected, complex and prone to systemic risk. Financial regulation has developed a systemic understanding of financial regulation in new practices of macroprudential financial regulation. This has been the regulatory shift from microprudential regulation to macroprudential regulation, defined as systemic regulation, with new regulatory assumptions, goals, institutions and regulatory tools (Crockett, 2000; Borio, 2003; Persaud, 2009; Brunnermeirer et al., 2009; Tucker, 2009; Berwell, 2013; Claessens and Evanoff, 2011; Tucker,2011; Blanchard et al., 2014; Haldane, 2014b; Borio, 2014; Akerlof et al., 2014; Freixas and Laeven, 2015). Financial regulation is now aware of the necessity and difficulties of regulating systemic risk and has developed innovative new counter-cyclical regulatory practices and tools for regulating systemic risk.

However, in this transformation in the understanding of the complexity of financial systems and systemic macroprudential regulation the focus has been on this understanding of systemic risk, and the regulation of systemic risk. This chapter argues for a new agenda for transformed financial regulation, moving beyond a central concern of systemic risk and the macroprudential regulation of systemic risk. The proposed new agenda, rather, looks to a central concern of the resilience of complex financial systems, and the regulation of the resilience of complex financial systems. Ensuring the resilience of financial systems is already one of the key concerns of macroprudential regulation alongside the management of systemic risk. Yet the phenomena of systemic resilience and the specific regulation for systemic resilience have not received the same depth of consideration as have systemic risk and the management of systemic risk. This chapter proposes a further transformation in financial regulation, building on the complexity theory framework for financial regulation, based on a complexity theory understanding of systemic resilience and a complexity jurisprudence understanding of managing this complexity systemic resilience.

The argument of the chapter is that the complexity theory concept of resilience as ecological resilience is now crucial to regulating complex financial systems and that complexity jurisprudence provides the new agenda for the regulation of complex financial systems for ecological resilience. The complexity concept of resilience is that of a dynamic far-from-equilibrium resilience, very different from a return-to-state, near-equilibrium concept of systemic resilience. In the new agenda for financial regulation complex financial systems will be understood as ecologically resilient systems that require new understandings and practices of regulating for ecological resilience. In the new resilience agenda financial regulation will increasingly turn to complexity jurisprudence so as to learn how to regulate complex financial systems for ecological resilience. Complexity jurisprudence is a rich resource for understandings and practices of managing and regulating for ecological resilience. This complexity jurisprudence understandings and practices include adaptive management, assisted self-organisation and second order self-reflexive management of the regulatory system itself. All three of these strategies for regulating complex adaptive systems for ecological resilience make up the new regulatory agenda for the further development of complex financial regulation.

Complexity theory framework for financial regulation: a brief history

A complexity theory framework for financial regulation has been gradually emerging since the late 1980s, first in complexity economics, increasingly in financial theory, and now in financial regulation itself. This complexity theory framework was well developed prior to the 2007/8 global financial systemic crisis, but the development of a complexity theory framework for financial regulation has become particularly intensively developed since the financial crisis.

Complexity economics has been at the core of complexity theory in its Santa Fe development from very early on, and complexity economics was established by 1988 (Anderson et al., 1988). Complexity economics is now a broad and rich field (Kauffman, 1995; Arthur et al., 1997; Beinhocker, 2007; Colander et al., 2008; Rosser, 2009; Fontana, 2010; Holt et al., 2010; Farmer, 2012b; Helbing and Balietti, 2010; Kirman, 2011; Omerod and Helbing, 2012; Helbing and Kirman, 2013; Helbing, 2013b; Colander and Kupers, 2014; Helbing, 2015a; Helbing, 2015b; Arthur, 2015). Complexity economics has developed a distinct approach to economics in a central concern on formation and change in the economy, self-organisation and emergence in economies, economies as complex adaptive systems, the role of attractors and path dependencies in economies, economic evolution, systemic economic collapse events and transformation in economies (Beinhocker, 2007; Rosser, 2009; Kirman, 2011; Colander and Kupers, 2014; Arthur, 2015). Strongly fitting within the concerns and styles of the economics generally identified as heterodox, complexity economics and the wider heterodox economic field, have grown considerably in confidence and broader consideration in the light of the conceptual failure of neoclassical economics in relation to the event of the global financial crisis (Colander et al., 2008). A complexity theory understanding of financial systems can, thus, be seen as one exemplary case of a broader complexity theory understanding of economics. Indeed, many academics working on developing aspects of a complexity theory of financial systems are also active in developing the more general complexity economics (Farmer et al., 2012a, 2012b; Kirman, 2011; Helbing, 2010, 2012).

An early complexity theory understanding of financial systems was Sornette's 2003 *Why Stock Markets Crash: Critical Events in Complex Financial Systems.* Sornette's central concern is to develop prediction models, particularly stock market crashes but in so doing draws specifically on features of Santa Fe complexity economics – far-from-equilibrium conditions, self-organisation and emergence; nonlinearity and positive feedbacks; extreme events in complex systems; power law distribution of extinction events; self-organised criticality – to develop a model of financial markets as evolving complex adaptive systems (Sornette, 2003 p. 130, 2004). In many respects Sornette's *Why Stock Markets Crash* remains a founding source on the development of a complexity theory understanding of financial systems and merits in-depth exploration as the field of complexity theory of financial systems consolidates.

A 2006 *New Directions in Systemic Risk* was the first collective consideration of complex systems theory and financial regulation (Kambli et al., 2007). This conference firmly introduced complexity theory as exemplary for better understanding, and therefore better regulating, global financial systems. The conference organisers were resolutely transdisciplinary, bringing together ecologists and engineers with financial professionals and regulators. The focus of the conference was systemic risk as a specific feature of complex systems and how exactly to understand the operation of systemic risk in multiple complex systems. In this, the clear overall framing of financial systems was that of the problematics of complex systems developed in complexity science and theory and that the understanding of complex financial systems should take freely from the understanding of dynamic complexity already experienced and built up in relation to ecosystems and dynamic engineering systems. This complexity theory framing of financial systems as complex adaptive systems was followed up May et al. (2008). In this the authors further develop complexity theory framework for understanding financial systems as complex adaptive systems, with ecology and the complexity theory of ecosystems positioned as a privileged basis for rethinking banking and financial systems and their regulation.

Andrew Haldane was the first central banker to foreground the potential importance of complex systems theory for understanding financial systems and financial regulation (2009a). Invoking the paradigmatic shift in understanding and regulation of financial systems as complex systems conceptualised in complexity theory and various complex system disciplines, Haldane particularly drew on the value of the network theory branch of complex systems thinking. In complex systems theory networks are understood to exhibit endogenous dynamics, patterns and structures, as general classes of networks. Haldane argued this understanding of network dynamics should be extended to the understanding and regulation of financial systems. Haldane's intervention was part of a major development in the rethinking of financial systems as complex systems and complexity theory at the Bank of England and other central banks (Haldane, 2009a; Landou, 2009; Arinaminpathy, Kapadia, May, 2010; May, 2012; Haldane, 2012;

Amar and Avgouleas, 2016; Hollow et al., 2016). This work explored a multitude of problems of contemporary financial systems in terms of complexity theory and complex systems theory, with work on network theory and financial systems (Haldane, 2009a; Gai et al., 2011; Anand et al., 2013; May, 2013; Battiston et al., 2013), systemic risk in complex financial systems (Kambli et al., 2007; Schwarcz, 2008; Helbing, 2010; May and Arinaminpathy, 2010; Gai and Kapadia, 2010; Utset, 2010; Anabtawi and Schwarcz, 2011; Gai et al., 2011; Haldane and May, 2011; Gai, 2013; Ellis et al., 2014; Mitts, 2015; Freixas and Laeven, 2015) and financial cycles in complex financial systems (Borio, 2014; Aikman et al., 2013). In a parallel trend of development in the post–financial crisis US financial regulation literature, a complex systems theory understanding of financial systems were increasingly invoked in order to explore problems of systemic risk in complex financial systems (Schwarcz, 2008; Utset, 2010; Anabtawi and Schwarcz, 2011), governance of financial systems as complex systems (Schwarcz, 2009; Chinen, 2011) and a complexity theory re-assessment of fundamentals of pre-crisis financial and economic theory (Cooper, 2011). In particular, clear agenda-setting proposals of complexity theory framework for rethinking understandings of complex financial systems were made (Zieden and Richardson, 2010; Johnson and Lux, 2011), with Baxter in a number of articles proposing and pursuing a clear agenda of employing Santa Fe complexity theory to fundamentally rethink the understanding of complex financial systems (Baxter, 2012, 2016).

More recently, the complexity theory agenda for rethinking financial systems has been pursued in work conducted at the Institute for New Economic Thinking in Oxford (www.inet.ox.ac.uk). Whilst pursuing a broad complexity theory framework for rethinking economics and pursuing further development of the complexity economics approach, the institute has produced and hosted considerable work on financial systems as complex systems and understood in terms of complexity theory (www.inet.ox.ac.uk/news/Financial-Regulation). In particular, Farmer has been foremost in pursuing a complexity theory understanding of financial systems and the global financial system (2012a; also Cincotti et al., 2012). Farmer et al. sets out a framework for the progressive mapping and exploration of the full range of the complexity theory of interconnectedness, self-organisation, emergence, self-organised criticality, innovation and systemic risk onto a transformed understanding of financial systems and the global financial system. The complexity theory paradigm for understanding financial systems as complex systems has been pressed again in 'Complexity Theory and Financial Regulation' (Battiston et al., 2016), with specific emphasis on bifurcation points and tipping points in complex financial systems and crisis prediction, complex network theory for conceptualising systemic feedbacks, contagions and agent-based modelling for exploring the self-organising and emergent dynamics and patterns of complex financial systems and the promise of behavioural economics for understanding the complexity of financial systems. The Battiston et al. article marks out the contemporary stage of development of the complexity theory paradigm for understanding financial systems and the global financial system. Many of the authors named in the Battiston et al. article are familiar in the overall post-crash development of the complexity theory understanding of financial systems (Hommes, Haldane, May, Farmer, Kirman, Cincotti, Omerod, Helbing), and complexity economics, more generally, and mark a now established and maturing complexity theory paradigm in financial systems and economics.

Whilst the focus in the literature has been on how complex systems theory can improve the understanding of the organisation and dynamics of complex financial systems, there have been strands in the financial regulation literature pointing in the direction of a new complexity agenda specifically for financial regulation. Lippe et al. (2015) have called for a new paradigm for financial regulation informed by complexity, aimed at regulating financial systems for adaptive capacity and the use of modern notions of design to create new tools and methods of regulating complexity (2015, p. 835). Cincotti et al. (2012) have also posited a complexity agenda for financial regulation, particularly focused on issues of operationalising complex financial regulation in technological and computational technologies. This consideration of complex systems theory and financial regulation is at an early stage.

The complexity concept of resilience: ecological resilience

In complexity theory complex systems are understood in a set of core concepts. Complex systems are open systems to matter–energy–information flows, far-from-equilibrium and in continuous transformation. These systems are defined not by the presence of a single state attractor but, rather, by multiple system attractors that provide for potential multiple states for the system to move on. These systems are interconnected systems with multiple dynamic interactions and new interconnections generating complex dynamic networks. Complex systems theory is concerned with processes of order out of chaos, self-organisation, emergence, adaptation and evolution. This class of systems as a consequence of these features come to be understood as exhibiting a number of distinctive system properties: nonlinearity, contagion, events on power law probability distributions and systemic risk (Waldrop, 1992).

Of these core complexity theory concepts out-of-equilibrium, continuous transformation, self-organisation and emergence, are particularly important to understanding the resilience of complex systems. In out-of-equilibrium systems there are immanent processes of self-organisation that autonomously generate structures in operations of order-out-of-chaos. These novel structures emerge as counter-entropic organisation in the open matter–energy–information flows of the complex system. In complexity theory these counter-entropic emergent structures are theorised as dissipative structures (Prigogine and Stengers, 2018) and self-organising criticality (Bak et al., 1987). Understanding dissipative structures and self-organising criticality, in turn, becomes the next crucial stage to understanding the complexity theory concept of systemic resilience. There is a tendency in complex systems theory to go straight to a consideration of a network of heterogeneous agents and to skip over mechanisms of emergent complexity. However, the resilience of complex systems – the resilience this chapter foregrounds – is the scaling up of precisely the operations of dissipative systems and self-organised criticality. Dissipative structures give dynamic organisation to a complex system, providing the system with complex organisation and complex information processing. This complex organisation is in continuous change, holding together dynamic organisation only through continuous transformation. This dissipative organisation in continuous transformation self-organises, emerges and operates, at a tenuous border between firmly structured network organisation and network organisation in chaos (self-organising criticality). In this edge-of-chaos zone of dissipative structures and self-organising criticality complex systems can tap immense forces of self-organisation and emergence, immense complex dynamic organisation, immense dynamic processes for computing and information processing, immense potentials of adaptation and evolution (Kauffman, 1995; Bak et al., 1987).

Given these core complexity theory concepts of how complex systems are understood to organise and operate, the complexity concept of systemic resilience becomes a specialised and novel understanding of system resilience, differing significantly from generic understandings of the concept of resilience. This novel and specialist understanding of system resilience is particularly accentuated in Holling's resilience theory. In Holling's first broad formulation of the concept of resilience he defined it as the ability of a complex system to cope with change and adapt (Holling, 1973). However, Holling then made a fundamental distinction within this broad concept of resilience, between engineering resilience and ecological resilience. Engineering resilience was the concept of resilience that had been developed in equilibrium-orientated ecology and was centred on the idea of the speed at which a system hit by sudden change could bounce back to its equilibrium single stable organisation and operation. By contrast, ecological resilience was the concept to capture the resilience of out-of-equilibrium and multiple steady-states complex adaptive systems as they dynamically interconnected and interacted with change, through change in their own dynamic organisation and operation. Ecological resilience is the ability of a complex system to organise and operate and to re-organise and adapt operations, whilst undergoing change. With an initial core idea of dynamical persistence in complex systems ecological resilience was the capacity of a system to maintain a regime of structures and processes whilst undergoing change and to absorb change whilst maintaining regime. Ecological resilience was, thus, the result of complex system immanent dynamic processes and structures of interconnections and interactions and was an emergent capacity of a complex system operating in a regime of complexity.

In subsequent development in resilience theory, by encompassing interconnected and interacting self-organisation and emergence, ecological resilience became aligned on core Santa Fe complexity theory concepts of order-out-of-chaos, self-organised criticality, the regime of complexity and dissipative structures. This ecological resilience was thus, also, a concept of a complex system continuously adapting through self-organisation, moving around its dynamic state space of attractors and exploring new fields of system attractors (Garmestani et al., 2009; Garmestani et al., 2014; Garmestani and Benson, 2013). Ecological resilience was, thus, far more complex than the presence of multiple attractors and regimes in a complex adaptive system undergoing change. In response to change ecological resilience is understood to allow complex adaptive systems enhanced adaptive capacity. Adaptive capacity is the capacity of a complex system to explore, to experiment, to innovate and to create in relation to its own self-organisation, emergence and transformations. Indeed, the initial idea of ecological resilience being an ability to maintain a regime expands to a wider idea of ecological resilience as an ability to explore regimes, to systemically transform and to create new regimes of self-organisation and emergence in new dissipative structures (Cheffin et al., 2016). Ecological resilience provides a complex adaptive system not only with abilities to mitigate in relation to change and to adapt in relation to change but also with an ability to venture out into an unknown future and transform the organisation and attractors of the complex adaptive system itself (Cheffin et al., 2016).

With this concept of ecological resilience there can be fundamental transformations in the way particular complex adaptive systems are understood and regulated. As reflected in the complexity framework for financial regulation, there is a broad and firm commitment that financial systems are complex adaptive systems. Financial systems are characterised as out-of-equilibrium, open matter–energy–information flows, self-organisation, emergence, nonlinearity, contagion and systemic risk. However, in the existing complexity theory on financial systems and financial regulation the focus has tended to be systemic risk in complex financial systems. To the extent that the concept of resilience in complex financial systems has been discussed in the literature it appears as simply the other side of systemic risk, and implicitly understood in terms of an engineering resilience. The central concern with systemic risk in complex systems is with only one aspect of the operation of complex systems: the propensity of complex systems to systemically and catastrophically collapse. The regulatory priority becomes the maintenance of systemic functioning and the vigilant monitoring and removing as much systemic risk from the financial system as possible. The idea of financial system resilience accordingly becomes accentuated as the ability of a financial system to minimise and mitigate systemic risks and maintain robust functioning. This resilience is the ability of a complex financial system to robustly maintain a single stable state and to return to that single stable space in response to systemic risk perturbations. This is engineering resilience.

Yet when financial systems are rather understood in terms of ecological resilience, very different conclusions and questions emerge in terms of broader understandings of financial system resilience and the regulation of complex financial system resilience compared with when financial systems are understood in terms of engineering resilience. There are, of course, vary different conclusions as to how to understand the organisation of complex financial systems in terms of continuous transformation, dissipative structures, self-organising criticality, edge-of-chaos organisation and systemic no-analogue future. There has to be, also, new conclusions drawn on the aim of financial regulation. The vital conclusion to draw from the concept of ecological resilience is that complex systems are at their most creative in an edge-of-chaos zone of dissipative organisation and self-organising critical-ity. Creativity becomes the most extraordinary capacity of complex systems with ecological resilience, with this capacity of creativity running through all the other extraordinary capacities of such poised complex systems: responsiveness, intel-ligence, adaptation and evolution. The problem, though, is that an ecologically resilient complex system operating at the edge-of-chaos provides all these extraor-dinary capacities only at the cost of living with high levels of systemic risk. Major questions then emerge of whether social organisation would want a financial system organised and operating in ecological resilience or whether it is preferable to have a financial system organised and operating in engineering resilience. These questions become particularly charged if it is accepted that the processes and mechanisms of ecologically resilient self-organisation and emergence can be both understood *and* managed. The argument is that regulators can both understand the organisation and operation of complex adaptive systems and potentially steer complex adaptive systems, either closer to ecological resilience or to engineering resilience.

The overall new conclusion from the ecological resilience understanding of complex financial systems is that ecological resilience is both necessary and desir-able, that there can be developed strategies for actively regulating and managing complex financial systems, that this regulation and management should be to hold the complex financial system in the regime of ecological resilience and that regulation and management should ride the emergent transforming future that financial regulation co-creates with the complex financial systems.

Complexity jurisprudence: regulating and managing for ecological resilience

Thus, from the perspective of the ecological resilience understanding of com-plex financial systems, the crucial question becomes how to regulate and manage complex systems for ecological resilience. Regulatory assumptions, goals, institu-tions and regulatory tools, developed in relation to the primary task of mitigating systemic risk were not designed for the management of ecological resilience. The task of regulating complex financial systems for ecological resilience is a funda-mentally different enterprise than that of regulating for engineering resilience. The task of regulating and managing for ecological resilience sets a new agenda for financial regulation.

The field of complexity jurisprudence has developed precisely in response to the question of regulating complex adaptive systems for resilience, developing primarily in the context of ecological resource management (Ruhl, 1997; Hel-bing, 2008, 2015b; Craig, 2010; Garmestani and Allen, 2014). Complex finan-cial regulation can find in complexity jurisprudence resources and practices for regulating complex financial systems for ecological resilience. Complexity juris-prudence is characterised by a number of common features and developed in three substantive endeavours.

Complexity jurisprudence has been influenced by a broad rejection of ortho-dox legality and regulation (Ruhl, 1995; Craig, 2010; Arnold and Gunderson,

2014a, 2014b). Complexity jurisprudence has sought to establish that the very way of thinking in orthodox legality cannot understand problems of complex adaptive systems or the complexity of ecological resilience, and orthodox legality consequently cannot deal with problems of complex adaptive systems or regulate the complexity of ecological resilience (Helbing, 2013a, p. 58; Garmestani et al., 2014, p. 365). Indeed, this rejection of orthodox legality can be seen as an aspect of complexity jurisprudence's rejection of the whole reductionist approach of attempting to control and constrain non-equilibrium complexity. Helbing assesses the reductionist approach in the following terms:

> [regulators] have so far focused their efforts on attempting "control complexity" from the top-down through many regulations, laws and enforcement institutions. While this approach has served us reasonably well for a long time, it is eventually coming to its limits.
>
> (Helbing, 2015b, l.4267)

Thus, complexity jurisprudence has set out to explore new ideas of the organisation, regulation and management of complex adaptive systems and to new approaches and practices for regulating and managing complex adaptive systems for ecological resilience.

Complexity theory understands how complex adaptive systems organise and operate in immanent processes of self-organisation, emergence, experimentation, adaptation and evolution. Complexity theory, furthermore, understands how complex adaptive systems immanently self-regulate and self-manage in ecological resilience in these processes. Complexity jurisprudence, therefore, allows new ways to theorise and practice legality, regulation and management of complex adaptive systems and the complexity of ecological resilience on the basis of *an understanding* of how social-ecological complex adaptive systems organise and operate in immanent processes of self-organisation, and *the reality* of the immanent processes of self-organisation and self-management in social-ecological complex adaptive systems and the complexity of ecological resilience. Complexity jurisprudence proceeds with the rethinking of the law, regulation and management of complex adaptive systems on the basis of an open-ended uncertain future of continuous change and transformation. Complexity jurisprudence develops as a new idea of what legality, regulation and management should do: manage for ecological resilience and the enhancement of ecological resilience on a transforming line of creation, innovation, emergence, adaptation and evolution. Complexity jurisprudence is thus a positing of a new idea of regulation and management of complex adaptive systems for ecological resilience: *managing through the immanent self-managing processes and mechanisms of the managed complex adaptive system* (adaptive management; assisted self-organisation) and *managing through the immanent self-managing processes and mechanisms of the managing complex adaptive system* ('legal mapping'; Arnold and Gunderson, 2014a; Ruhl and Katz, 2015). In this complexity jurisprudence is developing as a highly ambitious active management of the enhancement of ecological resilience (Ruhl,

1995, 1996, 1997; Ruhl and Ruhl, 1996; Craig, 2010; Arnold and Gunderson, 2014b; Ruhl and Katz, 2015).

We can see three main approaches to the problem of the regulation and management of complex adaptive systems in complexity jurisprudence: resilience theory's adaptive management, Helbing's assisted self-organisation and Ruhl and Katz's second-order self-reflexive management of regulatory systems. These three approaches are considered in turn.

Adaptive management

The resilience theory of adaptive management was developed as practices adequate to the complex problems of managing and governing complex adaptive systems for ecological resilience (Hollings, 1978; Gunderson and Holling, 2002; Gunderson et al., 2010; Walker and Salt, 2008). Adaptive management puts complexity theory and resilience theory into practice.

The core of the idea of adaptive management is that the creative processes of self-organisation, emergence, innovation and adaptation in the managed system are the index of the system's ecological resilience. Adaptive management aims to manage systems so that they organise and operate in ecological resilience and so are able to generate self-organisation, emergence, innovations and adaptations, so increasing the managed system's own adaptive capacity to manage itself (Holling, 1978; Gunderson and Holling, 2002). Adaptive management manages through managing interventions in managed system dynamic to provoke, trigger, inhibit, catalyse and harness managed system immanent processes of self-organisation, emergence, innovation and adaptation. These thresholds can be managed through altering rates of matter–energy–information flows across the system, driving it dynamically closer or further away from its critical thresholds.

Adaptive management can explore the problem space of the managed complex adaptive system, mapping the system's attractors and the shape of the system's basins of attraction. Adaptive management for resilience is to assess desirable and undesirable states of the system and to assess desirable directions for the managed system's trajectory of ecological resilience and adaptation (Gunderson and Holling, 2002). Thus, adaptive management aims to enhance the managed system's experimentation, self-organisation, emergence, innovation and adaptations, thereby enhancing the managed system's ecological resilience and, hence, then enhancing the managed system's adaptive capacity to self-manage for ecological resilience (Walker and Salt, 2008). In this way managed system's adaptive capacity is developed to manage its own ecological resilience, to mediate between potentially multiple system states and to manage transformations in ecological resilience (Garmestani et al., 2009; Garmestani and Benson, 2013).

Adaptive management proceeds in highly flexible regulation, a data-driven monitoring, measuring and mapping of the managed complex adaptive system and then practices of regulatory interventions and modifications into the managed system (Ruhl, 2004, 2005, 2011; Craig and Ruhl, 2014). Adaptive management is a procedural framework for learning about the managed complex adaptive system whilst that system is being managed, integrating learning into management processes and progressively learn to make more effective management interventions. This adaptive management is continuous: continuous learning, continuous experimentation, iterative feedback loops, continuous innovations and continuous adaptations (Garmestani et al., 2014, p. 349).

Adaptive management consists of adaptive planning and identifying regulatory goals, with iterative and evolving processes of regulation and management. In adaptive management decision-making is flexible and discretionary, and it is the construction of context regarding standards. Adaptive management, therefore, necessarily has to be experimental and innovative, which necessarily entail regulatory interventions that are trial and error. In contrast to present orthodox legality and regulation that is heavily front-ended, adaptive management is backended regulation where experimentation and change of approach are enabled and discretion to change decisions is retained (Ruhl, 2004). It is a regulatory approach designed to manage a system in the complex reality of nonlinearity, uncertainty and unpredictability. Practices of adaptive management are to be flexibly institutionalised, tending to distributed and polycentric organisation, multimodal in terms of regulatory approaches and tools, multi-scalar in terms of the levels at which management interventions can be targeted (Garmestani and Allen, 2014 p. 339).

Central to adaptive management is the role of monitoring of the managed complex adaptive system, focusing on specific variables, attributes and drivers of the managed complex adaptive system (Ruhl, 2011). It is these data about the dynamics of the managed complex adaptive system that allow adaptive management to then map and model these dynamics to better understand these dynamics. In adaptive management there is the mapping and modelling of the system multiple states thresholds, adaptive cycles and nested adaptive cycle dynamics (Gunderson and Holling, 2002). For adaptive management to be possible there will need to be very considerable volume of real-time data about the managed system available to the regulators.

Adaptive management, of course, adapts. Regulatory exploration of the managed complex adaptive system, regulatory experimentation and interventions, learning, monitoring and modelling the dynamics of the managed system, feedbacks and modification of regulation entail the adaptation in the management of the managed complex adaptive system. It is management that adapts to changing conditions through intervening into complex adaptive system dynamics, monitoring the consequences of the management interventions, and then feeding that information back into the planning and implementation of management intervention into managed complex adaptive systems.

Adaptive management seeks to craft guiding principles for human intervention to improve complex adaptive systems' ecological resilience and sustainability (Garmestani and Benson, 2013). Adaptive management develops by learning the design of mechanisms for moving and triggering thresholds and basins of attraction, mechanisms for tapping and harnessing systemic self-organisation and emergence, mechanisms for systemic accountability and conservation of social and ecological capital and mechanisms for continuously enhancing systemic eco-logical resilience, adaptive capacity and sustainability (Garmestani et al., 2009).

Assisted self-organisation

Dirk Helbing has also taken up the challenge of legally structuring, regulating and managing complex adaptive systems for creative, self-organising, emergent, adaptive resilience (Helbing, 2008, 2015a, 2015b). For Helbing, such an approach to regulating and managing complex adaptive systems cannot organise and operate in any orthodox legal and regulatory top-down structures but must instead organise and operate with bottom-up mechanisms and interventions of assisted self-organisation. For Helbing 'we need to step back from centralising top-down control and find new ways of letting the system work for us, based on distributed, "bottom-up" approaches' (Helbing, 2015b l.2294). This 'bottom-up' approach is to not fight and control the complexity of the regulated system but precisely, rather, to make use of the self-organisation of complex systems and the development of an assisted self-organisation approach to legally structuring, regulating and managing complex adaptive systems (Helbing, 2008, p. 7). This approach of assisted self-organisation is explicitly complexity theory–informed systemic regulation and management that knowingly draws on the immanent self-organisation and emergence present in complex adaptive systems. Helbing's idea here is that complex system immanent tendencies to self-organise and cre-ate dynamically resilient order that can be used and that we can learn to har-ness the underlying self-organising forces to our benefit (Helbing, 2013a, p. 54, 2015b l.314). Assisted self-organisation is an approach that seeks to draw a new approach to legally structuring, regulating and managing complex problems that differs from both top-down complexity control and from unmanaged bottom-up self-organisation (for example, markets). Thus, there is the necessity of manag-ing the self-organisation and emergence for the creation of desired structures, properties, functions and capacities and for the management of complex system enhanced innovation, resilience, learning and sustainability. Thus, in assisted self-organisation 'self organisation can be used to produce desirable outcomes, and this would enable us to create well ordered, effective, efficient and resilient sys-tems' (Helbing, 2015b, l.394).

This assisted self-organisation, based on a complexity theory understanding of complex systems and of how to best regulate and manage complex systems, operates through techniques of influencing complex systems and through an operationalisation of these techniques in information and computer technology. Helbing's proposal has been that through real-time big-data mapping, modelling and measuring of complex systems, assisted self-organisation management would know about a system's attractors, bifurcation points and cycles and the system's instantaneous position in relation to these key variables, and therefore through well-placed, careful interventions would be able to drive, steer, switch, tap, har-ness, call forth and transform, system dynamics. The techniques focus on multiple control variables, levels of system interconnectedness, matter–energy–information flow across the system levels, intervening in local rules of interaction, intervening in strengths of systemic interactions, the design of mechanisms to trigger attractors, thresholds, dynamic consistencies, managing cascades and contagion, systemic crises, stimulating diversity, enhancing systemic resilience and adaptability. The viability of such a complexity theory–informed regulation and management of complex system complexity in assisted self-organisation is, however, absolutely dependent on the development of a data-driven and data-mined real-time computer mapping and modelling of the regulated complex systems in high-dimensionality topology and massive sensor-data output arrays. This development is what Helbing refers to as a theoretical, computational, experimental and data-driven Planetary Nervous System (Helbing, 2013a, p. 57). This ICT (Information and Communication Technologies) platform would be operated by complex system observatories that would monitor and manage complex systems, developing new socioscopes for exploring real-time dynamics of complex systems and responding to system dynamics adaptively (FuturICT.org). The rolling-out of such a big-data internet of things has not proceeded smoothly but is something that could at some point come together (see FuturICT.org). This combination of complexity theory and technology in assisted self-organisation leads Helbing to the conclusion 'three hundred years after the principle of the invisible hand was postulated we can finally make it work' (Helbing, 2015b, l.2626).

Reflexive regulation self-management

A crucial feature of complexity jurisprudence is that it presents not only new regulatory resources for managing for ecological resilience but also a focus on the necessity of the self-reflexive management of the complexity of the regulatory system itself. In complexity jurisprudence, law and regulation must itself be understood as a complex adaptive system (Ruhl, 1995, 1996, 1997, 2013; Ruhl and Ruhl 1996; Tussey, 2005; Cherry, 2007, 2008; Kim and Mackey, 2014; Zhang and Schmidt, 2015; Ruhl and Katz, 2015). Law and regulation must be theorised and practised in terms of law and regulation's complexity, and as itself organised as a dynamic out-of-equilibrium complex adaptive system. This then raises the follow-on consideration of how best reflexively to legally structure, regulate and manage complex adaptive legal systems to enhance their self-organisation, resilience and adaptability (Ruhl and Katz, 2015). It is how complexity jurisprudence has reconceptualised law and regulation as complex adaptive systems organised and operating in the complexity regime. The shift to conceptualising, organising and operating law and regulation as complex and as a complex adaptive system has very profound consequences on how law and regulation is thereafter thought of, that law and regulation must then be understood to be transformed and the legal framework and regulation of the underlying complex adaptive system consequently transformed (Arnold and Gunderson, 2014a, 2014b). Complexity jurisprudence can then explore regime shift thresholds in legal complexity, engineering resilience and ecological resilience in legal complexity, adaptive cycles in legal complexity (Ruhl, 2011) and the multiple nested adaptive cycles in legal complexity (Ruhl, 2012; Arnold and Garmestani, 2014). This is to turn to the ecological resilience of the regulatory system itself and to its reflexive self-regulation for ecological resilience in adaptive management, assisted self-organisation, mapping, measuring and modelling of law and regulation, operationalised in ICT real-time platform (Ruhl and Katz, 2015).

Across these three approaches the outline of a broad complexity jurisprudence agenda for regulating complex adaptive systems for ecological resilience can be drawn out. The agenda is to regulate and manage a complex adaptive system *through* the complexity of the managed system (Chandler, 2014). Adaptive management and assisted self-organisation are key strategies for regulating and managing *through* complexity. This agenda involves active interventions into the organisation and operation of the managed system, though tentatively and subject to continual recursive assessment. The goal of regulatory intervention is to enhance the ecological resilience and adaptive capacity of the managed system. The regulatory agenda also includes the second-order self-reflexive regulation of the regulatory system itself, with the agenda for the managed system implemented on the regulating system itself (Ruhl and Katz, 2015). Across this agenda what becomes crucial is the monitoring of the complex adaptive system, and what has become clear is the quite astonishing scale and depth of the necessary monitoring systems in order to operationalise regulating for ecological resilience. Adaptive management, assisted self-organisation and self-reflexive regulation require immense real-time ICT-enabled monitoring and highly sophisticated computer modelling in order to regulate and manage for ecological resilience, and all three approaches strongly foreground the challenge that this places on the regulatory agenda.

The implications of regulating for ecological resilience for financial regulation would, thus, go to the need to develop new financial regulatory tools for regulating complex financial systems and to the scope and ambitions of the agenda of regulating financial systems for ecological resilience. Whilst the very notion of complex systems theory and the understanding of regulating complex adaptive systems mean that prescriptive accounts of what financial regulation for ecological resilience would look like, and what it should do, are not available, nonetheless a broad outline for the future of financial regulation can be sketched.

A financial regulation for ecological resilience would be macroprudential regulation, but it would develop new regulatory approaches and regulatory tools in the macroprudential management of ecological resilience. This would be for financial regulation to turn to existing approaches and tools of regulating complex systems that are well developed but as yet are untried in regulating complex financial systems. Financial regulation would need to move in the direction of adaptive management as core regulatory approach. As has been discussed earlier, adaptive management is well understood, and would move in the direction of existing regulatory trends of technologically enabled real-time monitoring of complex financial systems (Haldane, 2014a, 2015). Complex systems, such as financial systems, can be adaptively managed, and new regulatory tools of assisted self-organisation and reflexive self-management of legal and regulatory complexity could be developed for financial regulation for ecological resilience. Any notions of the top-down regulation of financial markets would be abandoned, replaced with bottom-up relational financial regulation that may indeed be rather quite interventionist (of which more later). Financial regulation would become somewhat experimental in its interventions and open to regulatory innovations.

The implications of regulating for ecological resilience for what financial regulation should do are secondary, though, to its implications for the agenda of financial regulation. Regulating for ecological resilience involves encompassing two regulatory approaches in one primary regulatory agenda and approach. Regulating for ecological resilience does act to mitigate complexities and systemic risks in regulated complex systems. Regulating for ecological resilience does seek to improve the ability of regulated systems to adapt to new regulatory goals and does seek to adapt to changes in the regulated systems. However, the primary regulatory agenda in regulating for ecological resilience is always ultimately – and sometimes urgently – for the transformation of the regulated system and the enhancement of the adaptive capacities of the regulated system (and, indeed, the transformation and enhanced adaptive capacities of the regulatory system itself). The primary agenda for regulating for ecological resilience is a transformative regulation that, at times, seeks to actively shift a complex adaptive system into alternative and inherently more desirable regime with respect to ecological resilience and adaptive capacity by altering the structures and processes that define the system (Cheffin et al., 2016). The ultimate agenda for regulating for ecological resilience is that the regulated system should be put in an ongoing systemic transformation, guided and steered by management and regulation, into ceaseless self-organising and emergent creative transformations that constitute the most desirable state for complex adaptive systems. Indeed, in regulating for ecological resilience the need for regulatory interventions to put the regulated system in transformation becomes urgent when the systemic ecological resilience of a regulated system has been eroded to the point of systemic collapse. This regulating for open-ended transformability is not only the ultimate agenda for regulating for ecological resilience; it is also its urgent one. Regulating for open-ended transformability 'is the capacity to create a fundamentally new system when ecological, social, economic and political conditions make the existing system untenable' (Walker and Salt, 2008 1.782). A complexity approach may only be able sketch out only a broad outline of the future of complex systems, but here with financial regulation the implications for financial regulation in undertaking regulation for ecological resilience (rather than engineering resilience) could not be more in relief and in contrast. One model is of complex financial regulation is that of the mitigation of systemic risk in a systemic engineering resilience that seeks to preserve the existing financial system at all costs (bailing out market externalities). The other model of complex financial regulation in regulating for ecological resilience is one that ultimately, and urgently, actively intervenes at a systemic level to put into transformation socially and economically undesirable and un-ecologically resilient financial systems to call forth an emergent radical transformation of the organisation and operation of those financial system. Regulating for ecological resilience is the financial regulatory agenda and practice that we should have had, but have not, in the wake of the 2008 global financial collapse.

Conclusion: complexity jurisprudence as management for ecological resilience

Moving from a central concern of managing systemic risk (and engineering resilience) to a central concern with systemic ecological resilience has profound implications for the understanding of complex financial systems and macroprudential financial regulation. Building on the existing complexity theory framework for financial regulation, placing systemic ecological resilience as the central concern sets a new agenda for understanding complex financial systems and financial regulation. The new agenda for financial regulation is for the regulation and management for ecological resilience and the enhancement of financial system adaptive capacity. Complexity jurisprudence has developed precisely as strategies for the structuring and managing of ecological resilience in complex adaptive systems and for the enhancement of adaptive capacity in far-from-equilibrium complex systems. Thus, the central concern with ecological resilience sets complexity jurisprudence as the new agenda for financial regulation. The new agenda of regulating financial systems for ecological resilience leads the further transformation of financial regulation towards complexity jurisprudence understandings and practices of adaptive management, assisted self-organisation and second order self-reflexive regulatory system regulation. The regulation of financial systems for ecological resilience in complexity jurisprudence further sets a new agenda for a combined complexity theory and computing programme for the operationalisation of a complexity jurisprudence complex financial regulation. A new agenda of regulating complex financial systems for ecological resilience in a complexity jurisprudence should follow the efforts of Helbing's ICT-enabled World Wide Nervous System, and Ruhl and Katz's ICT-enabled legal maps, to develop the central bankers' dream of a financial system regulatory dashboard and financial policy 'wind tunnel' simulator for real-time financial system management (Haldane, 2014a, 2015).

There is no doubt that this agenda for a further transformation of financial regulation for the enhancement of financial system ecological resilience is ambitious. In particular, the operationalisation of a real-time modelling of the complexity of global financial systems presents tremendous challenges given the scale and immense complexity of global financial systems. However, in complexity theory there are the theoretical resources for understanding the complexity and ecological resilience of complex systems, and in complexity jurisprudence there are existing strategies and practices for managing complex adaptive systems for ecological resilience. The ecological resilience paradigm for financial regulation promises much for the future management of complex financial systems, guided transformations and potentially enhanced financial system adaptive capacities in relation to social-ecological systems generally.

Bibliography

Aikman, D,, Haldane, A. and Nelson, B (2013), 'Curbing the Credit Cycle', 125 *The Economic Journal* 1072.

Akerlof, G. A., O. Blanchard, D. Romer and J. E. Stiglitz (2014), *What Have We Learned: Macroeconomic Policy After the Crisis* (Cambridge, MA: MIT Press).

Amar, D. and E. Avgouleas (eds.) (2016), *Reconceptualising Global Finance and Its Regulation* (Cambridge: Cambridge University Press).

Anabtawi, I. and S. Schwarcz (2011), 'Regulating Systemic Risk: Towards an Analytical Framework', 86 *Notre Dame Law Review* 1349.

Anand, K., P. Gai, S. Kapadia, S. Brennan and M. Wilson (2013), 'A Network Model of Financial System Resilience', 85 *Journal of Economic Behaviour & Organisation* 219.

Anderson, P, Arrow, K and D. Pines (eds.) (1988), *The Economy as an Evolving Complex System* (New York: Addison Wesley).

Arinaminpathy, M., S. Kapadia and S. May (2010), 'Size and Complexity in Model Financial Systems Bank of England Working Paper 465', 109(45) *Proceedings of the National Academy of Sciences* 18338.

Arnold, C. and L. Gunderson (2014a), 'Adaptive Law and Resilience', 43 *Environmental Law Reporter* 10426.

Arnold, C. and L. Gunderson (2014b), 'Adaptive Law', in A. Garmestani and C. Allen (eds.) *Social-Ecological Resilience and Law* (New York: Columbia University Press).

Arthur, B. (2015), *Complexity Economics* (Oxford: Oxford University Press).

Arthur, B., S. Durlauf and D. Lane (eds.) (1997), *The Economy as a Complex Adaptive System II* (New York: Addison Wesley).

Bak, P., C. Ting and K. Wiesenfeld (1987), 'Self-Organising Criticality', 59(4) *Physical Review of Letters* 381.

Battiston, S., J. D. Farmer, A. Flache, D. Garlaschelli, A. G. Haldane, H. Heesterbeek, C. Hommes, C. Jaeger, R. May and M. Scheffer (2016), 'Complexity Theory and Financial Regulation', 351 *Science* 6275, 818.

Battiston, S., G. Caldarelli, C. Georg, R. May and J. Stiglitz (2013), 'The Complexity of Derivatives Networks', 9 *Nature Physics* 123.

Baxter, L. (2012), 'Betting Big: Caution and Accountability in an Era of Large Banks and Complex Finance', 31 *Review of Banking and Finance Law* 765.

Baxter, L. (2012), 'Capture Nuances in Financial Regulation', 47 *Wake Forest Law Review* 537.

Baxter, L. (2016), 'Understanding the *Global* in Global Finance and Regulation', in R. Buckley, E. Avgouleas and D. W. Arner (eds.) *Reconceptualizing Global Finance and Its Regulation* (Cambridge: Cambridge University Press), 28.

Beinhocker, E. (2007), *The Origin of Wealth: Evolution, Complexity and the Radical Remaking of Economics* (London: Random House).

Berwell, R. (2013), *Macroprudential Policy* (London: Palgrave MacMillan).

Blanchard, O., G. Dell'Ariccia and P. Mauro (2014), 'Introduction: Rethinking Macro Policy II', in G. Akerlof et al. (eds.) *What Have We Learned: Macroeconomic Policy After the Crisis* (Cambridge, MA: MIT Press).

Borio, C. (2003), 'Towards a Macroprudential Framework for the Financial System and Regulation', 49(2) *CESifo Economic Studies* 181.

Borio, C. (2014), 'Macroprudential Policy and the Financial Crisis', in G. Akerlof, O. Blanchard, D. Romer and J. Stiglitz (eds.) *What Have We Learned: Macroeconomic Policy After the Crisis* (Cambridge, MA: MIT Press).

Brunnermeirer, M., A. Crocket, C. Goodhart, A. Persaud and H. Shin (2009), *The Fundamental Principles of Financial Regulation* (Geneva: International Centre for Monetary and Banking Studies).

Cheffin, B. C., A. S. Garmestani, L. H. Gunderson, M. H. Benson, D. G. Angeler, C. A. Arnold, B. Cosens, R. Kundis Craig, J. B. Ruhl and C. R. Allen (2016), 'Transformative Ecological Governance', 41 *Annual Review of Environment and Resources* 399.

Chandler, D. (2014), *Resilience: The Governance of Complexity* (Abingdon: Routledge).

Cherry, B. (2007), 'The Telecommunications Economy and Regulation as Co-Evolving Complex Adaptive Systems', 59 *Federal Communications Law Journal* 369.

Cherry, B. (2008), 'Institutional Governance for Essential Industries under Complexity' 17 CommLaw Conspectus L & Pol 1.

Chinen, M. (2011), 'Governing Complexity', in L. Boule (ed.) *Globalisation and Governance* (Cape Town: Siber Ink) 43.

Cincotti, S., D. Sornette, P. Treleaven, S. Battiston, G. Caldarelli, C. Hommes and A. Kirman (2012), 'A European Economic and Financial Exploratory', 214 *European Physical Journal of Special Topics* 361.

Claessens, S. and B. Evanoff (2011), *Macroprudential Regulatory Policies* (London: World Scientific).

Colander, D., H. Föllmer, A. Haas, M. Goldber, K. Juselius, A. Kirman, T. Lux and B. Sloth (2008), 'The Financial Crisis and the Systemic Failure of Academic Economics', Opinion Par 98th Dahlem Workshop.

Colander, D. and R. Kupers (2014), *Complexity and the Art of Public Policy* (Princeton, NJ: Princeton University Press).

Cooper, M. (2011), 'Complexity Theory After the Financial Crisis', 4(4) *Journal of Cultural Economy* 371.

Craig, R. (2010), 'Stationarity Is Dead: Long Live Transformation: Five Principles for Climate Change Adaptation Law', 31 *Harvard Environmental Law Review* 9.

Craig, R. and J. B. Ruhl (2014), 'Designing Administrative Law for Adaptive Management', 67 *Vanderbilt Law Review* 1.

Crockett, A. (2000), 'Marrying the Micro- and Macro-Prudential Dimensions of Financial Stability', Bank of International Settlements Eleventh International Conference of Bank Supervisors Basel <www.bis.org/speeches/sp000921.htm>, accessed 12 March 2018.

Ellis, L., A. Haldane and F. Moshirian (2014), 'Systemic Risk, Governance and Global Financial Stability', 45 *Journal of Banking and Finance* 175.

Farmer, J. D. et al. (2012a), 'A Complex Systems Approach to Constructing Better Models for Managing Financial Markets and the Economy', 214 *European Physical Journal of Special Topics* 295.

Farmer, J. D. et al. (2012b), 'Economics Needs to Treat the Economy as a Complex System' <www.inet.ox.ac.uk/library/view/595>, accessed 12 March 2018.

Fontana, M. (2010), 'The Santa Fe Perspective on Economics', 18(2) *History of Economic Ideas* 167.

Freixas, X. and L. Laeven (2015), *Systemic Risk, Crises, Macroprudential Regulation* (Cambridge, MA: MIT Press).

Gai, P. (2013), *Systemic Risk: The Dynamics of Modern Financial Systems* (Oxford: Oxford University Press).

Gai, P., A. Haldane and S. Kapadia (2011), 'Complexity, Concentration, and Contagion', 58(5) *Journal of Monetary Economics* 453.

Gai, P. and S. Kapadia (2010), 'Contagion in Financial Networks', Bank of England Working Paper No 383.

Garmestani, A. and C. Allen (eds.) (2014), *Social-Ecological Resilience and Law* (New York: Columbia University Press).

Garmestani, A., C. Allen and H. Cabezas (2009), 'Panarchy, Adaptive Management and Governance: Policy Options for Building Resilience', 87 *Nebraska Law Review* 1036.

Garmestani, A. S., C. R. Allen, J. B. Ruhl and C. S. Holling (2014), 'The Integration of Social-Ecological Resilience and Law', in A. Garmestani and C. Allen (eds.) *Social-Ecological Resilience and Law* (New York: Columbia University Press).

Garmestani, A. and M. Benson (2013), 'A Framework for Resilience-Based Governance of Social Ecological Systems', 18(1) *Ecology & Society* 9.

Gunderson, L., C. Allen and C. S. Holling (2010), *Foundations of Ecological Resilience* (London: Island Press).

Gunderson, L. and C. Holling (2002), *Panarchy: Understanding Transformation in Human and Natural Systems* (Washington, DC: Island Press).

Haldane, A. (2009a), 'Rethinking the Financial Network', Speech to Financial Student Association, Amsterdam, 24 April 2009.

Haldane, A. (2009b), 'Small Lessons from a Big Crisis', Remarks at the Federal Reserve Bank of Chicago, 45th Annual Conference, Reforming Financial Regulation, 8 May 2009, Chicago.

Haldane, A. (2012), 'The Dog and the frisbee', Speech Jackson Hole 31 August 2012 <www.bis.org/review/r120905a.pdf>, accessed 12 March 2018.

Haldane, A. (2014a), 'Managing Global Finance as a System', Maxwell Fry Annual Global Finance Lecture Birmingham University 23.10.14.

Haldane, A. (2014b), 'Macroeconomic Policy in Prospect', in G. Akerlof et al. (eds.) *What Have We Learned: Macroeconomic Policy After the Crisis* (Cambridge, MA: MIT Press).

Haldane, A. (2015), 'On Microscopes and Telescopes', Speech Lorentz Centre, Leiden 27.03.15.

Haldane, A. and R. May (2011), 'Systemic Risk in Banking Ecosystems', 469 *Nature* 351.

Helbing, D. (ed.) (2008), *Managing Complexity: Insights, Concepts, Applications* (Berlin: Springer).

Helbing, D. (2009), 'Managing Complexity in Socio-Economic Systems', 17 *European Review* 423.

Helbing, D. (2010), 'Systemic Risks in Society and Economics', International Risk Governance Council <www.irgc/IMG/pdf/Systemic_Risks_Helbing2.pdf>, accessed 12 March 2018.

Helbing, D. (ed.) (2012), *Social Self-Organisation: Agent Based Simulations to Study Emergent Social Behaviour* (Berlin: Springer).

Helbing, D. (2013a), 'Globally Networked Risk', 497 *Nature* 51 02.05.13.

Helbing, D. (2013b), 'Economics 2.0: The Natural Step Towards a Self-Regulating Participatory Market Society', 10(1) *Evolutionary and Institutional Economics Review* 3.

Helbing, D. (2015a), *Thinking Ahead: Essays on Big Data, Digital Revolution and Participatory Market Society* (Berlin: Springer, Kindle ed.).

Helbing, D. (2015b), *The Automation of Society is Next* (Kindle: CreateSpace Independent Publishing Platform).

Helbing, D. et al. 'FuturICT Participatory Computing for our Complex World' <www.futurict.eu/>, accessed 12 March 2018.

Helbing, D. and S. Balietti (2010), 'Fundamental and Real World Challenges in Economics', 76(10) *Science & Culture* 399.

Helbing, D. and A. Kirman (2013), 'Rethinking Economics Using Complexity Theory', 64 *Real World Economic Review* 1.

Holling, C. (1973), 'Resilience and Stability of Ecological Systems', 4 *Annual Review of Ecology and Systematics* 1.

Holling, C. (1978), *Adaptive Environmental Assessment and Management* (New Jersey: Blackburn Press).

Hollow, M., F. Akinbami and R. Michie (eds.) (2016), *Complexity, Crisis and the Evolution of the Financial System: Global Perspectives on American and British Banking* (London: Edward Elgar).

Holt, R., J. Rosser and D. Colander (2010), 'The Complexity Era in Economics', Middlebury Economics Discussion paper 10–01 <sandcat.middlebury.edu/econ/repec/mdl/ancoec/1001.pdf>

Johnson, N. and T. Lux (2011), 'Financial Systems: Ecology and Economics', 469 *Nature* 302.

Kambli, J., S. Weldman and N. Krishnan (2007), *New Directions for Systemic Risk* (New York: National Academy Press).

Kauffman, S. (1995), *At Home in the Universe: The Search for Laws of Self-Organisation and Complexity* (Oxford: Oxford University Press).

Kim, R. and B. Mackey (2014), 'International Environmental Law as a Complex Adaptive System', 14 *International Environmental Agreements: Politics, Law and Economics* 5.

Kirman, A. (2011), *Complexity Economics: Individual and Collective Rationality* (London: Routledge).

Landou, J. P. (2009), 'The Macroeconomy and Financial System in Normal Times and Times of Stress', Speech at conference Gouvieux-Chantilly 08.06.09.

Lippe, P., D. Katz and D. Jackson (2015), 'Legal by Design: A New Paradigm for Handling Complexity in Banking Regulation and Elsewhere in Law', 93 *Oregon Law Review* 833.

May, R. (2012), 'What Biology Can Teach Us About Banking', 26 *The Santa Fe Institute Bulletin* 32.

May, R. (2013), 'Networks and Webs in Ecosystems and Financial Systems', 371 *Philosophical Transactions of the Royal Society A* 20120376.

May, R. and N. Arinaminpathy (2010), 'Systemic Risk: The Dynamics of Model Banking Systems', 7(46) *Journal of Royal Society Interface* 823.

May, R., S. Levin and G. Sugihara (2008), 'Complex Systems: Ecology for Bankers', 451 *Nature* 893.

Mitts, J. (2015), 'Systemic Risk and Managerial Incentives in the Dodd-Frank Orderly Liquidation Authority', 1 *Journal of Financial Regulation* 54.

Omerod, P. and D. Helbing (2012), 'Back to the Drawing Board for Macroeconomics', in D. Coyle (ed.) *What's the Use of Economics: Teaching the Dismal Science After the Crash* (London: Publishing Partnership).

Persaud, A. (2009), 'Macroprudential Regulation: Fixing Fundamental Market (and Regulatory) Failure', World Bank Crisis Response: Public Policy for the Private Sector.

Prigogine, I and Stengers, I (2018), *Order Out of Chaos* (London; Verso Reprints).

Rosser, J. (ed.) (2009), *Handbook of Research on Complexity* (London: Edward Elgar).

Rhul, J. B, (1995), 'Complexity Theory as a Paradigm for the Dynamic Law-and-Society System: A Wake-Up Call for Legal Reductionism and the Modern Administrative State', 45(5) *Duke Law Journal* 849.

Rhul, J. B.(1996), 'The Fitness of Law: Using Complexity Theory to Describe the Evolution of Law and Society and Its Practical Meaning for Democracy', 49 *Vanderbilt Law Review* 1407.

Ruhl, J. B. (1997), 'Thinking of Environmental Law as a Complex Adaptive System: How to Clear Up the Environment by Making a Mess of Environmental Law', 34 *Houston Law Review* 933.

Ruhl, J. B. (2004), 'Taking Adaptive Management Seriously: A Case Study of the Endangered Species Act', 52 *University of Kansas Law Review* 1249.

Ruhl, J. B. (2005), 'Regulation by Adaptive Management – Is It Possible?' 7 *Minnesota Journal of Law, Science & Technology* 21.

Ruhl, J. B. (2011), 'General Design Principles for Resilience and Adaptive Capacity in Legal Systems – with Applications in Climate Change Adaptation', 89 *North Carolina Law Review* 1373.

Ruhl, J. B. (2012), 'Panarchy and the Law' 17(3) *Ecology & Society* 31.

Ruhl, J. B. (2013), 'Managing Systemic Risk in Legal Systems', 89 *Indiana Law Journal* 559.

Ruhl, J. B. and D. Katz (2015), 'Measuring, Monitoring and Managing Legal Complexity', 101 *Iowa Law Review* 191.

Ruhl, J. B. and Ruhl, H. (1996), 'The Arrow of Time in Modern Administrative Stares: Using Complexity Theory to Reveal the Diminishing Returns and Increasing Risks the Burgeoning of Law Poses to Society', 30 *UC Davis Law Review* 405.

Schwarcz, S. (2008), 'Systemic Risk', 97(1) *Georgetown Law Review* 193.

Schwarcz, S. (2009), 'Regulating Complexity in Financial Markets', 87(2) *Washington University Law Review* 211.

Sornette, D. (2003), *Why Stock Markets Crash: Critical Events in Complex Financial Systems* (Princeton, NJ: Princeton University Press).

Sornette, D. (2004), *Critical Phenomena in Natural Sciences: Chaos, Fractals. Self-Organisation and Disorder* (Berlin: Springer).

Tucker, P. (2009), 'The Debate on Financial Resilience: Macroprudential Instruments', Speech Barclays Annual Lecture October <www.bankofengland.co.uk/archive/Documents/historicpubs/speeches/2009/speech407.pdf>, accessed 12 March 2018.

Tucker, P. (2011), 'Macroprudential Policy: Building Financial Stability Institutions', Speech 14.04.11 Bank of England <www.bankofengland.co.uk/archive/Documents/historicpubs/speeches492.pdf>, accessed 12 March 2018.

Tussey, D. (2005), 'Music at the Edge of Chaos: A Complex Systems Perspective on File Sharing', 37 *Loyola University Law Review* 101.

Utset, M. (2010), 'Complex Financial Institutions and Systemic Risk', 45 *Georgia Law Review* 779.

Waldrop, M. (1992), *Complexity: The Emerging Science at the Edge of Order and Chaos* (New York: Touchstone).

Walker, B. and D. Salt (2008), *Resilience Thinking: Sustaining Ecosystems and People in a Changing World* (London: Island Press).

Zeidan, R. and K. Richardson (2010), 'Complexity Theory and the Financial Crisis: A Critical Review', 14(6) *Corporate Finance Review* 10.

Zhang, K. and A. Schmidt (2015), 'Thinking of Data Protection Law's Subject Matter as a Complex Adaptive System: A Heuristic Display', 31 *Computer Law & Security Review* 201.

Section V

Complexity and the ethics of law and legal practice

11 Nonlinearity, autonomy and resistant law

*Lucy Finchett-Maddock**

It can be a little difficult to plot a timeline of social centres when you're dealing outside of linear time.
                                    – Interviewee from rampART collective, 2009 in
                                                    Finchett-Maddock (2016, p. 168)

This chapter argues that informal and communal forms of law, such as that of social centres, occupy and enact a form of spatio-temporal 'nonlinear informality', as opposed to a reified linearity of state law that occurs as a result of institutionalising processes of private property. Complexity theory argues the existence of both linear and nonlinear systems, whether they be regarding time, networks or otherwise. Working in a complexity theory framework to describe the spatio-temporality of law, all forms of law are argued as nonlinear, dependent on the role of uncertainty within supposedly linear and nonlinear systems and the processes of entropy in the emergence of law. 'Supposedly' linear, as in order for state law to assert its authority, it must become institutionalised, crystallising material architectures, customs and symbols that we know and recognise to be law. Its *appearance* is argued as linear as a result of institutionalisation, enabled by the elixir of individual private property and linear time as the congenital basis of its authority. But linear institutionalisation does not account for the role of uncertainty (resistance or resistant laws) within the shaping of law and demonstrates state law's violent totalitarianism through institutionalising absolute time. Unofficial, informal, autonomous and semi-autonomous forms of law such as those expressed by social centres (described as 'social centre law') remain non-institutional and thus perform a kind of informal nonlinearity, expressed through *autonomy-as-practice* and *autonomy-as-placement*, highlighting the nonlinear nature of autonomy and the central role of spatio-temporality within law and its resistance. The piece argues it is important that lawyers and other thinkers understand the role of space and time in the practices of law, how forms of spatio-temporality shape the ideologies that determine law and how it is organised, in order to better understand the origin and trajectory of law, resistance and the world it shapes around us and the usefulness of complexity theory in demonstrating this.

Emergent Themes

Ten years have passed since Manuel DeLanda mapped his dynamics of materialism in the 2007 *A Thousand Years of Nonlinear History* (DeLanda, 2007); subsequently, the terminological focus of this chapter 'nonlinearity' has become a familiar presence in legal, social, ecological and artificial relations today, whether we are aware of it or not. DeLanda's account of history took us through geology, language and markets to explain how change in time, although appearing to happen in a linear fashion from one chronology to the next, actually moves in an indirect, baroque movement, catalysing as the world responds to unforeseeable events, changing and adapting as it shifts and morphs across space and temporalities. One society to the next can be at infinite levels of transformation at the same time as another, but no one society is more 'developed' than another. This is not the linear story of progress in history or the Social Darwinism with which we have become familiar. DeLanda highlighted how 'both classical thermodynamics and Darwinism admitted only one possible historical outcome, the reaching of thermal equilibrium or of the fittest design. In both cases, once this point was reached, historical processes ceased to count' (DeLanda, 2007, p. 15). Within the nonlinear account of history, however, history is open, multiple and unbounded, a chorus of stages, events, atrocities, celebrations, milestones and developments occurring simultaneously across the globe's ecologies and cultures. A nonlinear history does not discriminate through the binding of solely linear time and a dogma of progress.

So what is nonlinearity? Nonlinearity is a scientific term for the way complex and dynamic systems behave, where the outcome of a system cannot be reduced to its input, inferring a 'bottom-up' and unplanned motion, where something has been created in the process of a phenomenon changing and responding to its environment. Both linearity and nonlinearity are the results of these *emergent* patterns of organisation and 'complex adaptive behaviour', whereby systems form in response to the *uncertain* nature of their surroundings, creating irreducible structures from 'emergent properties', the whole greater than the sum of their parts, order out of chaos and the same vice versa. It is this element of uncertainty that can account for the apparent 'leaderlessness' of nonlinear phenomena, as it occurs in response to its environment and not the top-down decisions of something or someone organising. As DeLanda explained, it is not the planned results of human action but the unintended collective consequences that create the world around us (2007, p. 17). This science of nonlinearity (complexity science, which we will come to in more detail later) has become central in describing not only our natural ecosystems but also our human-made social, cultural, political and economic systems and how we organise and govern ourselves as a result. The bottom-up nature of nonlinearity is of core import to this journey into the world of a so-called resistant law of social centres (radical community centres that are often squatted), given the predominantly anti-authoritarian make-up of their participants, and the legal loophole within which they can exist (if occupied without the owner's permission). For the purposes of this chapter, and for reasons of clarity, given the vastness of the scientific and non-scientific literature around nonlinearity, the chapter focuses predominantly on the spatio-temporal relation of nonlinearity and how this grounds legal (and otherwise) forms of organisation as a result.

So how is this plurivocality so prevalent in our everyday lives? It is interesting to consider the extent to which the understanding and resultant harnessing of nonlinear organisation have been incorporated into scientific and technological practice, effecting and affecting the forms of technology and interfaces we use daily. Marketised understandings of nonlinearity now propel and shape our technological lives, such as through the use of algorithms, computational models to generate predictions of peoples' tastes and preferences, sold back to advertisers and companies for marketing purposes. Social networking sites such as Facebook are a prime example, where your data are used in a nonlinear manner, their ever alternating in response to its users whilst the interfaces themselves being fine illustrations of 'emergent' networks. From the personalised adverts we receive catalysed by our interactions with social networks, through to predictive texts and weather apps, to the unfathomable growth of the internet and immense virtual and real networking platforms themselves, we can see that Stephen Hawking's prediction of the 21st century being one of 'complexity' is not too far off being proven correct (Hawking, 1988, p. 273).

Complexity theory is the practical science that grounds explanations of nonlinearity and the forms of *emergence* that occur as a result of nonlinear processes, describing and predicting the network's relationships in all forms of life. The commercial co-optation example of complexity is interesting considering this chapter's non-marketised nature of nonlinear organisation expressed in the example of social centres and the resistant laws they perform. Yet at the same time what we are about to deliberate is the potential nonlinear choreography of everything, from law (whether to resistant or otherwise) to property relations to the commercial co-optation of nonlinear dynamics, neoliberalism and post-truth politics itself and back around to explaining the motherboard functions and nature of all dynamic and inert life.

You might ask how an intuitive text message function may in any way be connected to understanding law, or why this is of any use to the study of law and the communal forms of resistance described. For any scholar of complexity, the association would not be such a mystery. In *Protest, Property and the Commons: Performances of Law and Resistance* (2016), I argued that social centres create their own form of law, premised on the framework of complexity theory to explain communal behaviour, the organisation and resistant forms of legality created as a result.

The law was resistant on the very basis of its nonlinearity, whereby it evaded state law institutionalisation through remaining informal (bottom-up) and thus, I argued, nonlinear. This nonlinearity is expressed in the spatio-temporal practices of the centres, both in the form of time that the participants identified with (in examples such as the creation of their nonlinear timeline) and the philosophy and practices of autonomy attached to their spatio-temporal organisation. This chapter similarly hopes to explain the nonlinear nature of both state and non-state laws, state law's desire to appear organised in a linear manner and the usefulness of complexity in understanding relations of law and resistance, in general. My research led me to analyse the workings of the UK social centre scene, the portrayals of which offering the basis for the descriptions used in this work.

Ultimately, it is important that lawyers and other thinkers understand the role of space and time in the practices of law, how forms of spatio-temporality shape the ideologies that determine law and how it is organised and the usefulness of complexity theory in demonstrating this. Understandings of law (and property) are argued as supported by understandings of time, where through private property, law becomes institutionalised and linear, in distinction from its nonlinear origin of the commons. As a result, one form of organisation is discredited at the behest of another, that is nonlinear versions of time that support communal practices of leaderless networks silenced by absolute linear time that sits with the vision of capital, individual private property and progress. Social centres, and other such examples, demonstrate (within their limitations) that there may be other forms of social organisation that operate on communal, less hierarchical terms akin to nonlinear dynamics and the form of time that supports this.

First we shall introduce social centres and then look at the complexity framework and its relation to nonlinearity as a concept and framework in law. Next state law and how it seeks to appear organised in a linear manner and why, shall be considered. After that, legal pluralism is considered as an explication of informal and formal laws, the move of institutionalisation that occurs between them and the role of linearity and nonlinearity within this. We then turn to the ascribed social centre law to understand how it is performed and how the philosophical underpinnings of autonomism in social centres, and the practices that ensue, offer a description of nonlinearity and nonlinear law. Autonomy-as-practice and autonomy-as-placement are used to describe the innate role of time and space within the organisation of nonlinearity and the autonomous law produced as a result, as well as the moulding role of spatio-temporality in state law and what this may teach us about the nature of law, property and resistance overall.

Social centres

Social centres are communally run buildings that are either squatted, rented or owned. There are varying concerns that shape the make-up and activities within social centres, propelled by premises of community and politically based activity, creativity, inclusion and, most relevant for this discussion of social centres and nonlinearity, autonomy. Each centre operates according to its own agenda and thus has peculiar characteristics as moulded by their participants, the community surrounding them and the philosophy and politics with which they associate themselves. Some spaces see themselves as more community-driven, whereas others are more event and political meeting spaces. For instance, the Library House in Camberwell, London (now evicted), was an example of a social centre that did a lot of outreach work with the local community, whereas the 'rampART' space in Whitechapel (London) was more of a meeting space and one that held benefit events and fundraising nights. Social centres attract a pastiche of folk, from young to old, from all racial and ethnic backgrounds, genders and abilities. Despite this, the organised squatting scene of which social centres are part, has been criticised for the prevalence of its participants' privileged higher education backgrounds, where the spaces are seen as occupied less for pure need than more a social experiment. That said, the social centre and squatting scenes offer access to a rich tapestry of not just London and UK radicalism, but anti-authoritarianism and social centre traditions across the rest of Europe. They are places in which, according to the Social Centre Network (SCN) website, 'people can come together to create, conspire, communicate and offer a collective challenge against capitalism'. Within this research, the social centres of the UK have been of focus, specifically within London and Bristol between 2006 and 2011 when the project was undertaken (Finchett-Maddock, 2011).

The groups' form of organisation is markedly altered to, and to some extent, outside of state remit, although with some fundamental reliance on state Law in the UK, to allow the spaces to happen lawfully through 'squatters' rights' as per the Criminal Law Act 1977. This condoning of squatting by state law has been limited in recent years through its recategorisation as a criminal offence in residential property under the Legal Aid Sentencing and Punishment of Offenders Act s. 144; thus, this semi-autonomy from the state is only available now in commercial properties. Not all social centres are squatted, but the focus of my research became these squatted social centres because of their contestation of space, property and the law, through their own form of alternate social organisation.

A number of interviews were conducted as the basis of the research. I distinctly remember interviewing a member of the rampART collective back in 2009 whose insights were of great value to the task of understanding a potentially law-innovating energy emanating from social centres. I remember she drew a cartographic picture of the centres on a planetary scale, saying squats and social centres find themselves within the 'chinks of the world machine', using a famous quote from feminist science-fiction writer Alice Sheldon (better known as James Tiptree Jnr). I was so moved and inspired by this; how were these chinks of the world machine created and how important are those who find themselves in these apertures to the workings of this world machine? What alternative way of life was being offered within these spaces and what role did law and resistance have to play? Of course, not all these questions were answered, but one way of formulating a theory around these chinks, these openings, was to look to complexity theory and nonlinearity for help.

The dynamics of the centres are argued as nonlinear by their nature, as bottom-up horizontal structures that operate in reaction to, and as a result of, state structures and forms of individual property relations. Social centre participants directly critique the vertical hierarchies of law, individualism and capitalism, through contesting the spaces and organising themselves communally, in a radically different way to that of state law sees itself.

Complexity and the nonlinear

Complexity, as the grounding enquiry of this collection, is a theoretical framework explaining the networked relationships in all parts of life, where all material and ontological happenings occur as a result of their interface with their environment. This can be in an irreducible 'bottom-up' movement of cause and effect, with no presupposed order or unidirectional sequence, of which nonlinearity is an organising relation. The nonlinear nature of complex adaptive behaviour accounts for the element of unpredictability in events that may occur, their unfinished openness, as well as the leaderless, upended movement and change, where within systems their output is disproportionate to their input, creating their disordered characteristics.[1]

Beginning in the sixties, and yet having gathered momentum back the century before through the bio-mathematical science of thermodynamics as well as the philosophical thinking in Spinoza (2001) and Schelling (1989), work on complex adaptive systems and nonlinearity first determined itself as a science in its own right through the increasing pace of computer information and the development of cybernetics. Since then, explanations of the patterns in urban development, biological systems, cybernetic networks, law and social movements have all become subject to complexity theory (Johnson, 2001; Escobar, 2003; Urry, 2006; Byrne, 2005), the method and metaphor for understanding the underlying nonlinear causation of social activity, as well as describing the substance of all forms of life itself.

Complexity or *emergence* thus explains bottom-up behaviour 'when the actions of multiple agents interacting dynamically and following local rules rather than top-down commands result in some kind of visible macro-behaviour or structure' (Escobar, 2003, p. 351). Escobar applies an analogy of the 'swarm' that he paints so effectively, whereby sea life sometimes amasses to create a greater shape in order to protect themselves. He treats categories of self-organisation, nonlinearity and non-hierarchy as those that are not peculiarly the products of biology but can also be applied to the observance of social movement behaviour and social life, in general.[2]

What about complexity in relation to law? Philippopoulos-Mihaloupoulos (2015) talks about there being nothing outside of law, the law being part of every 'assemblage' of everything else that makes up ourselves and our reality around us. This moveable feast of atomical structuring and re-structuring demonstrates law's 'becoming', echoing Deleuze and Guattari (2004), Latour (2007), DeLanda (2007, 2011) and Johnson (2001), 'whereby the actions of multiple agents interacting dynamically and following local rules rather than top-down commands result in some kind of visible macro-behaviour or structure' (2001, p. 231). This body of Deleuzian-inspired work, as well as speculative realist thinking and Object Oriented Ontology (Meillassoux, 2008; Harman 2016; Levi Bryant et al., 2011), works on the same nexus of non-hierarchical, plateaued principles where even objects have agential force within networks, part and parcel of a shift from humanism to post-humanism through complexity within philosophical thinking (Finchett-Maddock, 2017).

Within legal theory specifically, Jamie Murray's Deleuzian 'emergent law' (Murray, 2008, p. 236) relays complexity lawyer J. B. Ruhl's work on constitutional law, chaos theory and the overarching study of law (Ruhl, 1996). Murray and Ruhl argue nonlinear dynamics and complexity as having considerable impetus to the study of law, highlighting how the element of uncertainty in complexity, the element of unpredictability and preponderance for 'entropy' of ornate scales, describes law's emergence, as opposed to the much-propounded reliability and supposedly foreseeable nature of juridical structures. This openness to systems of law being unfixed and *becoming* is a nod to emergent processes and its role in the process and as the product of law and its rules and procedures.

Underlying the study of complexity is a reliance on this scientific measurement and substance of *entropy*. Entropy is, amongst many other definitions, a scientific explanation of the relationship between order and chaos within dynamic systems and is a quantitative measure of the amount of disorder in a system (Arnheim, 1971, p. 8). The more entropy there is, the more there is chaos,[3] and thus, systems strive for order but move towards maximum disorder. Entropy is referred to in mathematics and biology in a myriad of terms, most notably within thermodynamics, complexity science or chaos theory, and can be divided into three broad contexts: information or complexity, the arrow of time and uncertainty. The more information there is in a system, the more entropy there is and the more complex it becomes. According to the second law of thermodynamics, entropy can only ever be supplemented to and not be reduced, giving credence to the argument that time can only ever go in one direction. Entropy, however, also accounts for nonlinearity, as a result of the emergent movement and interfacing of each new part of a burgeoning system, where *uncertainty* allows for the chaotic drive of entropy production – a chance for *nonlinear* time.[4]

Because of this consideration of entropy, complexity theory is altered from systems theory and its emphasis on the self-producing 'autopoeisis', found in the work of Luhmann (2012) and Maturana and Varela (1974), whereby unity and distinction between the system and background environment collide, denoting first-order systems and second-order systems with the alternating role of the boundary as central. And it is because of this element of uncertainty that systems theory and complexity theory are so different, as complexity can argue the emergent and open nature of systems because of the instrumental role of entropy. The saliency of complexity theory has been re-asserted by Thomas Webb, as an alternative to systems theory (2013), and distinguished as quite separate from the work of Luhmann *et al.* by its *becoming* nature, its usefulness in describing the relations between law and its other, and yet setting it apart from systems theory as a body of understanding which is not closed, reliant on adaptation and a constant re-drawing, reconfiguration of boundaries. Here lies the contingent role of entropy in allowing for uncertainty, nonlinearity and change within systems and systems thinking.

Similarly, Jamie Murray argues a complexity understanding of law's nonlinearity permits the search for "lost, hidden, local, bottom-up, emergent modes of legality, and for a new conceptual creativity in [legal] work" (2008, p. 227).

This emergent understanding of law describes not just state law processes but also those of non-state laws and non-legal movements and has been used extensively in the legal pluralism of sociologist and economist Boaventura de Sousa Santos (2004b, 2005a, 2005b). Santos describes a 'sociology of absences' and a 'sociology of emergence' where emergent thinking can allow us to see laws that emanate from below, reaching upwards the subaltern, as opposed to the top-down structures of law with which we have become so familiar (1998, 2003, 2004a). He also proposes a 'continuum of formalism' mapping the movement from formal to informal legalities that is useful for understanding both linear and nonlinear juridical forms (1997).

Law and linearity

The easiest way to understand nonlinearity is to outline the linear nature of state law (or at least its desire to be linear) and the juncture or point at which state law begins to 'institute' or formalise itself and to understand exactly why linearity and nonlinearity mean anything in relation to law and its resistance, in turn.

Linearity is a simple term that has a number of ascriptions, predominantly within history, aesthetics, politics, geography and the algorithmic spatio-temporal organisation grounding each and all of these. Linear projections of time have dominated conceptions of history and temporality, at the behest of other understandings (Bastian, 2014, p. 145), thus to suggest an altered stance of time is to critique linearity's absoluteness. This preponderance of temporal understandings is evocative of E. P. Thompson's work on the role of time in labour and the capitalist workplace (Thompson, 1967). Walter Benjamin's allegory of the 'angel of history' (1999) is probably the most famous and simultaneously arrestingly beautiful critique of linear conceptions of time and historicity and our association of linearity with evolutionary progress and thus the manner in which time is predominantly understood in a monocultural form, as de Sousa Santos would say (1999). Linear time and chronological history exemplify the Western liberal concern for progress in society through developments in technology, industry as a result of accumulation, colonialism and the primordial role of capital as 'time as money' and the progenitor of organisation and order within society. As Robinson and Twyman (2014, p. 53) explain through the progressive trajectory of Parliamentary process,

> [p]arliamentary time is inherently progressive; it presupposes constant development along a linear trajectory. To be progressive is therefore to be successful. It is to demonstrate the capacity to shape the future – or at least to anticipate it. The description of particular policies as 'progressive' carries the implication that they are inevitable; historical time moves on and we must move with it or be left behind. Those who do not progress can only decline.

Baudrillard similarly speaks of the congenital link between linearity and accumulation, the fixation of dominant forms of capital with perfection that leads to a total annihilation (1993). State law time seeks to be linear as it confers time as capital, time as property and time as progress with a march forwards at all costs and proclaims itself as absolute, necessitating the categorising of private property through limitations and registers, excluding all other notions of law, property and temporality as practices by resistance movements such as squatters of the social centres discussed in this piece.

This hegemony of linearity is repeated within science. According to a New-tonian conception of time that was stalwart until Einstein's theory of relativity, there is such a thing as 'absolute time' that has 'its own nature, [it] flows equably without relation to anything eternal [. . .] the flowing of absolute time is not liable to change' (de Sousa Santos, 2004a. p. 19). This differs from Einstein, whose arguments echo the possibilities of temporal nonlinearity, where time has no being outside of the system of its signifiers (de Sousa Santos, 2004a, p. 19). It can be stretched and shrunk and varies from system to system and constitutes space-time supportive of a nonlinear understanding of temporality by allowing for the possibility of time to happen spontaneously, out of uncertainty, as opposed to the linear straitjacket of forward-facing absoluteness. Time and space are four-dimensional and curved under the influence of mass, allowing such things as 'wormholes' where the past may catch up with the future, and time travel could be possible. Hawking echoes Einstein, stating,

> Space and time are now dynamic qualities: when a body moves, or a force acts, it affects the curvature of space and time – and in turn the structure of space-time affects the way in which bodies move and forces act.
>
> (1988, p. 33)

Linearity does not have to be all bad. Robinson and Twyman recount how conceptions of progress can be varied and thus open to misunderstanding (2014, pp. 51–67). Similarly, Keenan has critiqued the Torrens system of land registra-tion for freezing the linear histories of indigenous native title in Australia, remov-ing past entitlements through the colonial imposition of the common law in preference for formalised, registered settler claims (2017, p. 87).

Nevertheless, linearity in relation to this exposition of law is critiqued by the framework of nonlinear complex systems and an argument for the universality of nonlinearity through the prevalence of uncertainty in all things, exemplified by the changing processes of entropy. Arguably, when state law is described as linear, it is meant that state law *aspires* to be linear through institutionalisation and relies on this appearance of linearity, predictability and constancy to legitimate itself and the ideology of liberal capitalist progress that it actuates. The desired linear direc-tion of institutionalisation occurs through a forward-facing preoccupation with progress, creating institutions through forms of force, following Weberian and Foucauldian biopolitics of rationality and bureaucracy and forms of representation and vertical hierarchies as a result. Specifically speaking of the common law (which was supplemented to and imposed on other models across the world in order to find more capital, more private property, more legitimacy), power is removed from the direct domain of the people by the organisation and central monopolisation of force (the state), leading to the people as represented by the simulacra of democracy. Vertical hierarchy is formulated through the establishment of the institutions and organs of the state. State law happens as a result of this supposedly linear institutionalisation, which impresses as a linear progression of pre-institutional rules and procedures at grass-roots level becoming reified, much like the progressive connotation of parliamentary pace that Robinson and Twymon speak of.

How can we make such a distinct linearity of state law and yet have an argument for nonlinearity within law? Referring back to the complex adaptive behaviour that DeLanda spoke of within his alternative account of history, I argue that all forms of law, whether resistant or institutionalised, are seen as contingent of one another; their true relation is nonlinear (2011, p. 39). In this sense, it is argued that there cannot be a pure form of law but that law is always coloured with resistance, and resistance always casts the seeds of law.

This is inspired by the work of Margaret Davies, through her *proper* and *improper* of property outlining the creation of real and imagined distinctions between formal and informal law (Davies, 2007). Davies describes very clearly the way in which the etymological connotation of being *propertied* and *proper* is the ability to exclude others, stating, 'Positive law itself is also conceptually based upon an originating exclusion, decision, or splitting which establishes a realm of law and a realm of that which is other to law' (Davies, 2007, p. 31). Any 'pure' formation of law (she gives Kelsen's pure law as that which is a law free of foreign elements) will always disallow the 'impure', or that which muddies the sleek surface and constitution of the law. Davies reveals the existence of the improper within the proper realm of the law, as through repetition, it is never unique and thus loses all purity. It is thus 'iterable', a form of mimesis and performance and never peculiar to itself:

> In other words, and to simplify, the formal deconstructive argument is essentially that the proper must refer outside of itself to that which is common, and to its (improper) other. It is never itself, and is therefore a nonidentity, equally common and improper.
>
> (Davies, 2007, p. 31)

How can we assert a cogency between linearity and purity? You only need to look to the laws of traditional Western aesthetics to understand the demoted place of unfinishedness, messiness and unpropertied notions of art and the art world up until the 20th century (when that very messiness became contemporary art at highly saleable prices). Western aesthetics has asserted the need for form, composition and completeness, in a way that attests to a colonial striving of domination, all contaminated determinations to be removed and invisibilised in an enhancement and mechanisation of purity, institutionalisation and absoluteness.

State law thus arguably concerns itself with an aesthetics of purity, order and authority in order to legitimate itself, founded on a doctrine of private property that has to formalise relations in order for capital to exist and flow. Law seeks to assert its cogency and authority through its linear *praxis* of institutionalisation, which is essentially myth made fact, exerting the appearance of an inherent normativity. The same is to be said for law's understanding of linear time in that it is as positive and absolute as it sees itself, in order to support conceptions of *progress* through which individual property rights in the name of capital can flourish. To talk of positive law is tantamount to speaking of absolute time, where law is time, and the same in turn (Finchett-Maddock, 2016, p. 169).

It is true that in a world of the state being subsumed by the market, by these very processes being discussed – the impurity, mixedness, dependence of nonlinearity – that the state can only ever *aspire* to be linear. The congenital role of uncertainty is even more stark, as the world alters in response to itself at increasingly rapid rates on global levels, demonstrating a nonlinearity of magnitude. And yet state law continues to incorporate individualism and totalitarianism to its extreme resulting in the institution of neoliberalism, to the detriment of any openness and plurality.

Legal pluralism

What does differentiate alternative forms of resistance or other forms of law from state law's artificial linearity? A useful body of literature and tool for understanding the burgeoning relation between state law and other forms of law is that of legal pluralism (Griffiths, 1986; Tamanaha, 2000; Engle-Merry, 1988; Teubner, 1992). Legal pluralism describes laws that exist either entirely outside of state bounds, such as 'strong' forms of legal plurality within community dispute mechanisms that have not as yet been subsumed into the state law matrix[5] and thus remain informal or plural forms of laws and organisation such as *Sharia* law that have been recognised and formalised by state law that have become 'weak' in light of their corroboration with state law institutionalisation (Griffiths, 1986). Forms of legal pluralism are synonymous with colonial imposition (in the weak sense), where a settler law has superimposed itself on pre-existing forms of law, as was the circumstance of both common and civil law jurisdictions, and the resultant complexities of multicultural societal make-up both in the Southern and Northern Hemispheres. Informal forms of legal pluralism arguably exist in both Western and Southern spheres, through examples such as anti-authoritarian and anti-capitalist movements synonymised by the social centre groups that seek to resist the dominant law, as well as dispute resolution mechanisms that remain outside of state law formality, such as community customs, rules and regulations specific to an area or practice, with remedies and etiquettes found in arts, sports or traffic, as examples. Santos, for instance, is not speaking of the more postcolonial forms of legal pluralist work but refers to a spatio-temporal understanding of legal influence, where there are facets that interpenetrate in a given zone. He speaks of

> the conception of different legal spaces superimposed, interpenetrated, and mixed in our minds as much as in our actions, in occasions of qualitative

leaps or sweeping crises in our life trajectories as well as in the full routine of eventless everyday life.

(Santos, 1987, p. 287)

But how does an informal law become formalised? It is arguably through the linear processes of institutionalisation that place individual property rights and resource management at the heart of law, that these chaotic, informal pluralities become the order of centralised state law. A very useful means of explicating the movement from informal to formal law is found in Santos and his 'continuum of formalism', describing the processes of institutionalisation from popular forms of justice, protest and extra-state law dispute remedies to those forms of law being incorporated into state law proper. In his article 'The Law of the Oppressed: The Construction and Reproduction of Legality in Pasargada' (1987), he takes of the formation of legalities from a Southern setting which informs this understanding of *institutionalisation* of law. Santos' work is known to draw on Weberian critiques of rationality that question the edifice of institutionalisation, bureaucracy and individualism, in stark contrast to the *emergent* and *bottom-up law* of Pasargada of which he speaks.

Pasargada is a fictitious suburb of Rio de Janeiro; thus, Santos has named the legality that is created 'Pasargada Law' (Santos, 1997, p. 126). The law of Pasargada is a law that deviates; it is a vernacular dispute prevention and dispute settlement court of the 'Pasargada Residents Association' that is made as a result of social exclusion, and yet it lends and borrows from the dominant law (Santos, 1997, p. 100). It is created as a result of necessity, whereby the state system does not accommodate for the said community and therefore other methods of cohesion have been developed. The interclass legal pluralism of which Santos recalls is one that selectively borrows from the official legal system and accordingly occupies a position along a 'continuum of formalism' (de Sousa Santos, 1997, p. 90). The official law does not cater for the community because of their precarious housing status, where 'the strategy of legality tends to transform itself in the legality of the strategy' (1997, p. 104).

Although Santos does not overtly say that this continuum of formalism is linear, its illustration of the unidirectional from informal to formal through the inclusion and co-optation of state law mechanisms insinuates a linear movement of institutionalisation where enforcement, power and forms of representation start to shape the character of the resultant legality.

Taking a legal pluralist stance is to assert that legal innovation occurring outside of state law is not limited to being shaped by the influence of state law institutionalisation, but in most instances this does happen. It certainly does not have to look like state law in order for it to be a form of law, following from a strong legal pluralist conception of legal plurality, and it is this strong legal pluralist conception of legality that is so helpful in describing the informality, and indeed nonlinearity, of resistant laws such as those of the social centre scene. The law produced by social centres, for example, is altered from that of state law, primarily on the basis that it is not constituted or institutionalised; it is a direct form of action and therefore purely present, not representational.

Legal pluralism allows us to understand other forms of laws and compare formal laws with informal laws. The work of Boaventura de Sousa Santos is used to explicate a continuum of formalism, where we can see the move from informal to formal in juridical structures, and is helpful in showing a connection between informality/nonlinearity and formality/linearity. Using a complexity framework combined with legal pluralism, we are able understand how *all* forms of law and social organisation are subject to uncertainty thus nonlinearity, but in order for state law to appear different and unitary from other forms of law, it must *appear* to move in a linear direction. This linear movement is expressed in practices of material and spatial reality where law crystallises as an institution, as an authority, as something that is fixed and has a legitimacy, in an onward march of progress, order and property. This materialism is similarly based on the spatio-temporal organising forces of capital and liberal individualism whose reasoning relies on a linear trajectory of progress and time, thus highlighting the congenital and interlinked role of time and space within the organisation of both linear and nonlinear laws.

The point at which law becomes concerned with private property, I argue is the point at which nonlinear informal law becomes formalised, institutionalised law. Nonlinear law thus prioritises presence as opposed to representation, community as opposed to individualism, horizontal hierarchy as opposed to vertical institutionalisation, the dispersal of power as opposed to the monopoly of force; informality as opposed to formality. This makes nonlinear law entirely differential to state law – which is why the law we have been discussing as state law seeks to concern itself with a mythical linearity, to distinguish itself from its nonlinear nature, and seeks to deny pluralities of law that may take away from its authority and the legitimacy of neoliberal capitalism to which it gives agency. Laws arguably need to be collective by nature, including state law, but the practice of state law becomes fetishised with the process and product of the institution of individual property rights (as expressed through the monopoly of force, representation and hierarchy), to the detriment of its originary present and collective consciousness.

The complicity of law, resistance and non-laws determines whether to direct a continuum of formalism as expressed by the creation of state law or to remain in a 'nonlinear informality' as manifested by examples of laws of resistance and resistant property such as those arguably expressed within the social centre phenomena.

Social centre law

There are a number of revealingly intersectional ways in which the law of social centres displays characteristics of nonlinearity. To understand this, first, we can very briefly have an illustration of what the law purports to be like, and how. Using Tamanaha to assert that law is non-essentialist (2000) and therefore there are no exact definitions of what law 'should' be like, the combination of *re-occupation* and *enactment*[6] is used to describe some of the elements and practices of the law being recorded.

Re-occupation:

a   the *legal or illegal occupation* of the space requiring legal knowledge of squatters' rights, as well as the actual and physical entering into the space, checking the building, organising the security and control of the building, changing locks;

b   the *vernacular* nature of running the centres, from washing up to cleaning the toilets and cooking.

The re-enactment of the spaces:

c   the participants' ability to understand themselves as part of a wider movement, through the recording and *archiving* of their events linked to other social centres, squatting, housing, land and activist movements around the world, predominantly through a 'nonlinear timeline' as well the internet, flyering and amotive feelings of the participants;

d   their *self-management practices* of autonomy that put into *praxis* the philosophies of anarchism and autonomism, relying on leaderlessness, collective and consensus decision-making, hierarchies of skills as opposed to power, and the absence of monetary concern.

The processes and product of the law that were performed are altered from that of state law through its direct, informal, leaderless, anti-hierarchical characteristics, as opposed to the concern for linearity and formality with which the authority of state law preoccupies itself. Of course, social centres are not devoid of conflict and points at which this *theory* of their law can break down or not exist at all. The domination of the spaces by certain characters, genders, racial, ethnic backgrounds and hierarchies of different kinds occur, so this description of their 'law' obviously has its realities and must be presented within its limits.

How these practices and performances observed in the social centre scene create a form of nonlinearity isfirstly organisational, and secondly spatial-temporal. These two corollaries are interchangeable as expressing nonlinear informality, orienting the way social centres of the kind visited for the research,[7] situated their practices and performances of law, temporally and spatially organised. This is referred to as *autonomy-as-practice* and *autonomy-as-placement*, in turn. Autonomism is the philosophical organising force of the social centres discussed, with autonomy argued as being a form of nonlinearity. Autonomy-as-practice and autonomy-as-placement are used to describe the similarities between autonomy and nonlinearity, highlighting the agential spatio-temporality of both. Autonomy-as-practice relates to the form of social organisation, and autonomy-as-placement refers to the spatio-temporal distance the organisation takes from the state.

Autonomy-as-practice

Social centres specifically are known for their autonomist and anarchist leanings, forms of social organisation that directly repeat nonlinear relations expressed by complexity. Social centres' use of autonomous methods of organisation not only describe the distance they situate themselves from the state but also demonstrated the key elixir of time and space within the practices of the participants and the squats they occupied. The emergent, spontaneous and bottom-up nature of autonomy operates in an exact formation of nonlinearity, where autonomy is demonstrably a form of organisation as well as an expression of nonlinear spatio-temporality.

According to Pickerill and Chatterton, autonomy is a principle that concerns movements seeking freedom and connection beyond nation states, international financial institutions, global corporations and neo-liberalism (2006, p. 746). Accordingly, 'autonomy is a socio-spatial strategy, in which complex networks and relations are woven between many autonomous projects across time and space, with potential for trans-local solidarity networks' (2006, p. 732). The recurrent conceptions used by the social centres include those of a lack of central force of power, delineating vertical hierarchies as unnecessary and making redundant any position of leader and leadership. This is altered greatly from the false linear architectures of state law that we have been speaking of thus far.

Crucial is the notion of mutual aid, based on a trust in the goodwill of social organisation, thus rendering any coercive power as unnecessary. Social centres and squats self-organise themselves, through self-management, where they believe in a collective who decide on the initiatives and the rules of the centre, according to consensus. According to Chatterton and Hodkinson, self-management and the characteristic organisational traits of social centres and squats are horizontal formations of open discussion, shared labour and consensus channelled through to generate 'a "DiY politics" where participants create a "social commons" to rebuild service and welfare provision as the local state retreats' (2007, p. 211).[8] A characteristic of self-management and self-organisation is this very *leaderlessness* that is key in the self-organising and spontaneous nature of nonlinearity. This performance of self-organisation, or leaderlessness, through the self-management practices of mutual aid, *trust* and cooperation, are the key values in anarchist and autonomist thought. These are played out in *praxis* and those that operate as a performance, an immanency that relies on autonomy-as-practiced and placed. Other examples may be alternate commons-based, non-vertical hierarchies such other resistance movements, peer-to-peer networks (Dulong de Rosnay, 2016).

A further useful way of seeing autonomy's praxis intertwined within spatio-temporality is how social centres and squats see themselves in relation to the state, having some autonomy taken away through the law partially allowing squats to occur even in limited form, where their law is always cushioned and defined by the disappearing doctrine of squatters' rights that enables it. Squatting and social centre movements are a clear example of a movement that shifts up and down a continuum of formality or along nonlinear versions of informality, dependent upon the level of autonomy from the state. This is a crucial part of understanding social centres and their forms of organisation due to their proximity to the state and simultaneous seeking of autonomy, it still remaining a lawful activity if within a commercial premise (as per Criminal Law Act 1977 s.6; Legal Aid Sentencing and Punishment of Offenders Act 2012 s.144). The less illegal the spaces are (according to the state), the longer they are likely to occur given their acceptance/distance from the state and thus enact forms of 'semi-autonomy' (Falk Moore, 1973). As soon as a space is rented or owned, one can arguably see the processes of individual property occurring through the imposition of capital in the necessity to pay rent, accounting for materials and supplies for the centres, having to consider paying rates, as examples.

Autonomy-as-placement

Autonomy is thus by its nature a nonlinear socio-spatial strategy, as outlined by Chatterton, performing a moment or coordinate of presence or re-presence in proximity to (or within) the coercion of the state. The nonlinearity of autonomy also relates to its spatio-temporality, where nonlinear accounts of space and time inform autonomous and informal forms of law, just as linear accounts of time influence state law structures and formalities. It becomes clear here how semi-autonomy works at a temporal and spatial level, specifically through the guise of complexity theory.

The role of time and space combined in the nonlinear character of autonomy is congenital. This occurs through the presence of *uncertainty* which is key in both the autonomous and self-organising behaviour of social centres and their conception of time. Uncertainty, or what Meillassoux would term as 'hyperchaos' (Meillassoux, 2008), within nonlinear time speaks of the spontaneity of social centre law and can account for its collective unpredictability as well as its lack of formality, its transient nature always evading institutionalisation. It is here that nonlinearity and uncertainty come together as one within both social organisation and time itself, thus reasserting the central role space and time plays in social organisation, as well as law.

Not only that, the nonlinear nature of the autonomous informal laws of social centres as explicated in theories of emergence and complexity, are not just a body of theory describing all forms of organisation but are also a formula for understanding the basic movements of chaos, order and time itself based on processes of entropy mentioned earlier (and thus the underlying role of the fourth dimension – time and space). Linear time prefers the scientific explanation of the arrow of time and the gathering of entropy as the measurement of disorder within a system, with no room for the possibility, and reality, of nonlinearity. It's not to be forgotten that in entropy, in fact, supports the possibilities of *both* linear and nonlinear time. Prigogine highlights how theories of complexity and emergence open up the acceptance of different levels of time experienced by differing individuals, groups and tribes across the globe, highlighting 'social time', 'individual time', 'geographical time' (Prigogine, 1980, p. xiii). This experiencing of time as different between different groups confers with the idea of autonomy as practice and autonomy as placement, where social centre law's time is different to that of the state – the more institutionalisation, the more absolute time becomes; the more informal and collective the law, the more nonlinear the more heterogenous time becomes. The social centre participants' nonlinear timeline gives a very interesting indication of how they saw themselves in relation to their counterparts and that they did not see their progression in history as something occurring in a Darwinian unidirection.

Nonlinearity denotes there may be concurrent contingencies of time and thus law as opposed to an absolute, positive time, with plural ways of understanding a form of disordered thinking and ontology distinct from the control fetish of capital, much as legal pluralist thinking asserts the same over law. Whether all social centre law is nonlinear or not, it highlights the integral role of time in autonomy, and this spatio-temporality in our distance and relation to forms of authority, and the surrounding natural and non-natural environment. It further reasserts the superficial character of linearity, the synthetic nature of individualism, the artifice of private property, progress, capital and institutionalisation.

Conclusion

It is hoped this piece has demonstrated the import of considering the role of spatio-temporality within law, particularly scientific views of time, and the manner in which these versions of time are turned into ideology to support a given form of law, based on the spatio-temporal organising force of property. Linearity and nonlinearity described through complexity theory were posed to demonstrate a distinction between formal state law and its informal resistance, the preponderance and plurality of forms of law and social organisation.

Complexity was proposed as revealing how understandings of law and property are supported by understandings of time, where through private property, law becomes institutionalised and linear, in distinction from its nonlinear origin of the commons. As a result, one form of organisation is discredited at the behest of another, that is nonlinear versions of time that support communal practices of leaderless networks silenced by absolute linear time that sits with the vision of capital, individual private property and progress. As a result only the voice of individualism, hierarchy and institutionalisation is heard, creating the violence of totalitarianism.

By demonstrating this reification of linearity through using complexity theory to describe law and its resistance, it is hoped that movements such as those of social centres, and the autonomy-as-practice and autonomy-as-placement they instil, may not always be demoted as a mere *chink* in the world machine but understood and accepted as offering real, radical and alternate ways of being, law and property, despite all.

Notes

* Thank you so much to Steve Wheatley, Tom Webb, Jamie Murray for including this piece as part of their collection on complexity theory. Thank you very much also to Ting Xu and Sarah Blandy for offering the research environment in which this original piece was written, and Donald McGillivray, Anne Bottomley and Nathan Moore for their support and inspiration over the last few years with this spatio-temporal exposition on law, time, the communal and complexity. Thank you to Andres Guadamuz and Tarik Kochi for their invaluable comments on this piece.

1 Thank you very much to Andres Guadamuz for assistance in explaining the scientific basis of the relations of linearity and nonlinearity. See Guadamuz's excellent Creative Commons licenced 'Networks, Complexity and Internet Regulation' for an in-depth explanation of the science behind complexity (2011).

2 Emergence and assemblage is conspicuously found in the work of Deleuue and Guattari (2004), Latour (2007), DeLanda (2007) and Johnson (2001), with regard to cyberspace, urban studies and, of course, assemblage and actor/network theory and more directly in relation to law in the work of lawscaper Philippopoulos-Mihalopoulos (2015).

3 This is the second law of thermodynamics: that energy, although constant in amount, is subject to degradation and dissipation (Arnheim, 1971, p. 9).

4 For example, it is far less likely, in fact, almost infinitely unlikely, that a cliff could reverse the forces of erosion, but it is highly probable that erosion will cause a cliff to lose its order through the interaction with the order of the elements, forcing materials and rocks to fall and diminish the cliff. Nevertheless, there still exists that possibility of time moving in nonlinear directions in the face of the preponderance of irreversibility in time.

5 A fascinating example of strong legal pluralism was demonstrated in 2016 through former lawyer now artist Jack Tan's 'karaoke court', involving an exhibition and performance at a theatre in Hackney Wick, London, where those bringing a case to the court have to sing for a resolution to a given conflict in front of a real judge and jury, the most convincing performance being the successful party, with a legally binding contract thereafter. This method is similar to the Inuit method of dispute resolution (see Tan, 2016) and https://jacktan.wordpress.com/art-work/karaoke-court/, accessed 2 October 2016.

6 Re-occupation denotes the symbolic taking of space and the requiting of the sense of loss, a re-justification of property through its occupation and a configuration of spatial justice (Finchett-Maddock, 2016, pp. 92–119). The element of re-enactment implies not only the repetitive nature of the law but also its re-staging and archiving of the law of movements of the past, the present and the future, the retelling of a story where alternate conceptions of law are re-animated through the practices and actions (performances) of the social centre participants (Finchett-Maddock, 2016, pp. 92–119).

7 The research was undertaken as part of my PhD 'Observations of the Social Centre Scene: Archiving a Memory of the Commons', between 2006–2010, at Birkbeck Law School. It combined participant observation, interviews and theoretical investigations of social centres in London, Bristol and around the UK, including the rampART (Whitechapel), the Library House (Camberwell), 56a Infoshop (Elephant and Castle), and Kebele (Bristol), amongst others (Finchett-Maddock, 2016).

8 They argue that self-management is 'horizontality (without leaders); informality (no fixed executive roles); open discussion (where everyone has equal say); shared labour (no division between thinkers and doers or producers and consumers); and consensus (shared agreement by negotiation)' (2007, p. 211).

Bibliography

Arnheim, Rudolf (1971), *Entropy and Art: An Essay on Order and Disorder* (Berkeley: University of California Press).

Bastian, Michelle (2014), 'Time and Community: A Scoping Study', 23(2) *Time and Society* 137.

Baudrillard, Jean (1993), *Symbolic Exchange and Death* (London: Sage Publications).

Benjamin, Walter (1999), 'Theses on the Philosophy of History', in *Illuminations* (London: Pimlico).

Byrne, David (2005), 'Complexity, Configurations and Cases', 22(5) *Theory, Culture & Society* 95.

Chatterton, Paul and Stuart Hodkinson (2007), 'Autonomy in the City: Reflections on the Social Centers movement in the UK', 10(3) *City* 305.

Davies, Margaret (2007), *Property: Meanings, Histories, Theories* (London: Routledge Cavendish).

de Sousa Santos, Boaventura (1987), 'Law, a Map of Misreading: Toward a Postmodern Conception of Law', 14(3) *Journal of Law and Society* 297.

de Sousa Santos, Boaventura (1997), 'The Law of the Oppressed: The Construction and Reproduction of Legality in Pasargada', 12(1) *Law and Society Review* 5.

de Sousa Santos, Boaventura (1998), 'Oppositional Postmodernism and Globalisations', 23(1) *Law and Social Inquiry* 121.

de Sousa Santos, Boaventura (1999), 'The Fall of Angelus Novus: Beyond the Modern Game of Roots and Options' <www.eurozine.com/the-fall-of-the-angelus-novus/>, accessed April 2008.

de Sousa Santos, Boaventura (2003), 'Nuestra America, Reinventing a Subaltern Paradigm of Recognition and Redistribution', 18(2–3) *Theory, Culture and Society* 185.

de Sousa Santos, Boaventura (2004a), 'A Critique of Lazy Reason: Against the Waste of Experience', in Immanuel Wallerstein (ed.) *The Modern World-System in the Long Durée* (Colorado: Boulder).

de Sousa Santos, Boaventura (2004b) 'The World Social Forum: Toward a Counter-Hegemonic Globalisation', in Jai Sen, Anita Anand, Arturo Escobar and Pete Waterman (eds.) *World Social Forum: Challenging Empires* (New Delhi: The Viveka Foundation).

de Sousa Santos, Boaventura (2005b) *Towards a New Common Sense: Law, Science and Politics in the Paradigmatic Transition* (New York: Routledge).

de Sousa Santos, Boaventura and César Rodríguez-Garavito (eds.) (2005a), *Law and Globalization from Below: Towards a Cosmopolitan Legality* (Cambridge: Cambridge University Press).

DeLanda, Manuel (2007), *A Thousand Years of Nonlinear History* (Brooklyn: Zone Books).

DeLanda, Manuel (2011), 'Emergence, Causality and Realism', in L. Bryant, N. Srnicek and G. Harman (eds.) *The Speculative Turn: Continental Materialism and Realism* (Melbourne: Re-Press).

Dulong de Rosnay, Melanie (2016), 'Peer to Party: Occupy the Law', 21 *First Monday* 12.

Elder-Vass, David John (2006), 'The Theory of Emergence, Social Structure, and Human Agency', PhD Thesis, University of London, Birkbeck.

Engle Merry, Sally (1988), 'Legal Pluralism', 22(5) *Law and Society Review* 869.

Escobar, Arturo (2003), 'Actors, Networks, and New Knowledge Producers: Social Movements and the Paradigmatic Transition in the Sciences', in de Sousa Santos and Bouventura (ed.) *Conhecimento Prudente para una Vida decente: 'Um Discurso sobre as Ciêncas' Revisitad [Prudent Knowledge for a Decent Life: 'Discourse on the Sciences' Revisited]* (Porto: Afrontamento).

Falk Moore, Sally (1973), 'Law and Social Change: The Semi-Autonomous Social Field as an Appropriate Subject of Study', 7(4) *Law & Society Review* 719.

Finchett-Maddock, Lucy (2011), 'Observations of the London Social Centre Scene: Archiving a Memory of the Commons', Ph.D Thesis, Birkbeck School of Law, British Library.

Finchett-Maddock, Lucy (2016), *Protest, Property and the Commons: Performances of Law and Resistance* (London: Routledge).

Finchett-Maddock, Lucy (2018), 'Speculative Entropy: Dynamism, Hyperchaos and the Fourth Dimension in Environmental Law Practice', in Andreas Philippopoulos-Mihalopoulos and Victoria Brooks (eds.) *Handbook of Research Methods in Environmental Law* (London: Edward Elgar).

Gilles, Deleuze and Felix Guattari (2004), *A Thousand Plateaus: Capitalism and Schizophrenia* (London: Continuum).

Griffiths, John (1986), 'What Is Legal Pluralism?' 24 *Journal of Legal Pluralism* 1.

Guadamuz, Andres (2011), *Networks, Complexity and Internet Regulation* (Cheltenham: Edward Elgar).

Harman, Graham (2016), *Immaterialism: Objects and Social Theory* (New York: Polity Press).

Hawking, Stephen (1988), *A Brief History of Time* (London: Bantum).

Jane, Pickerill and Paul Chatterton (2006), 'Notes Towards Autonomous Geographies: Creation, Resistance and Self-management as Survival Tactics', 30 *Progress in Human Geography* 730.

Johnson, Stephen (2001), *Emergence* (Brooklyn: Scribner).

Keenan, Sarah (2017), 'Smoke, Curtains and Mirrors: The Production of Race Through Time and Title Registration', 28(1) *Law and Critique* 87.

Latour, Bruno (2007), *Reassembling the Social: An Introduction to Actor-Network Theory* (Oxford: Oxford University Press).

Levi Bryant, Nick Srnicek and Graham Harman (eds.) (2011), *The Speculative Turn: Continental Materialism and Realism* (Melbourne: Re-Press).

Luhmann, Niklas (2012), *An Introduction Systems Theory* (London: Polity Press).

MacKay, Robert (2008), 'Nonlinearity in Complexity Science', 21 *Nonlinearity* 273.

Maturana, Humberto and Francisco Varela (1974), 'Autopoiesis: The Organization of Living Systems, Its Characterization and a Model', 5 *Biosystems* 187.

Meillassoux, Quentin (2008), *After Finitude: An Essay on the Necessity of Contingency* (London: Continuum).

Murray, Jamie (2008), 'Complexity Theory & Socio-Legal Studies', 29 *Liverpool Law Review* 227.

Philippopoulos-Mihaloupoulos, Andreas (2015), *Spatial Justice: Body, Lawscape, Atmosphere, Space' Materiality and the Normative Series* (London: Routledge).

Pickerill, J. and P. Chatterton (2006), 'Notes Towards Autonomous Geographies', 30(6) *Progress in Human Geography* 730.

Prigogine, Isabelle (1980), *From Being to Becoming* (San Francisco: Freeman).

Robinson, Emily and Joe Twyman (2014), 'Speaking at Cross Purposes? The Rhetorical Problems of "Progressive" Politics', 12 *Political Studies Review* 51.

Ruhl, J. B. (1996), 'Complexity Theory as a Paradigm for the Dynamical Law-and-Society System: A Wake-Up Call for Legal Reductionism and the Modern Administrative State', 45(5) *Duke Law Journal* 849.

Schelling, F. W. J. Von (1989), *Ideas for a Philosophy of Nature* (Cambridge: Cambridge University Press).

Spinoza, Benedict (2001), *Ethics* (London: Wordsworth Editions).

Tamanaha, Brian (2000), 'A Non-Essentialist Version of Legal Pluralism', 27(2) *Journal of Law and Society* 296.

Tan, Jack (2016), *Voices from the Courts: A Collaboration Between Jack Tan and the Commnity Justice Centre* (Singapore: Darius OU).

Teubner Gunther (1992), 'The Two Faces of Janus: Rethinking Legal Pluralism', 13 *Cardozo Law Review* 1443.

Thompson, E. P (1967), 'Time, Work-Discipline and Industrial Capitalism', 38 *Past and Present* 56.

Urry, John (2006), 'Complexity', 23(2–3) *Theory, Culture & Society* 111.

Webb, Thomas (2013), 'Exploring System Boundaries', 24(2) *Law and Critique* 131.

12 Complexity and the normativity of law[1]

Minka Woermann

Complexity and the law

The preceding chapters have amply motivated the fact that the law is complex. Yet, what has also come to light is that there are several different interpretations of how we are to understand the complexity of law.

Broadly speaking, there are two outlooks on complexity (*cf.* Morin, 2007). The first is that complexity is a function of our knowledge and that complexity can ultimately be resolved through means of more data collection and better information processing. The second outlook – which forms the theoretical basis of this chapter – presents complexity as a particular view of ontology rather than a theory of causation (Byrne and Callagham, 2014). This view affirms the idea that the world is inherently complex because of organising processes and that complex behaviour is the result of the interrelations between components rather than the components themselves. No central organising principle exists in complex systems, with the consequence that complexity is dispersed throughout the system. An implication of this is that it is impossible to further compress complex systems without discounting some of the complexity (Cilliers, 1998).

From the preceding discussion, it follows that any codification of law (what, in complexity terms, is referred to as a model) will be incomplete. In other words, the realities that we seek to interpret through the lens of law will necessarily contain more complexities than the specific lens of law can account for. And yet, codifying law is necessary since we need to reduce the complexities in order to arrive at useful knowledge (i.e. considered judgment). As an illustration, consider the claim that no law can correspond with all dimensions of justice. Modelling justice according to a redistributive or compensatory framework of law, for example, will, in all probability, mean that we must forego a notion of justice based on formal equality, but it also means that we can then further the end of just compensation. The practical challenge that complexity presents us with is therefore not a problem of quantification (i.e. we cannot account for complexity because of the sheer number of variables at play). Rather, the challenge concerns the over-determinacy that characterises most of our social realities. Paul Cilliers (2004, p. 23) motivates this problem as follows within the context of justice:

> It is impossible to arrive at a complete and just description of society, not because we lack the intellectual resources, but because the demands made

on such a description are contradictory. To provide justice to someone will mean that somebody else is treated unjustly. One cannot begin to think about the problem of justice if one does not accept its impossibility.

The subject of justice will be returned to later. At this point, it is simply important to note that since we cannot fully and objectively codify our complex realities, our models – although useful in that they enable understanding and action – are necessarily limited and exclusionary, and the outcome of normative considerations (in addition to factual considerations). With regard to the latter point, the post-structural philosopher, Jacques Derrida (1988, p. 133; p. 134), notes that although the law is commonly framed as a ' "nonfiction standard discourse" [. . .] laws [. . .] in their very normality as in their normativity, entail something of the fictional'. He qualifies this statement by arguing that this does not mean that laws should be equated with, or understood as, novelesque fictions; merely that they rest on the same structural power as novels, that is choice and judgment (what I refer to as normativity in the broad sense) rather than brute fact. Laws (like all models) are also subject to further interpretation through application. This means that normativity pervades not only the establishment, but also the functioning, of law.

As concerns our understanding of complexity and law, we can derive the following two insights from the foregoing discussion. Firstly, the law is a model in light of which complex human actions are interpreted with the goal of promoting just ends; secondly, law itself functions as a complex system precisely *because* our models of law are incomplete (i.e. they can never simply correspond with justice). In practice, this means that law is dynamic; or, as stated by James Boyd White (2012, p. 1), the law is 'a form of life'. Understanding law as a form of life 'de-essentializes' (Derrida, 1988, p. 133) law to the extent that it opens the law to critical questioning. Derrida argues that in order to de-essentialise law, we should not lose sight of the fictional dimension of law. This can be achieved if we continue to scrutinise critically both the founding and functioning of law, the normalcy of law as such. This critical questioning foregrounds the ethics of law, which in this chapter should be understood as an ongoing and constitutive task, rather than as a system dictating right action (Woermann and Cilliers, 2012). As such, ethical considerations permeate the structural power that Derrida identifies as producing laws.

These two insights can be further re-inscribed in terms of systemic closure and openness. The codification of law through modelling is the process by which law is sanctioned, and by which it is given identity and autonomy. The codification thus marks a certain *closure* to alternative, competing views. Yet, as argued earlier, the law also operates dynamically and reflexively. This means that when understood as 'a form of life', the law must necessarily be viewed as an *open* system that stands in relation to the world and to other disciplines. In this view, the focus shifts from the law as a system to what happens when the system of law meets the world. However, one should be careful to construe openness and closure as a dichotomy. Rather, both views necessarily operate together in order to define the system as complex. In order to motivate this statement, I briefly turn to a second hallmark of complex systems (the first being that complex systems are irreducible), namely, that of organisational openness.

In *Method*, his five-volume magnum opus, the complexity theorist Edgar Morin (1992) argues that complex systems are both organisationally open and operationally closed. Organisational openness means that the organisation of the interrelations among elements, actions, or individuals serves to give rise to complex systems, as well as to relate these systems to their environments. Operational closure (or autopoiesis) means that complex systems function autonomously to the extent that they maintain their own internal processes and organisation. Morin (p. 133) describes the interplay between openness and closure as follows:

> active organizations of systems called open insure the exchanges, the transformations, which nourish and effect their own survival: the opening allows them to ceaselessly form and reform themselves; they are reformed by closing, by multiple loops, negative retroactions, recursive uninterrupted cycles. Thus the paradox imposes itself: an open system is opened in order to be closed, but is closed in order to be opened, and is closed once again by opening.

The interplay between systemic openness and closure is facilitated by the system's organisational activity, insofar as '[t]he closing of an "open system" is the loop itself [. . .] it is active reclosing which insures active opening, which insures its own closing [. . .] and this process is fundamentally organizational' (p. 133). Although Morin's analysis centres on living beings (whom he defines as constitutively open-closed), I contend that the system of law also functions as an open-closed system, to the extent that it maintains its internal integrity whilst engaging with the environment. What makes Morin's view of openness unique is that, unlike the traditional thermodynamic conception of openness, his view is not exclusively framed in terms of inputs and outputs (a view wherein the system itself serves as the connection point). Rather, in Morin's work, recursive organisation becomes the mechanism that serves to constitute the system as actively open insofar as the system is co-produced (and not only maintained) by its environment.

The question concerning the manner in which the system relates to its environment is also addressed in the literature on operationally closed systems. The systems theorist, Niklas Luhmann (1992), argues that a thermodynamic conception of opening construes the system's relation with the environment as a causal chain of inputs and outputs. In contrast, Luhmann frames this relation as one of structural coupling. What structural coupling and organisational openness have in common is that both concepts view the relations between the system and its environment as simultaneous and not causal. In terms of structural coupling, Luhmann maintains that the system must be coupled with its environment in order to perform its functions, but the environment does not contribute to the system's operations in much the same way as gravity enables walking but does not contribute to the motion of living bodies. In terms of organisational openness, the recursive loop between systemic openness and closure implies that the activity of the system is contingent on a constant communication and exchange with an environment that co-organises and co-produces the system, as opposed to just feeding an already existing system (as is the case under a thermodynamic conception of openness).

The difference between the autopoietic view of structural coupling and Morin's view of organisational openness, however, is that the former reduces the environment to a function of the system, whereas the latter does not. In this vein, Luhmann (p. 1432) argues that 'selections of information are always internally constructed'. In other words, selections of information can only be communicated within a given system if the information can be rendered intelligible by the norms, codes, or programmes of the system. In terms of the legal system, this means that the norms of the system produce and reproduce its operations, and external references can only be accommodated within the system to the extent that these references can be made to fit the legal framework. In a nutshell, 'the real operations which produce and reproduce [. . .] combinations [of internal and external references] are always internal operations' (p. 1431). This, Luhmann contends, is all that is meant by closure. Process and structure are therefore part of the same operation, and 'all operations gain their own unity by producing subsequent operations' (p. 1440). What this description affirms is that, from the point of view of operational closure, the system's operations are totalising.

In contrast, organisational openness implies that there is an opening in the system insofar as the environment enters the system in order to co-produce it. Morin (1992, p. 201) asserts that '[e]co-dependent beings have a double identity: an identity which sets them apart [from their environment], an identity of ecological belonging which attaches them to their environment'. The frontier is therefore not the boundary line between the system and the environment (as is the case in a thermodynamic conception of opening), but is rather the point that 'reveals the unity of the double identity, which is both distinction and belonging' (p. 201). The environment is therefore always both co-organising (in that it co-constitutes the system) and co-present. Translating this insight for the legal system, however, is challenging, and will – in part – constitute the focus of this analysis.

To this end, I investigate the manner in which the legal system establishes its autonomy by focusing on the founding and functioning of law. However, it will further be demonstrated that the structural power or organisational principle that serves to establish the law as a system in its own right (and which, over time, normalises law) also opens law to its environment, by ensuring that the activity of law is never arrested or concluded. This means that we should remain sensitive to transgressions of law and the transformative potential that transgression implies. Before turning to the analysis of the three 'moments' of law (i.e. the founding, functioning, and transformation of law), I first introduce Derrida's threefold framework of violence, as this framework can serve as a useful lens for understanding the law as open-closed and for demonstrating how the three "moments" of law are related to, and complicate, one another.

Normativity and the three 'moments' of law

Derrida's discussion of violence takes place in the context of language – the primary process by which we ascribe meaning to the world. Although Derrida does not refer to Ferdinand de Saussure's (1960) view of language in this specific argument, briefly drawing on his insights before presenting Derrida's views is instructive. Saussure argues that language is the means by which we divide up, order, and make sense of the conceptual plane. Furthermore, Saussure argues that this ordering is essentially arbitrary, since there are different divisions according to which meaning can be ascribed to stimuli (for example, Russians consider the colours 'blue' and 'sky blue' to be distinct colours rather than two shades of 'blue', as is common in English). It is this original naming of the world that Derrida (2016) refers to as arche-writing. He identifies arche-writing as the originary violence of language, 'which consists in inscribing within a difference, in classifying, in suspending the vocative absolute' (p. 121). The act of inscribing and classifying requires an originary cutting or splicing of the world, insofar as it divides the world into concepts. The first grunt for food, for example, bestows a differential meaning to the grunt, implying that we identify the meaning of the grunt on the basis of how it differs from all other sounds. Since the process of creating meaning is inherently characterised by a scission, it marks 'the loss of [. . .] self-presence' (p. 121).

The second level of violence seeks to cover over the originary violence by presenting itself as the very counter to violence. Elizabeth Grosz (1998, p. 193) describes second-level violence as a violence 'whose violence consists in the denial of violence'. Derrida (2016, p. 121) argues that this violence is 'reparatory [and] protective' and that it 'institut[es] the "moral"'. This level marks the violence of law, right, and reason, in which meaning becomes naturalised and 'truth' is created. The third level of violence, in turn, unmasks and de-naturalises meaning, and manifests (when it does) as empirical, mundane violence (i.e. as street violence or as violence between races, classes, genders, and political oppositions). Derrida (p. 121) argues that '[i]t is on this tertiary level, that of the empirical consciousness, that the common concept of violence (the system of the moral law and of transgression) whose possibility remains yet unthought, should no doubt be situated'. Derrida moreover argues that the third level of violence necessarily refers to the previous two levels in that it seeks to call into question the establishment and normalisation of our social and moral consciousness.

When one applies the preceding schema to law, it is clear that the foundation of law rests on a particular cut or scission that distinguishes the (il)legal from the a-legal (Lindahl, 2011), which both constitutes the law as such (level 1) by differentiating it from that which cannot be assimilated into its internal logic (the a-legal), and sanctions the force of law (level 2), by distinguishing the legal from the illegal.[2] Grosz (1998, p. 196) calls this the paradox of law, in that 'while [the law] orders and regulates, while it binds and harmonises, it must do so only through a cut, a hurt that is no longer, if ever, calculable as violence or a cut'. In order for the law to function, it must exercise force, but the very exercise of force is what also sanctions the law as right (*droit*), and as a major jurisprudence, thereby covering over the opening for alternative jurisprudences (level 2). On both level 1 and level 2, the law is essentially viewed as a closed system (level 1 concerns the establishment of the system of law, whereas level 2 concerns the autonomous functioning of law). Level 3 (which concerns radical transgressions, deconstructions, and transformations of law) marks the opening of law. Challenges to law highlight the very undecidability that permeates every complex life form and render these forms as open. However, since the law is inherently complex, it also means that the limits of law are felt in the very act of judgment (the system itself cannot therefore function as an entirely closed system). In other words, apart from explicit challenges to law, the system reveals its limits in the act of application. Grosz (pp. 196–197) affirms this view in arguing that '[t]he undecidable [. . .] is the very openness and uncertainty, the fragility and force of and in the act of judgement itself, the limit (as Drucilla Cornell (1992) puts it) of the law's legitimacy or intelligibility'. Therefore, and to repeat a previous premise, it is because the model of law is incomplete (in Derrida's terms, it cannot encapsulate the proper) that the law necessarily demands interpretation and judgment.

Derrida's three levels of violence thus operate together in order to both sanction the law's identity and authority, and to place this very identity and authority into question. What this implies is that we cannot analytically sever one type of violence from the others. Each judgment (i.e. each moment of law's functioning) institutes anew the scission between legal and illegal, and thereby confirms the inherent structure of law. However, judgment also demands a questioning in that a just decision cannot just reinstate a precedent but must also seek to critically engage the precedent in the context of a unique case, and in this process rewrites law.

Before continuing with the analysis, one note concerning the use of the term 'violence' is called for. Although, violence does much to highlight the fragile foundation on which the law (and all systems of meaning) rests, this term has been prone to abuse in recent years. Increasingly, we see (in both academic literature and common usages) a conceptual slippage that occurs between manifest violence and the violence that is inherent to the structural force of right, law, and reason (to meaning as such). This conceptual slippage dilutes the meaning of the term. However, since 'violence' is the term that is used by many of the theorists who direct this analysis, I will continue to employ the term but with the explicit caveat that – in the context of this analysis – level 1 violence should be understood as the structural normativity that defines any act of modelling, level 2 violence should be understood in terms of a normalising or moralising normativity that serves to entrench a given status quo, and level 3 violence should be understood as a transgressive normativity that seeks to challenge the limits of a given system in order to open up the system to new possibilities.

Normativity in the founding and functioning of law

From the very brief discussion offered earlier, it should now be clear that the founding of law – or the act of closure whereby the law is endowed with legitimacy – rests on a paradox, namely, that the origin of law, and hence of authority, is without foundation, or the law is founded on the basis of what Derrida (1992) describes as 'the "Mystical Foundation" of Authority' (which also constitutes the sub-title of his essay 'Force of Law'). Derrida characterises this mystical foundation as 'a violence without ground' (p. 14), in that the foundation cannot be justified on the basis of legal infrastructure. Rather, the law is inaugurated through virtue of a performative force which 'in itself is neither just nor unjust and that no justice and no previous law with its founding anterior moment could guarantee or contradict or invalidate' (p. 13). The paradox thus amounts to the fact that the founding distinction – which Derrida (p. 13) refers to as "interpretive violence" – between what constitutes legal and illegal actions (and moreover separates the law from that which is foreign to its internal logic; the a-legal) cannot itself be justified through an appeal to these categories.

Furthermore, the performative force whereby constitutional law is inaugurated through means of a declarative act ('We the people decree . . .') is based on a performative contradiction in that the constituting power (the people) only exist after the fact (once they have been declared so in law). As such, the sanctioning of a constitution is itself an unsanctionable act – both because the institution of law cannot be justified on legal grounds, and because the instituting agents are only defined as agents once the law is enacted. Stewart Motha (2013) – following Martin Loughlin and Neil Walker – refers to this as the 'paradox of constitutionalism', and this paradox begs the critical questions, '[W]ith what authority and in whose name is a constitution or constituted order created?' (p. 95).

Since the answers to these questions remain undecidable, law is founded and furthered on the basis of the 'as-if', which Motha (p. 94) defines as a 'consciously false, illusory knowledge, or legal fiction'. Theorising the 'as-if' has a long intellectual history, which dates back to Kant, who, in *The Critique of Pure Reason*, argues that we should proceed 'as if' the (moral) law was legitimately grounded in reason. The rich body of literature on social contract theory also provides robust arguments for the instantiation of a hypothetical social contract (that rational humans would be motivated to follow 'as if' they had explicitly sanctioned the contract). More recently (and as explained earlier in the chapter), Derrida has argued that the inevitability of the 'as-if' introduces a fictional element into legal thought.

The fictional basis of law (which is typified in reasoning that proceeds from the 'as-if') holds two implications: firstly, it is the generative power by which law comes into being. The law, in other words, is an operationally closed system precisely because we are willing to lend, what Motha (2013, p. 98) calls, 'our qualified and tentative consent' to these founding fictions (which are themselves normative, but which also inform the law's norms). However, and secondly, since the law is grounded in fiction as opposed to natural fact, the interpretative or formative violence of law should not be naturalised. The law becomes the authoritative power in determining which narratives are to be included and excluded from the judicial order, and our legal fictions should therefore be subjected to continuous critical reflection. Indeed, Motha (p. 97), following Jean-Luc Nancy, argues that making sense from 'invention, allusion or fiction' constitutes a form of ethical-political responsibility. The question thus arises as to how we are to respond to the paradox of constitutionalism.

Motha (p. 99) writes that responses to the paradox of constitutionalism 'either privilege the "political" or "legal" side of the binary between constituent and constituted power'. Whereas the former camp (represented by Hans Lindahl) presents a 'reflexive questioning of the conditions of collective selfhood' (p. 99), the latter camp (represented by Hans Kelsen) views the legal order 'as an expression of the unity of collective existence' (p. 99). According to Motha, Kelsen (1991) attributes collective subjectivity to the legal order, and as such frames this order in terms of a self-enclosed and unified norm that grounds collective existence. In contrast, Lindahl (2007, p. 21) understands the 'ontology of collective subjecthood' in terms of questioning and responding. The normative is thus framed in reflexive terms.

Although at face value, the political camp seems to be more sensitive to the normativity of law, Motha argues that the political camp – as with the legal camp – tends to gloss over the issue of violence (i.e. structural normativity) in its account of law. This is because despite continually reposing the question 'Who are we?', the question is always already asked from within a political horizon and, hence, from within an established form of life that is grounded in the legal order as such. In contrast to the constitutionalists, Giorgio Agamben (2005) attempts to escape the violence inherent to any political order of representation, thereby disavowing legally mediated forms of life. Motha, however, finds this strategy unconvincing, arguing that a robust theory should be able to account for both violence and normativity. Stated in terms of this analysis, a robust account should be able to engage with the structural implications arising from both the fictional basis of law and the normalising force of law.

Motha finds such an account in the work of Judith Butler, who contemplates the exclusions that define any legal or political order, whilst also recognising the normative force of law in constituting its subjects. With regard to the normative force of law, Butler (2009, p. 138) writes that 'any inquiry into [. . .] ontology requires that we consider another level at which the normative operates, namely, through norms that produce the idea of the human who is worthy of recognition and representation at all'. Yet, Butler is also sensitive to the fact that the formative norms are violent and that this violence also defines us (precisely because the law constitutes us as subjects). The critical question, according to Butler (p. 170), is thus whether 'one [can] work with such formative violence against certain violent outcomes and thus undergo a shift in the iteration of violence?' Butler's strategy, as summarised by Motha (2013, p. 101), is to acknowledge the conditions that give rise to a subject's or culture's violent production, whilst simultaneously 'assuming the responsibility of contesting its determining power'.

In order to contest the determining power of law, and hence to ensure that the form of life propagated by law is subject to critical scrutiny and transformation, one cannot merely stay within the system – as Motha claims the Constitutionalists do. However, Motha's reading of Lindahl's work as representative of an essentially closed constitutionalist frame is, I would contend, misleading. In the same source that Motha references, Lindahl (2007, pp. 23–24) explicitly states that

> responsiveness [within a constitutional framework] is radically finite because legislation does not merely integrate the strange [i.e. the outside of the legal system] into a legal order; it also always *neutralizes* strangeness, levelling down the extraordinary to a variation of the ordinary.

He further argues that '[t]o lose sight of this is to strip strangeness of its ambiguity, collapsing the threat posed by subversion into a mere opportunity for and celebration of legal change'. Like Butler, Lindahl acknowledges the subject's constitution in, and through, law whilst seeking strategies to overcome – or at least arrest – the manner in which law assimilates strangeness.

In another essay, Lindahl (2011) suggests one such strategy against the backdrop of Québec's application for secession from Canada. The application was rejected by the Supreme Court of Canada on the basis of the opinion that Québec had no right to unilateral succession, as rights demand reciprocal recognition under constitutional law. In other words, the appeal to such a right would only make sense within the very constitutional frame that the Québécois refuse to acknowledge in framing their application in terms of *unilateral* succession (i.e. without negotiation under constitutional terms). In the context of this case, Lindahl poses the question of whether constitutionalism is capable of responding to radical difference (or strangeness) without reducing it to a claim governed by the logic of the system. His suggestion is that an appropriate response in this regard may be to institute another 'novel unilateral act which suspends, *albeit partially*, the constitutional regimentation of reciprocity with a view to initiating political negotiations with those who want out' (p. 230). The suggestion, in other words, is to suspend the rules governing constitutional amendment in order to account for the interests of those who do not wish to be recognised under a given constitution.

As such, Lindahl's strategy constitutes an attempt to negotiate a-legality, in the sense that one recognises that 'there could be no *challenge* unless what is outside the order registers in some way within the order: a-*legality*.' But, 'there could be no *transgressive* challenge unless it calls into question, more or less radically, what the legal order frames as being (il)legal: *a*-legality'.[3] In other words, the transgressive challenge manifests as the imperative to contend with strangeness or that which lies outside of the system itself. In the context of the court's pronouncement on Québec's application for succession, this transgressive challenge did not register as the court argued from within the constitutional frame. What the court failed to see is that the Québécois separatists reject constitutional recognition as such, because such recognition would neutralise the 'politics of difference' that constitutes the very heart of the challenge that an application for unilateral secession demands.

The strategy for engaging with a-legality that is suggested by Lindahl (2011) constitutes a form of constitutionalism – what he calls 'para-constitutionalism' –

that allows one to respond to singularity or difference in an indirect manner, by temporarily suspending the constitutional paradigm that demands reciprocal recognition under its laws and by creating a new parallel system. This new system would, however, also be dictated by an internal code, albeit one that does not subscribe to the code of a given constitution. If this reading is correct, it implies that para-constitutionalism merely re-enacts the logic of constitutionalism, which, as should be clear at this juncture, is one that is dictated by operational closure insofar as environmental inputs are read and interpreted through the system's infrastructure (i.e. all inputs are rendered meaningful through virtue of being neutralised by the system's codes). The question that this interpretation begs is therefore how one is to account for the transgressive moment *within* constitutionalism that allows for recognising the need for a parallel system. In other words, wherein lies the impetus to transgress the constitutional system in the first place, if one cannot transcend the totalising logic of operational closure? Although Lindahl also addresses this question in his work, the response put forward in the next section is one that subscribes to the complexity logic of organisational openness.

Opening and closure in the functioning of law

Up to this point in the analysis, the functioning of law has primarily been described as constituting a conservative moment in that law affirms its identity and entrenches its founding violence through virtue of application (i.e. the law has been primarily framed as functioning as an operationally closed system). However, since legal systems are contingent on their own operations and are not grounded in natural fact, they are also open to challenges. Québec's application for unilateral secession is one example of such an external challenge, albeit that the application failed to usurp the legal infrastructure. As argued earlier, in order for the transgressive challenge that emanates from outside law, to register as a moment in law (and thus, in order for it to be heard), the law itself must carry within it the potential to reform. This is only possible if one can identify the organisational opening of law that initiates the deconstructive gesture or that positions a-legality (or strangeness) in the very heart of the legal system itself.

In his work on theoretical economies, Derrida (1982, pp. 26–27) calls the strangeness or difference that can manifest in the system the 'unnameable', but he is careful to qualify that this 'unnameable' should not be understood as an absolute outside (God, for example), but as

> the play which makes possible nominal effects, the relatively unitary and atomic structures that are called names, the chains of substitutions of names in which, for example, the nominal effect *différance*[4] is itself enmeshed, carried off, reinscribed, just as a false entry or a false exit is still part of the game, a function of the system.

Translated into the terms of the legal system, this citation implies that the operations of the legal system cannot be totalising in that interpretation and judgment always produce gaps in the system that cannot be accounted for by the system's apparatus. White (2012, p. 1) affirms the inherent openness that marks the functioning of law, in writing,

> what I shall say, in a phrase, is that law is not at heart an abstract system or scheme of rules, as we often think of it; nor is it a set of institutional arrangements that can be adequately described in a language of social science; rather it is an inherently unstable structure of thought and expression, built upon a distinct set of dynamic and dialogic tensions. It is not a set of rules at all, but a form of life. It is a process by which the old is made new, over and over again. If one is to talk about justice in the law, it must be in the light of this reality.

The complexity insight that the law is both open and closed complicates the strong dichotomy that White draws between the closure and openness of law. A productive reading of law requires that we think these two operations together (a reading that I shall again turn to at the end of this section). White's analysis of law as a fundamentally open system, is however illuminating, and for this reason I shall first briefly reproduce his argument here. White shifts the focus from law understood as a system to what happens when the system of law meets the world. It is in the confrontation between the system of law and the world that a number of openings in the functioning of law can be identified. Note that, on face value, this act of translating the world in terms of the legal system seems consistent with the insights from operational closure and structural coupling. However – and herein lies the crucial difference – before translation, interpretation, or judgment becomes legal fact, those agents acting in the legal system need to contend with these openings in order to further the work of translation. The process of law is thus open and subject to interpretation and judgment rather than being a mere mechanical process that translates the environment in terms of the system. White refers to these openings as 'tensions'. He identifies seven such tensions, which can be briefly described as follows.

Firstly, there exists a tension between ordinary language and legal language. White argues that the challenge introduced by this tension is an iteration of the more general challenge of translating experience into language as such. Successfully responding to this tension (i.e. helping the client to tell her story and translating this story into legal terms) necessitates that the lawyer is well versed in the art of language and the art of judgment. The second tension (which is related to the first) manifests as a result of the multiplicity of voices that define the legal system. This tension requires of lawyers and judges to be well versed in, and capable of moving between, these specialised languages. Thirdly, a tension exists between the substance of the case and legal procedures. Lawyers and judges must be skilled in simultaneously working in both tracks. Fourthly, the lawyers on opposing sides of the case must be able to demonstrate, and productively deal with, the range of possible meanings that legal texts hold. White argues that this tension breathes life and creativity into law, but it also creates a moral tension within lawyers, who – when faced with opposing arguments – need to question whether what they are doing is morally justifiable. This internal tension requires a high degree of maturity from lawyers and judges. Fifthly, the tension between rational, yet opposing, legal opinions results in a parallel tension in the mind of the judge, who must ultimately come to a judgment (which in itself is the outcome of the tension that exists between the judge's intuition and the ability to express her intuition in rational language and substantiated arguments). A sixth tension exists between past precedent and present judgment (which also creates something new for the future). The seventh, and final tension, exists between law and justice (which can also be described respectively as the tension between institutional justice and abstract justice or positive law and natural law). This tension requires the lawyer to navigate the unstable and dynamic relation that exists between the real and the ideal.

One could argue that all the other tensions are bound up in this final tension, inasmuch as the constraints and challenges presented by the real (as embodied in the legal system and the particularities of the case) must be handled with skill, judgment, and vision if lawyers and judges are to realise (what should be) their final goal, which is 'to bring into one field of vision the ideal and the real' (p. 12). White (p. 18) states as much in arguing,

> [t]o work well with the tensions that I describe is itself to achieve an important kind of justice; this kind of justice is no less important than abstract justice; indeed if abstract justice is to become real, and not merely abstract, it must itself be wisely, justly, and artfully located in the context of these tensions. It is not too much to say that in this process lies the only hope for justice in the law.

The subject of justice presents us with arguably one of the clearest examples of how autopoietic systems theory differs from an open complex systems theory. In autopoietic theory, the concept of justice contains no moral content since 'morality has no legal relevance – neither as code (good/bad, good/evil), nor in its specific evaluations' (Luhmann, 1992, p. 1429). Moral constraints can therefore only be incorporated by the system as legal constraints, which are produced by legal codes. Given this view, Luhmann (p. 1431) argues that '[t]he guarantee of "justice" is not the correspondence with external qualities or interests, but the consistency of internal operations recognising and distinguishing them'. In contrast to this view, Derrida (1992) defines justice as undeconstructible, which implies that justice does not have conceptual meaning (as all concepts are deconstructible). Derrida (p. 17) speaks of the '*act* of justice' (my italics) as something that 'must always concern singularity, individuals, irreplaceable groups and lives [. . .] in a unique situation'. As such, he writes that '[d]econstruction is justice' (p. 15) insofar as justice (as a type of empty, non-appropriable ideal or quasi-transcendental 'concept') acts as the impetus to challenge, transform, and transgress legal systems. Derrida argues that the interplay between the real and the ideal or between positive law and abstract justice can neither be ensured by a rule nor be the outcome of a purely private morality. Justice therefore protects law from operating merely as an operationally closed system, whereas law protects justice from becoming an arbitrary expression of subjective convictions or an abstract – but impotent – conversation about ideal ends. Otherwise stated, the law authorises justice and gives it position.

The difficulty, however, is that the relation between law and justice is governed by an aporia or logical disjunction (which is a stronger description than that of White's 'tension'). Derrida describes the aporia as follows: on the one hand, justice is 'a responsibility without limits, and so necessarily excessive, incalculable, before memory' (p. 19) (the ideal of justice is thus an empty ideal); on the other hand, the exercise of justice as law or right requires legitimacy or legality, stability and statute, calculation and coded prescription (p. 22). Despite the logical disjunction, a just decision requires both abstract justice and positive law or, as stated by Cilliers (2004, p. 23), '[t]o engage with the problem of justice in a philosophical way [. . .] entails entering into this dilemma, and, in a way, to accept both sides of it'. The practical question thus becomes how does one successfully engage with the dilemma?; or – in White's (2012, p. 18) terms – how does one interpret 'wisely, justly, and artfully'?

Derrida (1992) argues that just action, although requiring the rule of law, must necessarily transcend the law in the moment of judgment. In his words, '[n]o exercise of justice as law can be just unless there is a "fresh judgment"', which he understands in terms of 'responsible interpretation' (p. 23). Fresh judgment requires that we act through duty and not only in conformity with duty. In the latter case (which characterises a crude form of legal positivism), responsibility is shifted to the rule or law, and the outcome cannot be viewed as ethical because the judgment did not involve a decision. A decision, Derrida (1988, p. 116) writes, 'can only come into being in a space that exceeds the calculable program that would destroy all responsibility by transforming it into a programmable effect of determinate causes'. Thus, justice requires that we observe the law and consider the arguments of the case whilst recognising that neither the law nor the arguments will bring us to a decision. Rather, in making the decision, the judge accepts responsibility for proclaiming an outcome that could not have been determined in advance, and which is therefore not the product of a utilitarian and closed economy of directives and prescripts. In White's (2012, p. 16) terms,

> [law] is not a totalitarian system, closed and unlistening, but an open system, like a language, not only making creativity possible, but requiring it [. . .] Every case, every legal conversation, is an opportunity to exercise the lawyer's complex art of mind and imagination.

In summary, moral and political responsibility is instantiated in decision-making, whereby the law is re-evaluated and re-assumed each time that it is applied. A decision or judgment is therefore necessarily transgressive – the moment of the decision must transcend the system of rules and institutional arrangements that define the legal system in order to be a judgment (i.e. the moment of the decision must be open to strangeness). However, in transgressing the system of law in the moment of the judgment, the structural force of law is either suspended (As would be the case in Lindahl's suggestion of para-constitutionalism), or re-enacted. With regard to the latter. Cilliers (2004, p. 24) writes that ethical decision-making (i.e. judgment) requires that

> [w]e gather all possible information and consider all possible options, then make a decision *as if* we would expect it to be a universally valid decision while we realise that we could not consider all possible options, and that we have to be prepared to reconsider the choice.

The reference to the 'as-if' in the above citation highlights the fact that the 'mystical foundation of authority' is not confined to the instantiation of law, but is re-inscribed anew in every fresh judgment. After the decision is taken and the judgment passed, the moment of judgment is archived by the legal system and reified as an analysable object. In other words, the opening in law is closed again. Yet, to recall Morin's (1992, p. 133) words, the interplay between systemic openness and operational closure is ongoing in that 'an open system is opened in order to be closed, but is closed in order to be opened, and is closed once again by opening'. In the context of the legal system, this logic translates as the insight that the status of judgment is always provisional. Yet Derrida (1992, p. 28) warns that the fact '[t]hat justice exceeds law and calculation, that the unpresentable exceeds the determinable, cannot and should not serve as an alibi for staying out of juridico-political battles'. To the contrary, we must act and, moreover, recognise that each advance – each judgment – re-inscribes the field of law. It is this recognition that leads White (2012) to state that lawyers and judges do the work of poets, to the extent that every time they go to work, they engage with the tension between the ideal and the real, or the general and the particular.

Conclusion: transgression and the ethics of law

As argued in this chapter, the legal system translates (models) complex human behaviour in an attempt to promote just outcomes. However, because our models of law are limited, the law itself functions as a complex system. This means that although the law conserves its founding identity in the institutional arrangements and application of rules that define the legal system as such, the system cannot close in on itself because it cannot perfectly map the realities that it seeks to interpret. Law is an open form of life insofar as it engages, and is engaged by, the world and with the empty ideal of justice. Another way of putting this is to say that in complex systems – such as the law – signification cannot be fixed. The normalcy of law (law's conservative moment) is thus premised on the transformative potential of law (or the fact that it is open to its environment). Given this understanding, the ethics of law is an ongoing constitutive task, whereby we engage with the different – and conflicting – modes of normativity that characterise the law (which, in this chapter, have been defined as structural, normalising, and transgressive normativity).

This engagement constitutes a critical task, which is first and foremost concerned with how 'to become' rather than with what is currently the case (Pavlich,

2013). Such critique also does not endeavour to replace one system of thought with another, but spurs on reflection about new forms of meaning. In George Pavlich's (p. 39) words, the promise of critique 'lies not in illusory universal ideals; instead, it remains forever open to other ways of thinking and becoming. Deluded closures may breed despotic acts'. Michel Foucault (1994, p. 323) beautifully summarises the work of this kind of critique as follows:

> I can't help but dream of a kind of criticism that would try not to judge but to bring [. . .] an idea to life [. . .] It would multiply not judgements but signs of existence; it would summon them, drag them from their sleep. Perhaps it would invent them sometimes – all the better.

In light of the preceding analysis, it is clear that this type of critique is necessarily creative and transgressive in nature, in that '[i]t can never simply re-enforce that which is current, but [. . .] involves a violation of accepted or imposed boundaries' (Woermann and Cilliers, 2012, p. 453). However, 'in order to practise transgressivity responsibly, one must be modest enough to recognise the limitations of one's conceptual schema, and show a willingness to overcome these limitations' (p. 453). In terms of this analysis, this statement implies that it is because of the law's limitations and fragile foundations – in other words, it is because other legal forms are possible – that critique and transgression are possible and necessary. And it is this insight that fuels deconstruction, which Derrida (1999) argues is always undertaken in the service of ethical testimony.

Notes

1 I thank Hans Lindahl and an anonymous reviewer for their insightful comments on an earlier draft.
2 The functioning or normalcy of law is predicated on transgression (in the conservative sense), insofar as the 'possibility of transgression tells us immediately and indispensably about the structure of the act said to be normal as well as about the structure of law in general' (Derrida, 1988, p. 133). In other words, illegality belongs to the very structure of the law itself. This point is well summarised by Luhmann (1992, p. 1428), who states that '[i]f the question arises whether something is legal or illegal, the communication belongs to the legal system, and if not then not'.
3 From email communication.
4 Différance should be understood both as 'a regulated economy of difference and deferral' (Hurst, 2004, p. 254), wherein meaning is formed; as well as in the aneconomic sense as 'an expenditure without reserve, as the irreparable loss of presence [. . .] that apparently interrupts every economy' (Derrida, 1982, p. 19), wherein meaning is destroyed.

Bibliography

Agamben, Giorgio (2005), *The Time that Remains: A Commentary on the Letter to the Romans*, trans. Patricia Dailey (Stanford, CA: Stanford University Press).
Butler, Judith (2009), *Frames of War: When Is Life Grievable?* (London: Verso).

Byrne, David and Gillian Callaghan (2014), *Complexity Theory and Social Sciences: The State of the Art* (Oxon: Routledge).

Cilliers, Paul (1998), *Complexity and Postmodernism: Understanding Complex Systems* (London: Routledge).

Cilliers, Paul (2001), 'Boundaries, Hierarchies and Networks in Complex Systems', 5(2) *International Journal of Innovation Management* 135.

Cilliers, Paul (2004), 'Complexity, Ethics and Justice', 19(5) *Humanistiek* 19.

Cilliers, Paul (2005), 'Complexity, Deconstruction and Relativism', 22(5) *Theory, Culture & Society* 255.

Cornell, Drucilla (1992), *The Philosophy of the Limit* (New York: Routledge).

Derrida, Jacques (1982), 'Différance', trans. Alan Bass in *Margins of Philosophy* (Chicago: University of Chicago Press), 1.

Derrida, Jacques (1988), 'Afterword', trans. Samuel Weber in Gerald Graff (ed.) *Limited Inc* (Evanston: Northern Western University Press), 111.

Derrida, Jacques (1992), 'Force of Law: The "Mystical Foundation of Authority"', trans. Mary Quaintance, in Drucilla Cornell, Michel Rosenfeld and David Carlson (eds.) *Deconstruction and the Possibility of Justice* (New York: Routledge), 3.

Derrida, Jacques (1999), 'Hospitality, Justice and Responsibility: A Dialogue with Jacques Derrida', in Richard Kearney and M. Mark Dooley (eds.) *Questioning Ethics: Contemporary Debates in Philosophy* (London and New York: Routledge), 65.

Derrida, Jacques (2016), *Of Grammatology*, trans. Gayatri Spivak (Baltimore: John Hopkins University Press).

Foucault, Michel (1994), *Politics, Philosophy and Culture: Interviews and Other Writings, 1977–1984* (New York: Routledge).

Grosz, Elizabeth (1998), 'The Time of Violence: Deconstruction and Value', 2(2) *Cultural Values* 190.

Hurst, Andrea (2004), 'Derrida's Quasi-Transcendental Thinking', 23(3) *South African Journal of Philosophy* 244.

Kelsen, Hans (1991), *General Theory of Norms* (Oxford: Clarendon Press).

Lindahl, Hans (2007), 'Constituent Power and Reflexive Identity: Towards an Ontology of Collective Selfhood', in Martin Loughlin and Neil Walker (eds.) *The Paradox of Constitutionalism: Constituent Power and Constitutional Form* (Oxford: Oxford University Press), 9.

Lindahl, Hans (2011), 'Recognition as Domination: Constitutionalism, Reciprocity and the Problem of Singularity', in Neil Walker, Jo Shaw and Stephen Tierney (eds.) *Europe's Constitutional Mosaic* (Oxford: Hart Publishers), 205.

Luhmann, Niklas (1992), 'Operational Closure and Structural Coupling: The Differentiation of the Legal System', 13 *Cardozo Law Review* 1419.

Morin, Edgar (1992), *Method: Towards a Study of Humankind, vol. 1: The Nature of Nature*, trans. and intro J.L. Roland Bélanger (New York: Peter Lang).

Morin, Edgar (2007), 'Restricted Complexity, General Complexity', trans. Carlos Gershenson in Carlos Gershenson, Diederik Aerts and Bruce Edmonds (eds.) *Worldviews, Science and Us: Philosophy and Complexity* (Singapore: World Scientific), 5.

Motha, Stewart (2013), 'As If: Constitutional Narratives and 'Forms of Life', in Karin Van Marle and Stewart Motha (eds.) *Genres of Critique: Law, Aesthetics and Liminality* (Stellenbosch: SUN Press and STIAS), 91.

Pavlich, George (2013), 'Dissociative Grammar of Constitutional Critique?' in Karin Van Marle and Stewart Motha (eds.) *Genres of Critique: Law, Aesthetics and Liminality* (Stellenbosch: SUN Press and STIAS), 31.

Saussure, Ferdinand (1960), *Course in general linguistics*, Charles Bally and Albert Sechehaye (eds.), trans. Wade Baskin (London: Peter Owen).

White, James Boyd (2012), 'Justice in Tension: An Expression of Law and the Legal Mind', 9 *No Foundations: An Interdisciplinary Journal of Law and Justice* 1.

Woermann, Minka (2016), *Bridging Complexity and Post-Structuralism: Insights and Implications* (Switzerland: Springer International).

Woermann, Minka and Paul Cilliers (2012), 'The Ethics of Complexity and the Complexity of Ethics', 31(2) *The South African Journal of Philosophy* 447.

13 Regulating the practise of practice

On agency and entropy in legal ethics

Julian Webb

> We see that the intellect, so skillful in dealing with the inert, is awkward the moment it touches the living.
>
> – Henri Bergson (1998, p. 165)

This chapter seeks specifically to argue for the value of a complexity approach to legal ethics and professional regulation, as well as to make a more general contribution to understanding the complexity of law and regulation.

Put briefly, I propose that conventional thinking does not (and cannot) adequately account for the emergence of legal ethics from the *system of practice*. The context for social action is decisive, and context itself is shaped by the interplay of agents within a 'system', which, if we are to appreciate the full complexity of social practices like legal ethics, needs also to be understood intrinsically as the interplay between multiple agents and sub-systems. This framework should, in turn, provide a more psychologically and institutionally rich ground for normative debate about professional conduct and the allocation of responsibility, and a more adequately complex framework for regulation.

In presenting this argument, the chapter opens with a broad account of agency as currently represented in legal ethics and then sets out to construct a more adequately complex account of situated and organisationally based agency as (first) a descriptive rather than normative way forward for ethical theory and for regulation. This perspective implies that professional ethics are not, and cannot be, wholly normatively predetermined, as they are an emergent property of interactions within the 'system' of legal practice. It also suggests that the very complexity of legal ethics in practice itself makes a degree of ethical failure inevitable and even, in a narrowly functional sense (depressingly), 'necessary'. It follows that the primary challenge for regulation is to construct ethical climates that not just limit the risk of what we might call 'entropic' failure but also find ways to reduce the functional advantages of normatively undesirable behaviours.

The problem of agency and normative legal ethics

Agency in itself is a 'big', one might say complex, concept and one that goes to the heart of legal and other professional ethics. While it is, in some senses the subject of this chapter, this is not in any sense a philosophical disquisition on agency as such. Rather, I adopt a theoretically informed but ultimately descriptive, and functional, approach to the topic as part of a larger mission of understanding legal practice in more richly agentic terms. For present purposes, therefore, *agency* is defined, first, as the ability of individuals to have a personal effect or impact on the world. This recognises that individuals are, in a functional sense, 'agentic' whether they will it or not. Agency as willed or 'reasoned' action is, of course, important and central to most understandings of the term. Agency in that latter sense is commonly seen as central to human flourishing and exercises an important social and moral function since it enables us to control our own lives and pursue our life goals without being subjected to the domination of others (e.g., Gewirth, 1986, p. 288). In most philosophical accounts agency is thereby linked to a number of other thick concepts such as autonomy, free will, choice and responsibility. By contrast, this account, while it acknowledges the inherent normativity of agency, remains largely at a descriptive level and seeks to take seriously what is often described as the 'situationist' (Trevino, 1986; Zimbardo, 2007) challenge to accounts of (moral) agency, without itself engaging in further normative debate *about* agency.

This chapter thus proposes, as a starting point, that the theoretical focus on normative ethics, and specifically on the abstract principles and methods that guide and constrain the moral actor, has led to a neglect of the actual, situated *practise* of normativity itself. Let us now address why and how this has come to pass by considering, albeit broadly, both the value and the limits of existing accounts of professional agency.

In professional discourse, agency cannot be separated from another fundamental construct, namely, the notion of occupational *role*. I start this discussion from the position that role morality is both a plausible and potentially useful distinction, though it is not the end of the story.

In essence, role morality captures the idea that, in F.H. Bradley's famous phrase, 'my station and its duties' is shaped by a specific mix of external norms and behaviours and internalised attitudes and values. The emphasis on *role* here, importantly, focuses our attention on action and behaviour. One who merely thinks appropriately but acts otherwise therefore does not fulfil the fundamental functions of a role. *Morality* in this conjunction also specifically highlights three useful ideas. First, it reflects the fact that certain roles matter precisely because they involve some institutional commitment to advance specific social and moral goods.[1] Second, it reflects that the morality of the agent is, at least in part, determined by the morality of the role itself. Actors cannot readily be their own moral justification; that is simply too circular, and too agnostic about the normativity of roles in and of themselves, to be sustainable. Third, and more subtly, role morality also provides institutional grounds for selecting between otherwise conflicting social norms. It generally achieves this by instantiating moral *separatism*. Separatism 'holds that one aspect of morality, namely, [the professional's] specific role-based actions on behalf of his client or other valuable purposes of his profession, takes precedence over many other aspects of morality'(Gewirth, 1986, p. 283).

The propriety and extent of the separatist account underlie much of the debate about professional ethics, with theorists essentially adopting positions along a continuum between 'strong' and 'weak' versions of separatism (see, e.g., Radden, 2004). In legal ethics, this well-known debate is reflected in broadly two strands of thought. On one hand, there are those scholars who adhere to what is commonly called the 'standard conception' of the professional role. This sees the lawyer primarily as an advocate for his or her particular client's interests or entitlements (Pepper, 1986; Wendel, 2012). In this account, it is the law itself, and lawyers' (collective and individual) institutional obedience thereto (Dare, 2009; Wendel, 2012), that primarily sets the effective boundaries of lawyer agency. On the other hand, there are theorists who argue for a weaker version of role morality, which holds the role norms closer to the social expectations of 'common morality'. *Legal* agency in this version remains a proper constraint on action and should not be too readily set aside, but actors have a greater scope for individual *moral* agency through the exercise of 'moral activism' (Luban, 2007, 1988) or 'contextually-sensitive lawyering' (Nicolson and Webb, 1999; Simon, 1998) that addresses the harms that may otherwise be caused.

This positional, normative, debate has dominated legal ethics theory to a, perhaps, remarkable degree. I say remarkable because of the obvious, though not always well understood, limits of the dominant normative ethical debate. A number of these limits are relevant to the following discussion.

First, it is clear that there is no overarching version of moral theory that enables us to say categorically which ethical position is superior. The conflict between the role-constrained and the unencumbered self, as Luban (2005, p. 595) concludes, 'is not [one] that will ever go away'. Nor can we even say categorically that any one role morality is superior as an exercise in moral reasoning. Each may help determine where our ethical commitments lie and ensure that decisions are non-arbitrary. However, once the threshold of normative justifiability is satisfied, any one ethical theory is, from a moral reasoning perspective, as good as another (Miner and Petocz, 2003; Woolley, 2011).

Second, most normative models of the professional role assume a theory of agency that operates from a weak 'conception of sociality' (Rousse, 2016). There is a tendency in theory to assume that the key features of agency are all found within the human psyche, and failures of agency can therefore be ascribed primarily to an individual lack of sensitivity or inattention to the moral qualities of the situation, or poor reasoning, or sheer lack of moral 'backbone'.

Third, it follows, too, that in the conventional view, any multi-agent system is seen primarily as no more than an aggregation of such individual agents. In professional ethics and regulation this has been reflected in a tendency to regard lawyers, in David Luban's (2007, p. 237) memorable phrase, 'largely as self-contained decision-makers flying solo'. Regulatory and disciplinary norms have similarly tended to focus on dealing with transgression as a matter of individual failure – the classical 'bad apples' approach.

Both of these tendencies arguably reflect a more generalised underplaying of the situatedness of agency. Psychological studies of the power of the situation represent a significant challenge to standard conceptions of agency and responsibility (Mele and Shepherd, 2013; Shepherd, 2015), as Lieberman (2005, p. 746) observes:

> All of the most classic studies in the early days of social psychology demonstrated that situations can exert a powerful force over the actions of individuals. . . . If the power of the situation is the first principle of social psychology, a second principle is that people are largely unaware of the influence of situations on behaviour, whether it is their own or someone else's . . .

This takes us into much larger issues (in the interstices between social psychology and both moral philosophy and philosophy of mind) than can be properly considered here. Suffice it to say that one highly relevant implication may be that situational and relational factors operate, not as mere distractors from the 'proper' exercise of agency, but as prior constraints upon it. Moreover, from a situationist perspective, the significance of structural and organisational forces – 'bad barrels' and 'bad barrel-makers' – has, at least until relatively recently, been significantly underplayed in both ethical theory and practice.[2]

I propose instead that legal ethics theory would benefit if it were to start from a richer, more adequately complex vision of the *context* for ethical agency. This is, first, a descriptive task. The account of lawyers' ethical practice, I argue, needs to be both more completely 'agentic' *and* more situationally sensitive to the structural complexity of ethical decision-making. This is necessary to avoid the twin pitfalls of reducing the ethical account of lawyering to naïve realism and thin descriptivism (in which the role becomes simply what lawyers 'do'; Applbaum, 1999, p. 56) and producing a decontextualised normative ethics which may lack a proper degree of psychological and sociological realism (cp. Flanagan, 1991, p. 32ff).

Which complexity?

In arguing that we need a more adequately complex account of legal ethics as a practice, this chapter immediately begs a question as to how we should understand the 'adequately complex'. This is no small question, as there are multiple accounts of complexity and multiple approaches within the umbrella of complexity theory. Michael Lissack's (1999, p. 112) observation that complexity 'is less an organized, rigorous theory than a collection of ideas' still retains more than a grain of truth.

As the chapters in this collection indicate, approaches to social complexity range from the computational to the metaphorical and often draw on strong interconnections with cybernetics and systems theory. This paper also reflects that breadth, but nonetheless seeks to avoid a simple 'pic-n-mix' approach. It seeks to achieve internal coherence by three moves which position it as both an account of, and within, complexity theory.

First, the chapter takes a non-controversial approach to the concept of complexity itself. It thus starts from the widely held position that complex systems are those characterised by multiple interactions between multiple components or elements over time (Ladyman et al., 2013, pp. 35–6). Of course, this by itself does not take us far in understanding the distinctive nature of complexity. The chapter thus takes the further step of defining complex systems in terms of an established base of core concepts about which there is a high level of agreement in the literature. Thus, it assumes that complexity theory focusses primarily on the ways in which *order emerges as a property of self-organisation in dynamic, adaptive, and generally non-linear, systems*. These concepts have been substantively discussed in earlier literature on law and complexity theory (see, e.g., Ruhl, 2008; Webb, 2004, 2014), as well as in other contributions to this collection.

Second, the chapter locates its argument within the fundamental distinction between *general* and *restricted complexity* (Morin, 2007; Woermann and Cilliers, 2012). Restricted complexity assumes that systemic complexity can be simplified by a 'grand unifying theory' and thus reduced to a set of fundamental principles that apply trans-systemically. By contrast, general complexity views an object of study as a *complexus* – that which is 'woven together' – and argues that there is a quality to real-world complex systems that is in and of itself beyond the reach of any one formalist (reductionist) account to capture in all its properties. Whilst this chapter draws on literature from both theoretical positions, it ultimately positions itself as an account within general complexity.

Third, philosophically, the chapter is framed, phenomenologically, within a broadly realist-constructivist paradigm (Webb, 2006). Two central assumptions flow from this model:

a   Complexity in this sense is not a synonym for the merely complicated. A complex system is a qualitatively distinctive phenomenon. It is often said that systems are complex where the whole is more than the sum of the parts (Simon, 1962, p. 468). The critical point to note is that this statement is correct in ways that are *functionally non-trivial*.

b   It recognises that complex systems have qualities that are irreducibly distributed, contingent and dynamic. Complex systems theory thus contrasts with Parsonian systems theory, which assumes a high level of functional 'localisability' (i.e., that each system property should be isolatable to the actions of individual components or agents or the structural couplings amongst them).

In sum, the position adopted is that a complexity account is significant in emphasising the basic plasticity of social (including legal) structures, and the distributed, contingent and composite nature of social communication, and indeed of the knowledge that is the product of these systemic processes. Complexity approaches offer a uniquely useful, one might say 'evolutionary', perspective on normative adaptation and development. They particularly highlight the processual quality of systems, operating through a cyclical process of variation (disorder), selection and stabilisation (re-ordering). The mechanics of such processes are central to this chapter, and we therefore need, finally, in this section to address the way in which notions of agency and entropy do useful explanatory work in understanding the operation of complex systems.

Agency and system

This chapter offers a departure from those approaches to systems theory and complexity in law that (appear to) take the individual agent out of the equation. This is a criticism that has been particularly directed at Luhmann and autopoietic systems theory. Whether or not that critique is overstated (Paterson, 1995), I would suggest that autopoiesis in this regard has at least been a useful spur and corrective. By de-centring the individual and treating the operation of what Luhmann calls 'psychic systems' distinctly from communication in social systems, he has forced us to confront (again) the very real question of how much control the individual has over social processes. Nonetheless, the position adopted here is not autopoietic. This chapter takes an intermediate – cybernetic (Lippucci, 1998) – position on the role of agency in systems, rejecting both Luhmannian agency-scepticism and traditional, liberal idealism. Complexity does require us to assess and take proper account of the agents themselves – their diversity, localities, networks, and levels of interaction. However, it also acknowledges, not only that social systems possess properties that are not reducible to individual action but that such properties may also serve to suppress, amplify or otherwise distort agency effects. Conceptually, research on adaptive agents and agent-based systems, derived from the seminal work of theorists such as John Holland (Holland and Miller, 1991) and Robert Axelrod (1997, 1984), is useful in trying to capture some of that interplay between agency and structure. Three further, related points can be made about the approach to agency here adopted.

First, the cybernetic focus serves to reframe agency primarily as communicative action. As Wiener (1989, 1948) observes, communication is central both to individual and social learning and to the very construction of human society: 'community extends only so far as there extends an effectual transmission of information' (Wiener, 1948, p. 184). Communication is a product of agency and structure, reflecting agents' system location and levels of interaction. Moreover, it is suggested that system communications also need not be exclusively human but are often the product of interaction and co-functioning between networks of persons and objects. If, as John Law (1992, p. 380) observes, 'order is an effect generated by heterogeneous means', then we are also perhaps permitted to think of agency in heterogeneous terms.

Second, agency and social structure necessarily co-evolve. System adaptivity is, of course, affected by existing system structures – what Law (1999) calls the larger 'topology' of the system, that is, features such as spatiality, scale and density of connections demonstrated by a region or network. This serves to shape agent 'context' – essentially the (limited) information/stimuli that agents gain from their immediate environment – including, of course, other agents – and which creates a 'shadow of the adaptive future' (Axelrod, 1984; Cohen et al., 2001, p. 5). The system structure also influences agents' 'schematic preferences' and 'internal models' of the world, which both condition agent responses and allow them, to some degree, to 'look ahead' to the consequences of their actions (Holland, 1995). Thus, a complexity account can be said to describe agents as operating in ways that are both fundamentally relational and (broadly) rule-based in the sense that system operations can ultimately be rule-described. Agency, however, is more than mere rule-following, the very complexity and the dynamic quality of the system mean that agents are active and, to a degree, autonomous, elements of the system engaged in an evolving 'adaptive walk' (Arthur et al., 1997) across the system landscape. Agency thus understood involves both a playing by the rules and playing *with* the rules.

Third, in terms of understanding the internal operations of agents, it makes better scientific sense to link this complex, dynamic view of agency with the 'bounded rationality' model of human action rather than rational choice alternatives. The idea of bounded rationality, developed originally by Herbert Simon (1956), has gained increasing purchase in the social and psychological sciences as a way of describing the mechanisms of, and the cognitive and environmental, constraints on decision making under (normal) conditions of complexity and uncertainty. To focus on bounded rationality is to look, descriptively, at how individuals cope when knowledge, resources and time are all limited, and reasoning processes are, at best, stochastic but often more broadly based on trial and error heuristics or 'rules of thumb' (Axelrod, 1997; Gigerenzer et al., 2002). Crucially for our purposes, the idea of bounded rationality has, over the last decade or so, also influenced research in organisational and behavioural ethics, where we see a growing interest in accounts of 'bounded ethicality'(Bazerman and Moore, 2013; Chugh and Kern, 2016; De Cremer et al., 2010).

Entropy, organisation and adaptation

Entropy is a significant feature in understanding the (self-)organisation and adaptive operation of both closed and open systems. Following Bateson (2002, p. 211) and Bailey (2001, pp. 55–7), I define entropy relatively broadly as the degree of disorder or uncertainty present in a system. This broad approach makes sense in the context of social systems, where what we are dealing with is 'social entropy' (Bailey, 2001, 1990) rather than 'pure' energy.

Thinking about entropy helps us identify a number of key qualities of complex systems. First, it highlights that entropic decay is a key risk faced by complex systems (Mobus and Kalton, 2015, pp. 445–6). In rather crude mechanical terms, this is the tendency of system components to become less-efficient over time or to wear out. Complex systems have a capacity for self-repair, though the extent of that capacity is, in the biomechanical world at least, invariably limited. Social systems may be more resilient. There is at least some evidence that social entropy does not necessarily follow the second law of thermodynamics, and in some contexts net entropy may actually decrease over time (Bailey, 2001, p. 55); however, we need also to recognise a possibly countervailing insight from regulatory theory, which points to the extent to which system responses to potential 'entropic decay' may also add to the informational build-up and complexity of the system. As Robert Kagan observes, the natural corollary to Murphy's Law is that 'regulation grows' (Kagan, 1989, p. 89).

Second, thinking about entropy highlights that we need to think about information itself is a 'measure of organisation' (Wiener, 1989, p. 21). The functionality of a system is thus fundamentally a process quality: Can the system manage uncertainty through the process of successful communication? This points us to the way in which information is both a vital resource and a problem for the functioning of a system. Complex systems are thus inherently prone to information overload (Mobus and Kalton, 2015, pp. 441–2). Continued functional efficiency thus depends on the system's ability to be selective in its storage and use of information. In practice, these processes of selection/deselection will likely reflect the functional utility or disutility of certain information. Information that is not repeatedly selected will tend to be 'forgotten' and lost to the system.

Third, thinking about entropy offers some useful insights into the function and limits of regulation. Three key examples may be useful here. (1) Sustainable social systems require a capacity for uncertainty reduction through system selections that are at the same time *choice reductions*. In a social system, it is regulation that commonly performs this function, not only through formal rules but also through layers of governance. Again, an analogy can be made with biomechanical systems (Kugler and Turvey, 2016), where a 'macro' system of itself plays a fundamentally important coordinating role, responding to changes in entropy at a micro (multi-agent) level through coupled levels of dynamic activity. Metaphorically (albeit in a closed system analogy) we might say the macro-system acts rather like the regulator to an agentic boiler. (2) Regulation begets entropy; complexity itself is a function of the range of controls and decision alternatives created by the system (Palmer, 2012, p. 98). Any attempt at regulation changes system conditions, with consequences that are, under conditions of complexity, inherently unpredictable. In simple terms, the outcomes of efforts at regulatory steering are therefore intrinsically contingent and uncertain. (3) Whilst entropy creates risk, it also performs a useful function. It is part of what drives the dynamism and hence the evolution of a system. Too much entropy reduction thus creates systemic risks of stasis and decay. This insight may be a particular challenge for regulation, since entropy reduction in and of itself can readily take on a pseudo-normative quality:

> entropy is generally considered as a measure of the ability to predict the next state of a system. If the next state is highly predictable, then entropy is considered to be low and vice versa; consequently, a system that presents low entropy is considered to be organized and, by deduction, *desirable*.
>
> (Mavrofides et al., 2011, p. 354)

With these qualities of complexity in mind, we therefore turn our attention back to the 'system' of ethics in legal practice and the role of agency in that system.

Ethical decision-making in legal practice: a cybernetic model

A cybernetic model of legal ethics focuses on the process of ethical decision making (Figure 13.1) as its starting point. The model presented here draws on sources from a number of overlapping literature that contribute to complexity thinking: from cybernetics itself (Mobus and Kalton, 2015; Wiener, 1989); evolutionary theory, particularly as pertaining to organisation and governance theory (Axelrod, 1986, 1997; Deakin, 2011; Zumbansen and Calliess, 2011); and the relatively recent 'behavioural turn' across organisational and ethical theory (Bommer et al., 1987; De Cremer and Vandekerckhove, 2016; Palmer, 2012), including legal ethics (Perlman, 2015; Robbennolt and Sternlight, 2013; Webb, 2019). The model recognises that ethical decision-making does not take place in isolation. Ethical decisions are set in a specific transactional context and are shaped by larger organisational and cultural norms, values and beliefs. It differs from early ethical decision theory, however, in the extent to which it treats the effects of these contextual factors on decision makers as *situational and impermanent* rather than general and semi-permanent (cp. Palmer, 2012, p. 95). The extent to which this makes ethical decision making intrinsically unstable and unpredictable is a key, and somewhat unsettling insight, of a complexity approach. In the following sub-sections I highlight many of the reasons for that instability.

Legal processes are dynamic sub-systems

The potential for situational variation is highlighted in the very structure of the model. The micro-unit of analysis is a single process: decision making in a specific organisation, with respect to an individual legal case or transaction. Each organisation (law firm, in-house legal function) in this model thus constitutes a system (within the larger environment of legal practice), and each process operation represents a throughput of the organisation and, functionally speaking, is best conceived of as a sub-system of the organisation. Such processes are the locus within which agency operates; they are, in effect, the building blocks on which the multi-agent ethical system is built. Such operations can be conceived of as broadly linear in time, though features of the process may be strictly nonlinear and recursive, as we shall observe.

The schema presented focuses, at the centre, on five elements of the decision-making process. The first two, information gathering and processing, may appear to be the least controversial and are often framed (e.g., in legal training literature) as a rational, objective process. The reality, however, is that they take place within a relational context where tensions may well exist between conflicting objectives and values and are inevitably subject to the cognitive limitations of the decision-makers. Ethical issues, moreover, may arise at almost any point: in gathering information, advising on the client's objectives, assessing whether and how information is used and selecting between alternatives.

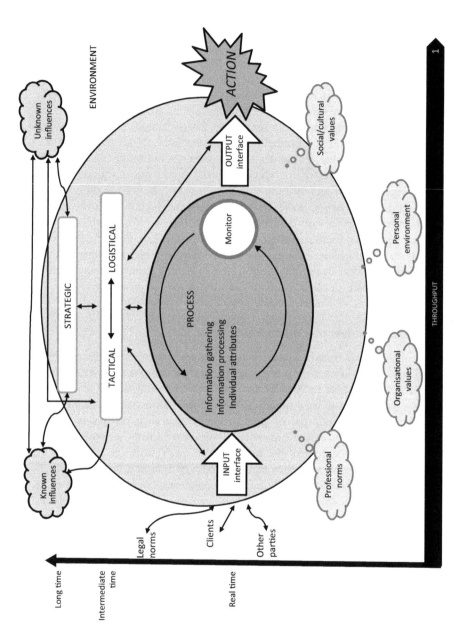

Figure 13.1 A cybernetic model of ethical decision-making

Individual and work-group attributes are ob viously material. Classically, regulatory frameworks have tended to subscribe to what I have already characterised as a 'bad apples' theory of professional misconduct, leading to resources being channelled primarily into some mix of psychologically crude pre-admission screening of character and fitness (Bartlett and Haller, 2013; Levin et al., 2013) and an individualised sanctions-based model of professional discipline. This focus on rogue individuals is not entirely misplaced: failures of 'character' and 'personality flaws' undoubtedly account for some of the more egregious misconduct, but the bad-apple approach remains simplistic and somewhat unfocused. First, as Woolley and Wendel (2010) observe, moral decision-making is shaped at the individual level by a complex mix of moral 'dispositions', personality traits and affective states, as well as cognitive capabilities and biases. Legal ethics theory (and, I would add, regulation) has generally, by contrast, taken quite a thin view of human character and, at least arguably, may even lack sufficient psychological realism about the desirable mix of traits and dispositions we want to see in our lawyers.[3]

Second, in any event (as Woolley and Wendel also acknowledge), research in behavioural ethics indicates that organisational culture is a much bigger determinant of conduct than are the personal characteristics of the agent. Behavioural research highlights the complex role of institutional structures, systems, processes, attitudes and values in shaping the 'ethical culture' of a workplace, recognising that organisations, and even sub-groups within organisations, constitute an important normative setting in their own right (see, e.g., De Cremer and Vandekerckhove, 2016; Treviño et al., 2006; Victor and Cullen, 1988). A complexity analysis suggests that these factors may well interact at different levels and in different ways within the system. This is reflected in the division in Figure 13.1 between in-process dynamics and wider organisational dynamics which are shaped both by the system and by the wider environment.

At a process level, teamwork and group dynamics may exercise considerable importance in understanding lawyers' ethics and their limits, particularly given the increasingly large role that teams play in legal work (Murray and Fortinberry, 2016; Rogers, 2017). This is a dimension of legal practice that has been relatively under-researched so far, though work in other domains highlights a range of ethical risks (including task fragmentation, 'groupthink' and conformity biases), and potential benefits (scope for enhanced ethical sensitivity, group ethical identity formation, and capacity for moral action) – Rogers (2017). Recent research does point to lawyers as somewhat reluctant and instrumental team players (Gardner and Valentine, 2015). This could limit some risks (e.g. groupthink and over-identification with a negative group ethic) but exacerbate others. For example, it may point to a reluctance amongst individuals to take ethical leadership or to develop effectiveness in that role. In-group bias and techniques of moral disengagement may also actively limit the scope for ethical agency. Bandura's (2002, 2001) work on social cognitive theory particularly points to ways in which conduct may be 'cognitively reconstrued' to make harms acceptable. Thus, techniques adopted include, for example, portraying the conduct as serving a larger moral or social purpose (e.g., through the deployment of 'role morality' itself – Kuhn, 2009, p. 691), diffusing responsibility or distancing the act from one's own agency (cp. discussions in Bauman, 2000, pp. 161–3, 1994, p. 7, and Luban et al., 1992) of the way responsibility may simply 'float' between actors in organisations), or simply disguising misconduct with euphemisms.

'Technology', in a broad, economic sense (Stephen, 2013, p. 128), and technological innovation are an increasingly important part of the law industry. Process re-engineering has become a significant feature of change in the workings of the legal services market and its regulation (Stephen, 2013, pp. 127–41; Susskind, 2010). The extent to which new technologies change or increase specific ethical risks remains moot and difficult to generalise. Undoubtedly, some technologies, such as the use of machine learning in automating legal advice may well create new and specific regulatory challenges. Less sophisticated technology changes, such as outsourcing, the use of contract lawyers, or increased reliance on a paralegal workforce may, at first sight, create only qualitatively different ethical (management) problems, particularly of risk assessment and supervision, from more traditional practices. However, their ethical consequences may also run deeper and be more subtle than that suggests, For example, changed processes may risk further stratifying and diffusing ethical responsibilities, or may potentially help fragment the sense of collective professional identity that forms part of the environment of legal practice (Mather, 2011).

Organisational and environmental effects

Each process is also shaped by the particular interplay of organisational and wider environmental effects. These may provide additional resources to the system, or act as a cause of (ethical) entropy. Key environmental forces identified by the cloud shapes in Figure 13.1 are generalised professional norms and regulations, organisational, and personal and wider cultural values.

Professional norms

Accounting for the role of professional norms in shaping and directing legal and moral agency is not at all easy. A complexity perspective can help us in this task, not least because it invites us to re-consider aspects of the relationship between professional codes and agency. There has been a tendency in the legal ethics literature (other than in the practice of narrow, doctrinal scholarship) to overlook or understate the codes: conceptually they are remote from the higher concerns of ethical theory and, practically, of rather limited assistance as an aid to ethical decision-making. Ethical rules, however framed, tend to get a bad press (see, eg, Boon, 2010, 2016; Nicolson, 1998). They are either too lacking in aspiration or too hortatory in nature; they tend to be a mishmash of rules and principles, short on coherence and perhaps displaying at best complacency, if not confusion, over the profession's core values. There is, moreover, empirical evidence that the codes themselves are relatively under-utilised. Lawyers, trained as they are in the dark arts of rule manipulation, are also quite capable of adopting modes of 'creative compliance' in respect of their own rules. They seem frequently to treat ethical matters (by default) as merely tactical or strategic decisions (Moorhead and Hinchly, 2015; Vaughan and Oakley, 2016; Wilkinson et al., 2000). At the same time, regulators tend to focus on the more straightforward breaches, particularly financial misconduct, and avoiding disciplining matters on which there is moral dissensus, or where 'thick concepts' (like competence) make core duties difficult to define and enforce in all but the most egregious cases (Woolley, 2012; cp. Haller, 2001). All of these suggest that we should not take the rules too seriously.

Arguably, however, such a conclusion also risks missing the mark. Codes, in short, do matter, even if that is sometimes as much for what they don't say as for what they do. Whatever their limitations, they nonetheless constitute part of the normative and risk framework of the profession. The growth of an internal compliance function in law firms may, in fact, be one feature of organisational structure that helps ensure the formal norms have continuing resonance (Kirkland, 2008; Parker et al., 2008). We see professionals (individually and collectively) placing considerable store by the need for and importance of professional ethics and standards in both professional formation and practice. A striking feature of the Legal Education and Training Review (LETR) research data, for example, was the clear priority given by individual legal service providers to legal ethics over other knowledge areas in legal education and training (Webb et al., 2013, paras. 2.67–2.73). Professional bodies (or their regulators) frequently invest resources in the work of code (re)design and implementation. Similarly, when influential organisations or groups of practitioners are confronted by tensions between their ethical obligations and (perceived) business imperatives, they do not ignore the problem but act as 'institutional entrepreneurs' (Flood, 2011, p. 511), committing not inconsiderable energies to stabilising any potentially conflicting normative and client-led expectations (see Loughrey, 2011; Rogers et al., 2017). Ethics in this light is more than window dressing.

In more specifically systemic terms, the codes thus provide both structural (e.g. business organisation, trust accounting and other aspects of 'ethical infrastructure') and symbolic (encodings of values, principles and relational norms) information to the system and thereby influence the agency of actors, even if only to a limited degree. They also do important work of boundary definition and maintenance. Most obviously, they instantiate and embed the fundamental coding of conduct as ethical/unethical from within the system, and (despite problems of uneven and under-enforcement) some at least are ignored at one's peril. Less obviously, perhaps, they also play a part in the general normalisation of role and the ends of representation: thick concepts like the duty to the court and loyalty to clients provide powerful, albeit problematic, heuristics. The codes are also a part of what generates 'jurisdiction' in the larger sociological sense used by Andrew Abbott (1988, 1986), which serves to differentiate actors within and between professions and may also generate ethical tensions and regulatory competition between occupational groups (see Muzio et al., 2016). In both these informational and boundary maintenance functions the codes can perhaps be said to exercise their own agency, in the sense of 'a performative stabilisation of relational networks' (Law, 2002).

Personal, social and organisational values

The other areas of influence highlighted are the range of organisational, personal and social values and their institutionalisation as 'logics' of the organisation. The influence of such factors is widely acknowledged: as noted already, a considerable body of work, particularly in business ethics, has highlighted the importance of embedded and institutionalised organisational values and 'culture' in influencing ethical behaviour.

By contrast, the role of personal values and attributes is a more difficult and, in some respects, neglected question. Bommer et al. (1987) point to a range of research that suggests peer and family environment and individual demographic, moral and personality attributes should not be discounted. Work by Evans and Palermo (Evans and Palermo, 2005; Palermo and Evans, 2005) on personal values and ethical decision-making by law students also offers some relatively decontextualised and therefore tentative evidence of the impact that personal values may have on professional attitudes and possibly behaviour. However, such analyses are contestable, in that relatively little work has sought to reflect or simulate real-world decision-making. There are important questions raised, particularly by more recent research, as to the primacy of individual or group identity and individually or group-based norms, in organisational decision-making (see, e.g., Tremblay and McMorrow, 2011, pp. 571–2). This chapter therefore focuses primarily on organisational norms and values.

Within each organisation, variables such as hierarchy and leadership style, relational models, peer support, reward systems, ethical infrastructure and trust climate (Fortney, 2005; Giessner and van Quaquebeke, 2010; Parker et al., 2008; Steinbauer et al., 2014; Treviño et al., 2006) are all capable of influencing the individual agent's moral motivation and behaviour. It is notable that, in the legal practice context, regulatory development of 'ethical infrastructure' has been patchy to non-existent in most jurisdictions, thereby potentially amplifying the extent to which organisational factors will be a material variable in shaping ethical decision-making. Theoretical work on social networks also demonstrates how individual's social and business connections (including client and investor relationships) may serve to create or constrain, in often complex ways, opportunities for unethical conduct (Brass et al., 1998; Palmer and Moore, 2016). Much of this theoretical work, however, still lacks application and rigorous empirical testing.

In system terms, we can think of organisational culture and values as working through an interplay of information flows and control mechanisms, influenced by the specific temporal dynamics of that system. The relative weighting given to normative and other values and objectives is a material variable. A focus on non-normative goals, such as billing targets, client satisfaction, etc., while of economic importance, can be ethically problematic if it leads to a process known as 'ethical fading'. Ethical fading occurs when non-normative objectives override normative considerations, so that they fade into the background, and are no longer part of the conversation (Prentice, 2015). In the language of complexity this could be described as a form of entropic decay: certain ethical precepts lack (relative and perceived) functional utility and thus tend to be downgraded or forgotten in system operations. Of particular interest here may be questions regarding the flow and quality of normative information – including whether the conduct in question is regarded as de facto normative or counter-normative within the organisation. Relationships and hierarchical structure, including individual location and authority in the hierarchy can also be important structural considerations, not least because they may reflect power dynamics that shape both information flows and effectiveness of control systems. The high levels of autonomy granted to more senior lawyers; the power and status associated with (successful) partners and organisational tolerance for 'lone ranger' and non-collegial behaviour may, for example, all contribute to an environment in which the risks of some forms of lawyer misconduct increase (Abel, 2011; Regan, 2009).

Locating power and control systemically

The question of control also brings our attention to other related but specific functionalities highlighted in Figure 13.1. In cybernetic terms, control mechanisms and resources must be deployed within each system to maintain operability. While some element of monitoring may take place from within a process (e.g., as a matter of decision-checking by the responsible individual, or within the team undertaking the work), larger control mechanisms are better seen as part of the system, not just the process. This reflects reasonable assumptions about their relational and often spatial-temporal positioning relative to specific processes and will reflect the fact that, in systems that have many operational processes, control functions will tend to co-evolve into sub-systems of their own. Co-evolution, itself, creates new coordination problems for the organisation and the emergence of new control mechanisms.

What do we know about organisation and control in law firms? Over the last 20-odd years, a growing body of literature, most of it emanating from university business schools, has sought to address the distinctiveness of the professional partnership as a form of business organisation.[4] It identifies the professional firm as one distinguished particularly by norms of collegiality, a strong alignment between membership and ownership of the business, and a reliance often on the (charismatic – in the Weberian sense) leadership of a managing or (increasingly) chief executive partner as *primus inter pares*. Strategic decision-making under this model has often tended to be conservative, mimetic and inward-looking (Beaton and Kaschner, 2016). The law industry has, on the whole, not been renowned for either its strategic creativity or for its willingness to invest in business skills or for research and development – though this is possibly changing under the conditions of competing in a fully mature market (Beaton and Kaschner, 2016). Traditionally, systems of organisation and control within this structure have been relatively informal and non-bureaucratic, relying heavily on getting the 'right people' and socialising them into a specific culture and way of working (Moorhead, 2014, p. 459).

This classic partnership model has come under pressure both from within and without, characterised variously as a rise in 'managerialism' and bureaucratisation (Sommerlad, 1995), a turn to 'defensive professionalism' (Muzio and Ackroyd, 2005) and to a 'managed professional business' model (Lawrence et al., 2012). As Greenwood et al. (2017, pp. 117–8) observe, the normative consequences of such changes remain under-researched. The archetypal form of the professional service firm was assumed (certainly by practitioners and, to some extent, by organisational theorists) as in and of itself a guarantor of ethical standards through its commitment to classical 'social trustee' professionalism. The naivety of this view has been made apparent by the wave of audit failures in accountancy and the involvement of lawyers and their firms in a number of high profile scandals, including Enron, the collapse of Lehman Brothers, the conduct of 'Big Tobacco' litigation, and the *News of the World* phone hacks (see, e.g., Moorhead, 2014; Parker and Evans, 2014, pp. 119–21, 331–4). The extent to which certain forms of ethical failure are 'built in' to the design of law firms is therefore, in this context, a highly pertinent question.

The issue of control opens up another dimension of (operational) agency. As the earlier discussion of situationism indicated, abstract accounts of agency may be considered flawed because they separate the ability to act from the norms and relations of power that shape the decision-making context. A realistic account of agency must, by contrast, treat it as a situated concept and inseparable from the analysis of power (McNay, 2016, p. 41) and, particularly, the structures of power embedded within professional institutions. Neo-institutionalist analysis by writers such as Clegg (1989; Deroy and Clegg, 2015), Lawrence (2008; Lawrence et al., 2012) and others (see, e.g., Becker-Ritterspach and Blazejewski, 2016) treats organisations as networks of both localised ('episodic') and more structural, hidden, systemic power relations and usefully segues into complexity thinking. Since such power relations form part of the conservational tendency of existing organisation forms and practices (Lawrence et al., 2012; cp. Mather, 2011) they are central to understanding not just change processes in the profession but also the social construction of institutional agency and control itself.

In sum: agency in complexity

The model presented here aims to provide an overarching account of lawyers' working processes as complex rather than merely complicated. It presents a description of the context for ethical decision-making which is contextually rich and, I would argue, at least closer to adequately complex description. It highlights how ethical decision-making is individualised and collective: it is transactional, systemically embedded and situationally contingent. In overall terms, it points to the interplay of agency and structure and the importance of multiple value systems and priorities – at the level of the profession, the organisation and the individual. It points thereby to human agency as both an operation in individual consciousness and in collective action, as part of the complexity of interactions in the social world and therefore as a critical part of the difficulty of formulating causal explanations (and hence of allocating responsibility). Unlike some sociological accounts, it seeks to take rules seriously but not too seriously. It recognises that agency is both expressly and tacitly rule-described and, to some degree, *rule-constrained*. In so doing, it recognises that the 'rules' include not just the formal, collective ethical norms of the profession but also the formal and tacit norms that operate at the firm level, and individual standards (e.g., the personal limits lawyers may set on representation – Vaughan and Oakley, 2016). This interplay and contestation of norms of itself mean that the nature and extent of functional agency is an important, and often contested, or at least negotiated, question in practice. This brings us back to the question of entropy posited in the title of this chapter.

Ethical failure: talking about entropy in legal ethics

The model presented in this chapter has 'conceive[d] of agents and practices as contingent assemblages in the making rather than as performed entities' (Tsoukas and Dooley, 2011, p. 732). It follows that this in and of itself warrants some rethinking of the nature and likelihood of ethical failure in practice and consequent responses. In this final section, I start that process by using a conversation around entropy, reflexively, to reframe our understanding of ethical failure, its consequences and the possible cures.

It will be recalled that we defined entropy broadly in the second part of this chapter as a degree of (decision-making) uncertainty and ultimately functional disorder. This would tend to translate in the present context primarily to a proliferation of unethical behaviour, which could lead to a loss of professional legitimacy, and loss of market privileges, and thereby threaten system stability. The operative system here cannot, as our discussion shows, be considered the ethical (sub-)system in isolation but the larger system of practice. Consequently, system entropy and ethical failure are not synonymous. Entropy is certainly wider than the notion of ethical failure, and not all ethical failures (if we think of ethical failure in terms of a broad range of misconduct) are necessarily entropic.

Nonetheless, the preceding analysis points to the fact that some degree of 'ethical entropy' is inevitable within modern practice systems. Complexity is a property as well as a description of the system, and it is therefore neither oxymoronic nor overstating the point to highlight the fundamental situational complexity of the practice context as a key explanation of ethical entropy. More specifically, we can highlight (at least indicatively) the interplay of the following agentic and structural factors as important in constructing the ethical climate and infrastructure of an organisation, and with it the risk of entropy.

Authority structures and group dynamics: law firms tend to be hierarchical institutions, operating according to sometimes conflicting principles and institutional logics. Like any form of organisation there will be a degree of internal politics and organisational power play, and this can have an impact, somewhat differently from organisation to organisation, on ethical decision-making and the tolerance for unethical behaviours. Hierarchy also often implies subordination, which brings its own forms of ethical disempowerment (Perlman, 2007).

Normative pluralism, contingency and complexity: these are inevitably part of the climate of professional workplaces, particularly as, in many jurisdictions, the volume and complexity of regulation have increased. Insofar as it makes sense to think of legal practice as a rule-constrained activity, we must also acknowledge that every playing by the rules entails some playing with the rules. This does not necessarily imply bad faith. Rule application always presumes, first, an ability to identify and frame the issues as ethical. There may be subtle environmental and organisational influences at play, which shape those interpretations, and permit or even encourage a degree of ethical fading. The formal norms may actually contribute to this effect: some of the biggest conflicts (e.g. around hourly billing practices) may barely be identified as ethical problems in formal discourse. Ethical norms, where they are explicit, can be relatively open-textured and subject to significantly different interpretations, particularly at the margins. Less innocently, strategies of creative compliance may be adopted to deal with client demands and/or personal cognitive dissonance. Ethical resources, such as commentaries and helplines are, moreover, sometimes of least help in the hard ethical cases where they seem most needed.

Role rationalisation: As noted, the idea of professional role, and its associated role morality, has both descriptive and normative value. However, drawing on work in behavioural ethics, including notions of moral fading and moral disengagement, we can see that role is intrinsically double-edged. Its moral content may be under-determined and insufficiently recognised in the heat of practice. It can operate as a rather crude heuristic which may be used to rationalise actions, under conditions of uncertainty, that would otherwise be deemed unethical (see also Hall and Holmes, 2008). The common tendency to think of the legal professional role in 'client-first' terms (notwithstanding overriding duties to the administration of justice) and use that to discount countervailing responsibilities to the court, opponents and third parties is a prime example of the way in which a broad sense of role morality has potentially unethical consequences.

Moral distance: the ideas of 'moral distance' and 'floating responsibility' have both been used by Bauman (2000, 1994) to explore the impact of social context on ethical behaviour. Moral distance describes the notion that people's ethical concerns for others depend on (relational) proximity. Proximity is ethically double-edged; it valorises relationship, on one hand, but highlights that, as distance increases, it becomes easier to act in unethical ways. Moral distance matters because it is embedded in a 'client-first' ethic and in the tendency of codes to underplay duties to non-clients or vulnerable others (Nicolson, 1998; Nicolson and Webb, 1999). Following Bauman's formulation, it can be regarded as a form of moral disengagement and, *contra* at least some theories of partisanship (e.g., Fried, 1975; Markovits, 2011), a problematic quality of 'client-first' role morality. Floating responsibility, as discussed earlier, can also be seen as a consequence of the greater moral distance created by bureaucratic and distributed systems of work.

Network location: in complexity terms, ethical practice is fundamentally relational and can be conceived of (in multi-agent terms) as a network of conversations. One's temporal and spatial location in the network may therefore also help determine one's ethical capacity and risk. A number of questions might usefully be asked in determining network location: Is the lawyer embedded in or marginal to a given organisational environment? To what extent are there specialised structures for (self-)regulation and ethical oversight in her organisation? Is the working environment one where ethical conversations are available or even encouraged? Is she working in a context where there is some normalisation of deviant norms, or is it one which is otherwise particularly vulnerable to ethical fading?

At the same time, there are a number of features of the ethical sub-system, and particularly of regulation, which have the potential to be entropy-reducing.

First, systematisation *aggregates and consolidates* information that would otherwise be dispersed across multiple agents. The role of professional associations, regulatory bodies and legal service provider organisations themselves, in providing a form of both data fusion and coordinated dissemination, cannot be discounted. These functions may be achieved through, for example, professional codes, ethics opinions, helplines, localised (organisation-level) standards, risk-management tools and 'playbooks' and through training.

Second, codes, standards and guidance can act as important choice-reduction tools, thereby assisting with problems of decision-making, information selection and possibly overload. Their efficiency in this regard is, however, moot, and there are important questions that still need to be asked regarding the most efficient form and focus of such tools. Ethical standard-making in most jurisdictions is, it is submitted, relatively amateurish and remains a contested and still somewhat experimental activity (cp. Boon, 2016). At the very least, I would argue, there is a need for those who design the rules to catch-up with the growing cognitive and behavioural science literature. If we were to take situationism and bounded ethicality seriously, what kinds of norms and guidance would we produce?

A third, and perhaps a somewhat counter-intuitive, point also needs to be made here. The resilience and adaptability of the professional model also suggest that the system as a whole is (and speaking in purely functional terms, needs to be) relatively tolerant of ethical failure. This, perhaps rather uncomfortable conclusion follows from two inherent qualities of the system of practice: the contingency of ethical standards, already discussed, and the system's self-organising character. The latter point requires some explanation. As a self-organising system, legal practice must embrace sufficient mechanisms for entropy reduction (including the reduction of 'ethical entropy') to at least maintain equilibrium. Formal ethical standards are, as we have seen, part of that entropy-reducing toolkit. However, other, less ethical techniques may also play a part, for example, information asymmetry between lawyers and clients, complexity and uncertainty around professional standards, disincentives to enforcement, and other techniques that make opaque what lawyers actually do. The *functional efficiency* of practices that are unethical, or at least of (more) dubious ethicality (than the formal rules), is a fundamental challenge for normative ethics and one that is perhaps not sufficiently acknowledged.

Conclusion: thinking about regulation under conditions of complexity

This chapter has sought to offer a plausible view of legal practice as a complex system and to identify a range of key features operating within that system. It has particularly sought thereby to 'complexify' our understanding of agency by drawing together insights from systems theory, organisational theory and behavioural ethics. This suggests that context is, in a sense, everything. Context actually *shapes* the fundamental nature of the decision-making process. Consequently, an adequately complex view of agency must take situationism seriously. Situatedness is a necessary condition of agency in the real world, rather than simply a *post hoc* problem for agency. A complexity approach also thereby highlights the agentic interplay of persons and structures in the system.

Whilst such a 'situationist' narrative may seem primarily to offer despondency about the scope for both genuine agency, and effective regulation, more optimistically, complexity theory may provide access to a richer understanding of the setting within which ethical agency is framed, and thereby support better normative decision-making under the everyday, multi-faceted, conditions of uncertainty that constitute professional life. The potential for creating adequately complex regulation is a central part of that normative endeavour. I conclude, therefore by offering some brief and necessarily tentative conclusions regarding regulation.

First, regulation must plan for failure. At least some of the value of a complexity approach, I suggest, rather paradoxically perhaps, lies in its capacity to normalise rather than pathologise ethical failure. In simple terms, if the practise of practice is itself always contingent, and 'in the making', then ethical failure must always already be immanent as a possibility of practice. Regulators therefore need both to engage in regulatory conversations about how much ethical entropy is normatively acceptable/permissible and to develop a breadth of vision and fleetness that is often lacking from existing models. The latter could include moving away from individual duties and sanctions to more co-evolutionary, design-based and educative solutions than are currently commonplace.

Second, regulation needs to take professional organisation and structure far more seriously. This is hardly a new insight. Theorists writing on the ethical climate and infrastructure of law firms since the 1990s, including Parker (2008, 2010), Chambliss and Wilkins (2002) and others have stressed the importance of embedding ethical education, conversation and oversight at the organisational level. Complexity theory, given its fundamentally communicational and relational view of the world, adds another conceptual perspective from which calls for ethical infrastructure make sense. Computational studies of complexity demonstrate that agents are unique and idiosyncratic because they learn from their own localised experiences. This diversity creates aggregate behaviour that cannot readily be explained when agents are treated as homogenous and creates important challenges for regulation. Cybernetic models usefully highlight the facts that (1) as systems become more complex, structural (as opposed to agent) coordination becomes functionally more efficient and (2) control mechanisms need to match the complexity of their environment in order to manage the number of operations and degree of functional differentiation in the system (Mobus and Kalton, 2015, pp. 404–24). This latter point, in particular, may present regulators with a resourcing challenge.

Third, there is a case from complexity to rethink the relationship between role and rule. As we have seen, the idea of role morality dominates ethical theory and shapes much of our thinking about agency, yet it often appears tangential (at best) in the formulation of rules. Whilst role captures something both theoretically and intuitively plausible about the professional function, that is not always reflected by the practical formulation of discrete principles and duties. Indeed, empirical studies suggest that practitioners often have more ready recourse to a (*de facto* normative) conception of role than they do to the formal ethical rules (Moorhead and Hinchly, 2015; Vaughan and Oakley, 2016). This suggests insufficient attention has been paid to the way in which social roles themselves emerge from the dynamics of the system. As a first step we need a richer understanding of the 'real' norms and boundaries in practice: what are the fundamental differences between the formal norms of lawyering and what I have elsewhere called 'street-level morality'?[5]

Related to this is a fourth observation: a complexity account, by taking bounded ethicality seriously, also begs some interesting questions about the form and effectiveness of ethical rules. We need to take rules, in their multiple formal and informal, personal and collective manifestations seriously. In functional terms, there is at least an argument that the conventional form of regulation is suboptimal. From a complexity perspective conventional, positivistic, accounts of legal ethics arguably pay insufficient attention to Schauer's claim (2015) that the internalisation of law, absent coercion, sanction, or other external incentives, is far rarer than is often supposed. Regulatory practice could also better address the role of intuition and 'fast thinking' in decision-making when engaging with rule design. Can we better develop rules so that they correspond more closely to the form of 'fast and frugal' heuristics (see, e.g., Gigerenzer et al., 2002) favoured by bounded rationality research? Do we need a more proceduralist turn in regulation to slow down thought and monitor key steps? Should regulators embrace more 'nudge' (Thaler and Sunstein, 2009) technologies and design solutions? These are large, and sometimes ethically challenging, questions in and of themselves.

Finally, complexity accounts offer rather mixed messages about the potential benefits of regulatory action. Complexity should make us more modest in our expectations of regulatory steering but, drawing on cybernetic thinking, also highlights the potential for relatively small interventions to create large reductions in entropy (Wiener, 1989, p. 39). Targeting matters, and for that we do need more research and a greater understanding of the form, risks and benefits of (ethical) complexity itself.

Notes

1 Thus, it may be said that professions provide a societal mechanism for institutionalising collective responsibility to those in need (Alexandra and Miller, 2009) or for enabling access to certain goods of citizenship (Pepper, 1986). As an early progenitor of systems theory argued, professions are functionally distinguished to enhance or 'augment' the quality of life (Spencer, 1898).
2 Compare Chambliss and Wilkins (2002), Parker et al. (2008).
3 However, see Luban (2010).
4 See the excellent summation in Greenwood et al. (2017).
5 This is the subject of a paper in progress, an earlier version of which was presented under the title 'Understanding the (New) Moral Economy of Regulating Lawyers' at the Fifth Australia and New Zealand Legal Ethics Colloquium (ANZLEC5) at Monash University, Melbourne in December 2015.

Bibliography

Abbott, A. (1986), 'Jurisdictional Conflicts: A New Approach to the Development of the Legal Professions', 11 *Law & Social Inquiry* 187.

Abbott, A. (1988), *The System of Professions: An Essay on the Division of Expert Labor* (Chicago: University of Chicago Press).

Abel, R. L. (2011), *Lawyers on Trial: Understanding Ethical Misconduct* (New York: Oxford University Press).

Alexandra, A. and S. Miller (2009), 'Ethical Theory, "Common Morality," and Professional Obligations', 30 *Theoretical Medicine and Bioethics* 69.

Applbaum, A. I. (1999), *Ethics for Adversaries: The Morality of Roles in Public and Professional Life* (Princeton, NJ: Princeton University Press.).

Arthur, W. B., S. N. Durlauf and D. A. Lane (eds.) (1997), *The Economy as an Evolving Complex System II* (Boulder, CO: Westview Press.).

Axelrod, R. (1984), *The Evolution of Cooperation* (New York: Basic Books).

Axelrod, R. (1986), 'An Evolutionary Approach to Norms', 80 *American Political Science Review* 1095.

Axelrod, R. (1997), *The Complexity of Cooperation: Agent-Based Models of Competition and Collaboration* (Princeton, NJ: Princeton University Press).

Bailey, K. D. (1990), *Social Entropy Theory* (Albany, NY: State University of New York Press).

Bailey, K. D. (2001), 'Towards Unifying Science: Applying Concepts Across Disciplinary Boundaries', 18 *Systems Research and Behavioral Science* 41.

Bandura, A. (2001), 'Social Cognitive Theory: An Agentic Perspective', 52 *Annual Review of Psychology* 1.

Bandura, A. (2002), 'Selective Moral Disengagement in the Exercise of Moral Agency', 31 *Journal of Moral Education* 101.

Bartlett, F. and L. Haller (2013), 'Disclosing Lawyers: Questioning Law and Process in the Admission of Australian Lawyers', 41 *Federal Law Review* 227.

Bateson, G. (2002), *Mind and Nature: A Necessary Unity* (Cresskill, NJ: Hampton Press).

Bauman, Z. (1994), *Alone Again: Ethics After Certainty* (London: Demos).

Bauman, Z. (2000), *Modernity and the Holocaust* (Itahca, NY: Cornell University Press).

Bazerman, M. H. and D. A. Moore (2013), *Judgment in Managerial Decision Making* (New York: Wiley-Blackwell, 8th ed.).

Beaton, G. and I. Kaschner (2016), *Remaking Law Firms: Why and How* (Chicago, IL: American Bar Association).

Becker-Ritterspach, F. A. A. and S. Blazejewski (2016), 'Understanding Organizational Behaviour in Multinational Corporations (MNCs) from a Micropolitical Perspective: a Stratified Analytical Framework', in F. A. A. Becker-Ritterspach, S. Blazejewski, C. Dörrenbächer and M. Geppert (eds.) *Micropolitics in the Multinational Corporation: Foundations, Applications and New Directions* (Cambridge: Cambridge University Press), 185.

Bergson, H. (1998), *Creative Evolution* (Mineola, NY: Dover Press).

Bommer, M., C. Gratto, J. Gravander and M. Tuttle (1987), 'A Behavioral Model of Ethical and Unethical Decision Making', 6 *Journal of Business Ethics* 265.

Boon, A. (2010), 'Professionalism Under the Legal Services Act 2007', 17 *International Journal of the Legal Profession* 195.

Boon, A. (2016), 'The Legal Professions' New Handbooks: Narratives, Standards and Values', 19 *Legal Ethics* 207.

Brass, D. J., K. D. Butterfield and B. C. Skaggs (1998), 'Relationships and Unethical Behavior: A Social Network Perspective', 23 *Academy of Management Review* 14.

Chambliss, E. and D. B. Wilkins (2002), 'Promoting Effective Ethical Infrastructure in Large Law Firms: A Call for Research and Reporting', 30 *Hofstra Law Review* 691.

Chugh, D. and M. C. Kern (2016), 'A Dynamic and Cyclical Model of Bounded Ethicality', 36 *Research in Organizational Behavior* 85.

Clegg, S. (1989), *Frameworks of Power* (London: Sage Publications).

Cohen, M. D., R. L. Riolo and R. Axelrod (2001), 'The Role of Social Structure in the Maintenance of Cooperative Regimes', 13 *Rationality and Society* 5.

Dare, T. (2009), *The Counsel of Rogues? A Defence of the Standard Conception of the Lawyer's Role* (Farnham, UK: Ashgate).

De Cremer, D., A. E. Tenbrunsel and M. van Dijke (2010), 'Regulating Ethical Failures: Insights from Psychology', 95 *Journal of Business Ethics* 1.

De Cremer, D. and W. Vandekerckhove (2016), 'Managing Unethical Behavior in Organizations: The Need for a Behavioral Business Ethics Approach', 23(3) *Journal of Management & Organization* 1.

Deakin, S. (2011), 'Legal Evolution: Integrating Economic and Systemic Approaches', 7 *Review of Law & Economics* 659.

Deroy, X. and S. Clegg (2015), 'Back in the USSR: Introducing Recursive Contingency into Institutional Theory', 36 *Organization Studies* 73.

Evans, A. and J. Palermo (2005), 'Zero Impact: Are Lawyers' Values Affected by Law School?' 8 *Legal Ethics* 240.

Flanagan, O. J. (1991), *Varieties of Moral Personality: Ethics and Psychological Realism* (Cambridge, MA: Harvard University Press).

Flood, J. (2011), 'The Re-Landscaping of the Legal Profession: Large Law Firms and Professional Re-Regulation', 59 *Current Sociology* 507.

Fortney, S. S. (2005), 'The Billable Hours Derby: Empirical Data on the Problems and Pressure Points', 393 *Fordham Urban Law Journal* 171.

Fried, C. (1975), 'The Lawyer as Friend: The Moral Foundations of the Lawyer-Client Relation', 85 *Yale Law Journal* 1060.

Gardner, H. K. and M. Valentine (2015), 'Collaboration Among Highly Autonomous Professionals: Costs, Benefits, and Future Research Directions', in S. R. Thye and E. J. Lawler (eds.) *Advances in Group Processes* (London: Emerald Publishing), 209.

Gewirth, A. (1986), 'Professional Ethics: The Separatist Thesis', 96 *Ethics* 282.

Giessner, S. and van Quaquebeke, N. (2010), 'Using a Relational Models Perspective to Understand Normatively Appropriate Conduct in Ethical Leadership', 95 *Journal of Business Ethics* 43.

Gigerenzer, G., R. Selten and Dahlem Workshop (eds.) (2002), *Bounded Rationality: The Adaptive Toolbox* (Cambridge, MA: MIT Press).

Greenwood, R., C. R. Hinings and R. Prakash (2017), '25 Years of the Professional Partnership (P2) Form: Time to Foreground Its Social Purpose and Herald the P3', 4(2) *Journal of Professions and Organization* 112.

Hall, K. and V. Holmes (2008), 'The Power of Rationalisation to Influence Lawyers' Decisions to Act Unethically', 11 *Legal Ethics* 137.

Haller, L. (2001), 'Solicitors' Disciplinary Hearings in Queensland 1930–2000: A Statistical Analysis', 13 *Bond Law Review* 1.

Holland, J. H. (1995), *Hidden Order: How Adaptation Builds Complexity* (New York: Basic Books, 1st ed.).

Holland, J. H. and J. H. Miller (1991), 'Artificial Adaptive Agents in Economic Theory', 81 *The American Economic Review* 365.

Kagan, R. A. (1989), 'Understanding Regulatory Enforcement', 11 *Law & Policy* 89.

Kirkland, K. (2008), 'Ethical Infrastructures and de Facto Ethical Norms at Work in Large US Law Firms: The Role of Ethics Counsel', 11 *Legal Ethics* 181.

Kugler, P. N. and M. T. Turvey (2016), *Information, Natural Law, and the Self-Assembly of Rhythmic Movement* (London: Routledge).

Kuhn, T. (2009), 'Positioning Lawyers: Discursive Resources, Professional Ethics and Identification', 16 *Organization* 681.

Ladyman, J., J. Lambert and K. Wiesner (2013), 'What Is a Complex System?' 3 *European Journal for Philosophy of Science* 33.

Law, J. (1992), 'Notes on the Theory of the Actor-Network: Ordering, Strategy, and Heterogeneity', 5 *Systems Practice* 379.

Law, J. (1999), 'After ANT, Complexity, Naming and Topology', in J. Law and J. Hassard (eds.) *Actor Network Theory and After* (Oxford: Blackwell and Sociological Review).

Law, J. (2002), 'Objects and Spaces', 19 *Theory, Culture & Society* 91.

Lawrence, T. B. (2008), 'Power, Institutions and Organizations', in R. Greenwood, C. Oliver, R. Suddaby and K. Sahlin-Andersson (eds.) *The Sage Handbook of Organizational Institutionalism* (London: Sage Publications), 170.

Lawrence, T. B., N. Malhotra and T. Morris (2012), 'Episodic and Systemic Power in the Transformation of Professional Service Firms', 49 *Journal of Management Studies* 102.

Levin, L. C., C. Zozula and P. Siegelman (2013), 'The Questionable Character of the Bar's Character and Fitness Inquiry', 40 *Law & Social Inquiry* 51.

Lieberman, M. D. (2005), 'Principles, Processes, and Puzzles of Social Cognition: An Introduction for the Special Issue on Social Cognitive Neuroscience', 28 *NeuroImage* 745.

Lippucci, A. (1998), 'Cybernetic Legal Analysis and Human Agency', 4 *Res Publica* 77.

Lissack, M. R. (1999), "Complexity: The Science, Its Vocabulary, and Its Relation to Organizations", 1 *Emergence* 110.

Loughrey, J. (2011), 'Large Law Firms, Sophisticated Clients, and the Regulation of Conflicts of Interest in England and Wales', 14 *Legal Ethics* 215.

Luban, D. (1988), *Lawyers and Justice: An Ethical Study* (Princeton, NJ: Princeton University Press).

Luban, D. (2005), 'Professional Ethics', in R. G. Frey and C. Heath Wellman (eds.) *A Companion to Applied Ethics* (Oxford: Wiley-Blackwell).

Luban, D. (2007), *Legal Ethics and Human Dignity* (Cambridge: Cambridge University Press).

Luban, D. (2010), 'How Must a Lawyer Be? A Response to Woolley and Wendel', 23 *Georgetown Journal of Legal Ethics* 1101.

Luban, D., A. Strudler and D. Wasserman (1992), 'Moral Responsibility in the Age of Bureaucracy', 90 *Michigan Law Review* 2348.

Markovits, D. (2011), *A Modern Legal Ethics* (Princeton NJ: Princeton University Press, 1st ed.).

Mather, L. (2011), 'How and Why Do Lawyers Misbehave? Lawyers, Discipline and Collegial Control', in S. L. Cummings (ed.) *The Paradox of Professionalism: Lawyers and the Possibility of Justice* (Cambridge: Cambridge University Press), 109.

Mavrofides, T., A. Kameas, D. Papageorgiou and A. Los (2011), 'On the Entropy of Social Systems: A Revision of the Concepts of Entropy and Energy in the Social Context: On the Entropy of Social Systems', 28 *Systems Research and Behavioral Science* 353.

McNay, L. (2016), 'Agency', in L. J. Disch and M. E. Hawkesworth (eds.) *The Oxford Handbook of Feminist Theory* (Oxford: Oxford University Press).

Mele, A. R. and J. Shepherd (2013), 'Situationism and Agency', 1 *Journal of Practical Ethics* 62.

Miner, M. and A. Petocz (2003), 'Moral Theory in Ethical Decision Makinig: Problems, Clarifications and Recommendations from a Psychological Perspective', 42 *Journal of Business Ethics* 11.

Mobus, G. E. and M. C. Kalton (2015), Cybernetics: The Role of Information and Computation in Systems, in: *Principles of Systems Science* (New York: Springer), 359.

Moorhead, R. (2014), 'Precarious Professionalism: Some Empirical and Behavioural Perspectives on Lawyers', 67 *Current Legal Problems* 447.

Moorhead, R. and V. Hinchly (2015), 'Professional Minimalism? The Ethical Consciousness of Commercial Lawyers', 42 *Journal of Law and Society* 387.

Morin, E. (2007), 'Restricted Complexity, General Complexity', in C. Gershenson, D. Aerts and B. Edmonds (eds.) *Worldviews, Science and Us: Philosophy and Complexity* (London: World Scientific), 5.

Murray, B. and A. Fortinberry (2016), *Leading the Future: The Human Science of Law Firm Strategy and Leadership* (London: ARK Group).

Muzio, D. and S. Ackroyd (2005), 'On the Consequences of Defensive Professionalism: Recent Changes in the Legal Labour Process', 32 *Journal of Law and Society* 615.

Muzio, D., J. Faulconbridge, C. Gabbioneta and R. Greenwood (2016), 'Bad Apples, Bad Barrels and Bad Cellars: A "boundaries" Perspective on Professional Misconduct', in D. Palmer, K. Smith-Crowe and R. Greenwood (eds.) *Organizational Wrongdoing: Key Perspectives and New Directions* (Cambridge: Cambridge University Press), 141.

Nicolson, D. (1998), 'Mapping Professional Legal Ethics: The Form and Focus of the Codes', 1 *Legal Ethics* 51.

Nicolson, D. and J. Webb (1999), *Professional Legal Ethics: Critical Interrogations* (Oxford: Oxford University Press).

Palermo, J. and A. Evans (2005), 'Australian Law Students' Values: How They Impact on Ethical Behaviour', 15 *Legal Education Review* 1.

Palmer, D. (2012), *Normal Organizational Wrongdoing: A Critical Analysis of Theories of Misconduct in and by Organizations* (New York: Oxford University Press, 1st ed.).

Palmer, D. and C. Moore (2016), 'Social Networks and Organizational Wrongdoing in Context', in D. Palmer, K. Smith-Crowe and R. Greenwood (eds.) *Organizational Wrongdoing: Key Perspectives and New Directions* (Cambridge: Cambridge University Press), 203.

Parker, C. and A. Evans (2014), *Inside Lawyers' Ethics* (Melbourne: Cambridge University Press, 2nd ed.).

Parker, C., A. Evans, L. Haller, S. Le Mire and R. Mortensen (2008), 'The Ethical Infrastructure of Legal Practice in Large Law Firms: Values, Policy and Behaviour', 31 *University of New South Wales Law Journal* 158.

Parker, C., T. Gordon and S. Marks (2010). 'Regulating Law Firm Ethics Management: An Empirical Assessment of an Innovation in Regulation of the Legal Profession in New South Wales', 37 *Journal of Law & Society* 466.

Paterson, J. (1995), 'Who Is Zenon Bankowski Talking to? The Person in the Sight of Autopoiesis', 8 *Ratio Juris* 212.

Pepper, S. L. (1986), 'The Lawyer's Amoral Ethical Role: A Defense, a Problem, and Some Possibilities', 11 *American Bar Foundation Research Journal* 613.

Perlman, A. M. (2007), 'Unethical Obedience by Subordinate Attorneys: Lessons from Social Psychology', 36 *Hofstra Law Review* 451.

Perlman, A. M. (2015), 'A Behavioral Theory of Legal Ethics', 90 *Indiana Law Journal* 1639.

Prentice, R. A. (2015), 'Behavioral Ethics: Can It Help Lawyers (and Others) Be Their Best Selves?' 29 *Notre Dame Journal of Law, Ethics & Public Policy* 35.

Radden, J. (2004), 'The Debate Continues: Unique Ethics for Psychiatry', 38 *Australian and New Zealand Journal of Psychiatry* 115.

Regan, M. C. (2009), *Eat What You Kill: The Fall of a Wall Street Lawyer* (Ann Arbor: University of Michigan Press).

Robbennolt, J. K. and J. R. Sternlight (2013), 'Behavioral Legal Ethics', 45 *Arizona State Law Journal* 1107.

Rogers, J. (2017), 'Since Lawyers Work in Teams, We Must Focus on Team Ethics', in R. Levy, M. O'Brien, S. Rice, P. Ridge and M. Thornton (eds.) *New Directions for Law in Australia* (Acton: ANU Press).

Rogers, J., D. Kingsford-Smith and J. Chellew (2017), 'The Large Professional Service Firm: A New Force in the Regulative Bargain', 40 *University of New South Wales Law Journal* 218.

Rousse, B. S. (2016), 'Heidegger, Sociality, and Human Agency', 24 *European Journal of Philosophy* 417.

Ruhl, J. B. (2008), 'Law's Complexity: A Primer', 24 *Georgia State University Law Review* 885.

Schauer, F. F. (2015), *The Force of Law* (Cambridge, MA: Harvard University Press).

Shepherd, J. (2015), 'Scientific Challenges to Free Will and Moral Responsibility', 10 *Philosophy Compass* 197.

Simon, H. A. (1956), 'Rational Choice and the Structure of Environments', 63 *Psychological Review* 129.

Simon, H. A. (1962), 'The Architecture of Complexity', 106 *Proceedings of the American Philosophical Society* 467.

Simon, W. H. (1998), *The Practice of Justice: A Theory of Lawyers 'ethics'* (Cambridge, MA: Harvard University Press).

Sommerlad, H. (1995), 'Managerialism and the Legal Profession: A New Professional Paradigm', 2 *International Journal of the Legal Profession* 159.

Spencer, H. (1898), *The Principles of Sociology* (New York: Appleton).

Steinbauer, R., R. W. Renn, R. R. Taylor and P. K. Njoroge (2014), 'Ethical Leadership and Followers' Moral Judgment: The Role of Followers' Perceived Accountability and Self-Leadership', 120 *Journal of Business Ethics* 381.

Stephen, F. (2013), *Lawyers, Markets and Regulation* (Cheltenham: Edward Elgar).

Susskind, R. (2010), *The End of Lawyers? Rethinking the Nature of Legal Services* (Oxford: Oxford University Press).

Thaler, R. H. and C. R. Sunstein (2009), *Nudge: Improving Decisions About Health, Wealth and Happiness* (London: Penguin Books).

Tremblay, P. R. and J. A. McMorrow (2011), 'Lawyers and the New Institutionalism', 9 *University of St Thomas Law Journal* 568.

Trevino, L. K. (1986), 'Ethical Decision Making in Organizations: A Person-Situation Interactionist Model', 11 *The Academy of Management Review* 601.

Treviño, L. K., G. R. Weaver and S. J. Reynolds (2006), 'Behavioral Ethics in Organizations: A Review', 32 *Journal of Management* 951.

Tsoukas, H. and K. J. Dooley (2011), 'Introduction to the Special Issue: Towards the Ecological Style: Embracing Complexity in Organizational Research', 32 *Organization Studies* 729.

Vaughan, S. and E. Oakley (2016), ' "Gorilla exceptions" and the Ethically Apathetic Corporate Lawyer', 19 *Legal Ethics* 50.

Victor, B. and J. B. Cullen (1988), 'The Organizational Bases of Ethical Work Climates', 33 *Administrative Science Quarterly* 101.

Webb, J. (2004), 'Turf Wars and Market Control: Competition and Complexity in the Market for Legal Services', 11 *International Journal of the Legal Profession* 81.

Webb, J. (2006), 'When "Law and Sociology" Is Not Enough: Transdisciplinarity and the Problem of Complexity', in M. D. A. Freeman (ed.) *Law and Sociology, Current Legal Issues* (Oxford: Oxford University Press), 90.

Webb, J. (forthcoming), 'Behavioural Legal Ethics', in M. Sellers and S. Kirste (eds.) *Encyclopedia of the Philosophy of Law and Social Philosophy* (Springer Verlag).

Webb, J., J. Ching, P. Maharg and A. Sherr (2013) Setting Standards: The Future of Legal Services Education and Training Regulation in England and Wales, *Coventry: Legal Education and Training Review*, <http://www.letr.org.uk/report>

Webb, T. E. (2014), 'Tracing an Outline of Legal Complexity', 27 *Ratio Juris* 477.

Wendel, W. B. (2012), *Lawyers and Fidelity to Law* (Princeton, NJ: Princeton University Press).

Wiener, N. (1948), *Cybernetics: Or Control and Communication in the Animal and the Machine* (New York: Technology Press and John Wiley & Sons).

Wiener, N. (1989), *The Human Use of Human Beings: Cybernetics and Society* (London: Free Association).

Wilkinson, M. A., C. Walker and P. Mercer (2000), 'Do Codes of Ethics Actually Shape Legal Practice?' 45 *McGill Law Journal* 645.

Woermann, M. and P. Cilliers (2012), 'The Ethics of Complexity and the Complexity of Ethics', 31 *South African Journal of Philosophy* 447.

Woolley, A. (2011), 'Intuition and Theory in Legal Ethics Teaching', 9 *University of St Thomas Law Journal* 285.

Woolley, A. (2012), 'Regulation in Practice: The 'ethical economy' of Lawyer Regulation in Canada and a Case Study in Lawyer Deviance', 15 *Legal Ethics* 243.

Woolley, A. and W. B. Wendel (2010), 'Legal Ethics and Moral Character', 23 *Georgetown Journal of Legal Ethics* 1065.

Zimbardo, P. (2007), *The Lucifer Effect: How Good People Turn Evil* (London: Rider).

Zumbansen, P. and G-P. Calliess (eds.) (2011), *Law, Economics and Evolutionary Theory* (Cheltenham: Edward Elgar).

Index

Note: Page numbers in *italic* indicate a figure and page numbers in **bold** indicate a table on the corresponding page.

Milton Keynes UK
Ingram Content Group UK Ltd.
UKHW040446071024
449327UK00020B/1039